FORECASTER!

Battling the Weather Odds in Peace and War

by

Theodore L. "Ted" Cogut

Copyright © 2001
by Theodore L. Cogut

Library of Congress Control Number: 2001012345

ISBN 0-9675347-2-0

Published by

Mining History
P.O. Box 1319
Thatcher, AZ 85552

This book may not be reproduced in any media, including electronic, or used as the basis for a television series or motion picture without the written permission of the copyright owner.

Dedication

*Though concerned almost exclusively with the meteorologists of the US Army and US Air Force, this book is dedicated to the men and women of whatever meteorological service, including the US Navy, the US Coast Guard, the Marine Corps and the civilian weathermen of this nation, who took it upon themselves to engage the often uncertain science of meteorology and in the process carried the flag to meteorology's bright tomorrow. But more than that, this book is also dedicated to **all** the men and women of the US military who served on land and on sea and whose fidelity to this great nation could not be compromised; including in particular the American servicemen and servicewomen who served in the Vietnam War, fighting a competent and determined military enemy while being vigorously opposed by the anti-US-military protesters at home.*

CONTENTS

	page
Preface	*i*
Introduction	*iii*
Timeline of Army/Air Force Weather Services	*v*
Acknowledgments	*vii*

Chapters
1:	Into the Wild Blue Yonder	1
2:	I Take Meteorology	9
3:	Chanute, the Home of the Weather Schools	17
4:	WETT	27
5:	From Tinker to St. Louis to Grand Central	37
6:	8th Weather	49
7:	A Weather Job at Last — Sort of	57
8:	Hitchhiking with the AAF	65
9:	Family Life in Massachusetts and a Christmas Surprise	77
10:	Air Training Command's Weather Forecaster Course	85
11:	Castles in the Air	93
12:	The President's Plane is Running Out of Fuel	101
13:	A Lonely Job	115
14:	Brown Shoe Air Force	123
15:	Jet Streams and Other Things — We Call Off the Bet	137
16:	Saudi Arabia — No Tree, No Grass on Ground to Spy	153
17:	Single-station Analysis and Other Trials	169
18:	Fighter Pilots, Flying Saucers and One Lucky Forecast	185
19:	Weather Gets in Your Blood	201
20:	Shark Oil and Spider Webs	215
21:	We Found Paradise, Almost	233
22:	Peace Is Our Profession	251
23:	The Clime Center and the Terror of Cheverly	265
24:	1st Air Cavalry	275
25:	300 NVA at 1,000 Meters...or 1,000 NVA at 300 Meters	289
26:	A Well-traveled Cat	303
27:	XXIVth Corps Artillery	309
Epilogue		335
Abbreviations		336
Index		337

Preface

As we were about to publish this book, terrorists attacked our country at the World Trade Center towers and the Pentagon. At a time like that, it did not seem that a book telling one man's story of the development of the modern weather services was an appropriate endeavor. And so it was set aside. But then, less than a month after those atrocities, certain persons began protesting our efforts to retaliate. That kind of conduct, at a time when bodies were still being pulled out of the rubble, was simply unbelievable, bringing to mind in a flashback kind of way the anti-military protesters who held sway in America while I was in Vietnam. And it brought to mind my fervently held hope that what had happened to our fighting men in Vietnam, as a result of the actions of those that I had then begun to call the North Vietnamese Army Auxiliary, the anti-US-military protesters, should not be allowed to happen again. In that way, this book is quite appropriate for the times, for it also is a story of that war.

But this book was written primarily for those who really want to know what it was like to deal with weather problems before satellites, computers and sophisticated radars came on the scene. At a reunion of weather people I met a former weatherman who served with me in the days when each weather station was solely responsible for plotting and analyzing *all* of its weather charts and from them, with no help from a centralized weather facility (as there is now), among other meteorological variables made precisely stated forecasts of cloud cover, cloud heights, visibility and weather. "You know," he said, "we didn't really know very much about forecasting in those days." I had to differ with him. Too much time had passed since we had fought the weather battles together. He didn't remember what we did. I reminded him that the very fact we had no radar worthy of the name, no satellites and no computers, made us work much harder on the forecasts. Today's TV weatherperson would shrink if faced with making such forecasts.

Weather has always been crucial to successful military operations. The commander who took due notice of the weather was the one most likely to succeed. Like Napoleon, Hitler's failure to conquer Russia was largely due to inattention to the effects of coming weather changes. With his troops in summer uniforms suffering from frostbite and engine oils thickening in the cold, Hitler's armored columns began to bog down as they faced their first Russian winter. Conversely, the 6 June 1944 World War II D-day date, the successful jump-off for the all-out effort to liberate Europe, was selected by General Eisenhower to coincide with an expected good-weather window. In December of that year, at the Battle of the Bulge, for weeks a fiercely effective German counterattack in the cold and snow of the Ardennes halted the Allies' seemingly unstoppable push into Germany. The Germans took full advantage of low ceilings and low visibilities which initially prevented the Americans from employing effective battlefield air support. Earlier in the war, for their upstream look for fronts or pressure systems that would later be affecting weather on the continent, the Germans established weather stations on Greenland's east coast. They also received weather information from their submarines engaged in the Battle of the Atlantic.[1] On our side, the Army Air Force Weather Service provided weather support for the greatest air armadas in history, the B-17 and B-24 heavy bombers that pounded targets deep in Germany. An F-51 fighter pilot told this writer that contrails produced in the 1,000-bomber raid in which he participated, which was supported with probably 500 fighters, were so numerous as to practically create a cirrus overcast. During that World War II era, not much thought was given to the similarity of the weather duties of the US military weathermen and their adversaries. In the fact that they were to provide for more efficient battleground operations which could also bring a personal familiarity with the hazards of battle, even if not charged with the primary task of shooting at the enemy, their jobs really were quite similar. In one of the last letters to be received from the surrounded German Sixth Army at the pivotal battle of Stalingrad in January of 1943, a German weatherman wrote to his wife saying he and three others measured temperatures and humidity, and recorded cloud ceilings and visibilities, adding that he knew so little about war that he had not so much as fired his pistol.[2] He would soon be killed, or captured with the strong likelihood of dying in captivity.

---

[1] Dan van der Vat, *The Atlantic Campaign, World War II's Great Struggle at Sea*, Harper & Row, Publishers, 10 East 53rd Street, New York, NY 10022, p.371.

[2] For the complete and poignant letter see: "So this is what the end looks like," *Volume One, The American Heritage Picture History of World War II*, American Heritage Publishing Co., Inc., p. 298

Introduction

In books about military men, it has been popular to write about one who began as a private soldier and ended as a beribboned general. As an exception to that practice, this is the story of a soldier of a respected but not particularly exalted rank, and of his family. He is a special kind of military man — a soldier and a meteorologist. His story traces the advances in meteorology of the post-Word War II to Vietnam era, an era of unparalleled achievement in meteorology, a time when military meteorologists led the way.

There was a time in the Army Air Force of the United States when some meteorologists held the rank of warrant officer. Having honed their skills as sergeants, to many pilots, to enlisted weathermen, and to fellow officers, these weather forecasters often were the ones to be consulted during critical flying operations or when meteorological situations were particularly complex.

When the Department of the Air Force was created from the Army Air Force on September 18, 1947, and for some time thereafter, the warrant officers, the former AAF weather forecasters, continued to function in their important roles, this time in the USAF. Near the end of the post-world War II decade, however, warrant officers in all fields in the USAF were considered superfluous and the rank was retired.

A small number of meteorological warrant officers did survive. These were the "Metro Warrants" of the Army Artillery. In 1965, as America began a serious participation in the Vietnam War, there were only 40 of them. They would contribute significantly to the Free World effort in Vietnam. Now, they too are gone. Not many who knew the meteorological warrants could understand why they went the way of the dinosaurs.

Beginning as a private in the US Army Air Force and ending in the Artillery as a warrant officer, the meteorologist in this story made thousands of weather forecasts — at a transport base for the Berlin Airlift, at fighter and bomber bases, for commercial airliners in Saudi Arabia and Bermuda. He was an environmental analyst in Washington, DC, taught meteorology, and made meteorological artillery trajectory corrections on the battlefields of Vietnam, serving Vietnam tours with the famed 1st Cavalry Division and XXIV Corps.

The author, at age one, held by his mother, Mary, with father, Louis, and sister, Virginia, in front of the house in which he was born that was built by his parents in Royal Oak Township (later incorporated as Hazel Park), Michigan. In about a year, with the onset of the Great Depression, this home would be lost and the family would move a hundred miles north to a farm near Caro in the Thumb section of Lower Michigan.

Timeline of Army/Air Force Weather Services — Precursor Organizations, Related Events[*]

1870   President Grant signs a bill creating a national weather service to be run by the Army. The Signal Service is assigned the task.

1890   The Army's meteorological equipment and meteorological personnel are transferred to the newly created US Weather Bureau of the Department of Agriculture. US weathermen are now civilians.

1917   Air Service is created out of the Signal Corps' Air Section. Weather operations return to the Army with creation of the meteorological section of the Signal Corps. Army weathermen serve in World War I.

1926   Out of the Air Service, the Army Air Corps is organized.

1937   On July 1, Army Air Corps takes over from Signal Corps responsibility for the weather service. A school for enlisted forecasters opens at Patterson Field, Ohio. Weathermen are now in the Army Air Corps Weather Service.

1939   A weather observer school opens at Scott Field, Illinois.

1940   The observer and forecaster schools are consolidated and moved to Chanute Field, Illinois.

1941   On December 7, Japanese naval air and sea forces attack Hawaii, bringing the US into World War II. Army weathermen will serve in all theaters of the war, hitting the invasion beaches with the first waves of American troops.

---

[*] Weather history information for the early years was taken from *A History of the United States Weather Bureau* by Donald R. Whitnah, University of Illinois Press, Urbana, Illinois, 1965, and from *Air Weather Service: a Brief History* by William E. Nawyn and Rita M. Markus, 1991.

Timeline continued

1941   Army Air Force is created. The Army Air Corps retains its identity within the AAF.

1945   AAF Weather Service became a worldwide command, reporting directly to Commanding General, AAF.

1946   On 13 March, the AAF Weather Service is assigned to ATC and its name is changed to Air Weather Service.

1947   In June, the Army Air Corps is dissolved. In September, the AAF is reorganized as the independent Air Force. Weathermen retain their AWS name.

1948   The Soviets blockade Berlin, precipitating the Berlin Airlift. Military Air Transport Service is created from ATC. AWS is now under MATS.

1950   US and United Nations forces respond to the North Korean invasion of South Korea. AWS personnel serve in Korea as do Army ballistic meteorologists. AWS weathermen serve in observing, forecasting and weather recon roles. Army weathermen operate electronic atmospheric sounding systems, providing artillery-trajectory corrections.

1957   Rudimentary cloud pictures via satellite and computer-made prognostic charts appear. Much-improved weather satellites (Tiros) appear in 1958. More reliable computer progs follow.

1965   Division-size US forces enter Vietnam War. As in Korea, Air Force and Army weathermen are involved. Army and Marine Corps ballistic meteorologists train at the Meteorology Division, Ft. Sill, Oklahoma. Later in the war, Army Meteorological Quality Control teams provide technical supervision of Army ballistic meteorologists in Vietnam.

1966   Military Airlift Command is created from MATS. AWS weathermen are now under MAC.

1991   Air Weather Service is a field agency reporting to Air Staff.

Acknowledgments

As this has been written from personal experiences of some 20 odd years in the military weather services, no bibliography is presented. However, the reader wanting a detailed account of military weather history will be well advised to see the excellent *America's Weather Warriors, 1814-1985* by Bates and Fuller, Texas A&M University Press, 1986, and the compact: *Air Weather Service: a Brief History,* with authors, in separate editions, Nawyn and Markus, Office of AWS Special Study, 1991; and Nolan and Murphy, AFWA History Office Special Study, 2000. For the earliest accounts of military weather, one should consult: *A History of the United States Weather Bureau* by Donald R. Whitnah, University of Illinois Press, 1961.

I am indebted to a number of my friends who, in conversations, supplied details that they remembered from 30 to 50 years in the past, confirming my memory of certain events. Any errors that might have eluded their capture are mine alone. Near the end of my military service I happened to have been assigned to participation in the Vietnam War. No one who served in that war could have returned without having developed somewhat strong opinions regarding it. I am no exception. The opinions I bring forth in this book regarding the Vietnam War, therefore, are mine alone and no one I mention in this book should be thought responsible for what I have written about it.

In addition to the persons mentioned above, I am also indebted to all my other comrades-in-arms of the Air Force and of the Army who served with me and with whom I could not have the privilege of recent conversation. Foremost among them were my commanders, first-rate gentlemen all, and in particular the late Colonel Donald J. Wolfe who, at what would become a significant turning point, practically insisted I return to the military. Most of these associates were meteorologists. To have had the privilege of engaging in the meteorological battles with those professionals was an experience not many were fortunate enough to have. Friendly lightning had brushed you, and you wondered why you were the one so favored.

But most of all, I am indebted to my family and especially my immediate family: my wife, Marie, and our daughters, Leta Mach, Willa Swartz, and Pamela Bryant, who made it possible for me to have a most interesting and rewarding career in the weather services of both the Army and the Air Force, even when it seemed more sensible, financially, to have opted for some other line of work. And I must not forget William C. Conger, my co-author of the *History of Arizona's Clifton-Morenci Mining District,* and his wife, Joan, who, unknown to them, did a lot toward keeping me on course through their questions regarding my progress with this book.

Jacket Photo and Photo Credits

Jacket photo is of the author following a weather balloon with a theodolite at Dhahran, Saudi Arabia in 1951. Most of the photos in this book were taken by the author. Where there are exceptions, a credit line appears below the picture. In some photos of aircraft, items that appeared in the original picture but were not essential to the story were electronically removed to avoid clutter.

CHAPTER 1

Into the Wild Blue Yonder

"You're taking meteorology, aren't you?" Andy asked. He had just stepped out of the cubicle where he was assigned to meteorology training in the Army Air Corps which then still existed under its parent organization, the Army Air Force. Technically, we were in both, but we gradually began to refer to our organization as the Force, a grander sort of title. The Corps, in any event was dissolved in the next year. Andy, I suspected, might simply have wanted someone else to share his fate should that job turn out not to his liking. Andy's question might also have been meant as a lead-in to some of the light-hearted banter we had pulled on each other since that day in early February 1946 when we first met. That was on the frozen shore of Lake Michigan at Ft. Sheridan, Illinois. From there an Army bus had taken us to Chicago where we were loaded into a troop train.

On calm nights back on the farm, when sound carried well, I liked to listen to the plaintive echoes of whistles on speeding steam trains that were too far away to be seen even in daytime. For me and my sisters and brothers, the sound evoked pictures of elegantly dressed ladies and gentlemen, riding in private rail cars. Settled into plush upholstered seats, they held crystal glasses of champagne, rolled over rails smoothed by expert gandy dancers, chatted of this and that, and anticipated a thrilling rendezvous in some distant exotic place. That was the image I carried as I stepped up into the rail car.

But our troop train consisted of wear-beaten cars that looked as if they had first seen service, and tough service at that, during the Civil War. We slipped this way and that on hard, shiny benches, oak seats worn smooth by the posteriors of thousands of bodies that once were, like us, on their way to somewhere. This steam locomotive pulled its cars at crawl speed, constantly gathering and losing momentum while belching soot that drifted back through our open windows. Someone suggested the windows be shut to keep the soot from settling, like black snowflakes, on our neat olive drab uniforms. Except for one or two keepers of the windows, who were not fully aware they were now in the Army and no longer free to do as they wished, that was quickly done. A slight threat of force, and the few recalcitrants fell into line. But on this "milk run" where nearly every town, no matter how small, had to be stopped at, we *did* relax discipline enough to raise the windows at the station stops to be enthusiastically greeted by folks who were just following their routine of meeting the "boys" as troop trains arrived.

They told us about Jimmy or Tom, in the ETO (European Theater of Operations) or the Pacific theater, who they expected to soon see arriving on a train such as ours. The shooting war had ended only some five months earlier. In about a month, President Truman would declare an end to the Period of National Emergency, so this was still officially that unpredictable time known as World War II. The greeters didn't know we were recruits and not returning war heroes who deserved to be handed candy bars, homemade cookies and soft drinks to supplement the troop train's colorless box lunches. We rocked on to the sunny South. The temperature rose. Sweat streamed down on those woolen uniforms. Someone suggested we throw discipline to the wind, raise all windows to the cooling slipstream, sooty that it was. That done, we sang "Off we go into the wild blue yonder," repeated "Nothing can stop the Army Air Corps," joked about our lives now that *we were in the Army.* Eventually, we reached Biloxi, Mississippi and its Keesler Field.

Now, after three months of basic training, on this May day at Keesler it seemed my friend Andy might have thought of a way to get back at me for the jokes I had pulled on him. This one might be one that couldn't be topped. We usually took pains to avoid any display of sincerity, but I had to admit he seemed sincere in recommending I take meteorology. If I went for it, I'd be tied up in a military job from which there'd be no escaping for the three long years of our enlistments. Andy was not his real name. Like many others with his last name, with some of his associates in the Army Air Force who felt unfulfilled if they didn't attach a nickname to the newcomer, Donald Clinton Anderson partially lost the use of his first name. This also happened to me. My parents did indeed call me "Theodore." But those of my classmates from immigrant families liked to call me "T-Door" (in search of the Old Country's fertile soil, a number of Ukrainian immigrants had settled on nearby farms); and George Abke, my boyhood pal of German extraction who lived on the state highway that also ran past our farm, called me "Thee." My Army associates further shortened my name to "Ted" and, much later, in Saudi Arabia, Air Force friends called me "X" as they noted the similarity of my last name to the band leader Xavier Cugat. (But my last name is pronounced quite differently, with the "o" and "u" in "Cogut" as in "bold" and "but.") As Andy's middle name conjured up a dignified De Witt Clinton building the Erie canal, and we recruits were starving for dignity, I often called him Donald Clinton but, as I will refer to him, I thought of him as Andy. I've used his actual name. A few others in this book are represented by aliases.

Andy and I were Michigan boys. He came from Durand, on the highway between Lansing and Flint. I was from the great motor city of Detroit but was raised on a farm near Caro, in the "thumb" of the Michigan "mitten." Both Caro and Durand were small towns, though I liked to claim Caro would take the honors for being small, which would not now be so true. I had graduated from a high school in Detroit, but always thought of Caro as my hometown (even though we didn't live in town, but on a farm a short distance from it). Caro, Michigan, with its small-town charm and serene pastoral surroundings, was the place where the prerequisites for what I was going to be in my future adult world were pleasantly brought to the fore. The town had only two stoplights. You would barely have shifted into second gear after going through the stoplight at the east end of town when it was time to brake for the one near the west end. The Strand Theater was located at that west-end stoplight. Its octagon door windows, that are still there, had been made by my ninth-grade shop teacher sometime before we entered World War II. At the Strand, for eleven cents, I saw the first Caro showing of *Gone with the Wind*. When I was a child, horses were still used around Caro in field work, and were often seen on the roads, mostly in the process of pulling a grateful motorist's car out of the mud. When I was ten, it was my job to lead Prince, the big black horse, down the rows of corn as Dad steered the cultivator behind him. Horses then were a common part of everyday life and were frequently mentioned in conversation. Caro in those days was often referred to affectionately as the proverbial "one horse town."

Although I hadn't the slightest inkling of it then, to have been raised on a farm near Caro, Michigan was great training for a person who'd be a meteorologist. If it was Saturday afternoon, you went to town and while waiting at the Rexall drug store for the Sunday *Detroit News* to arrive on the truck from Detroit, picked up vital weather data from farmers who gathered there. I picked up a number of gems that way. One example: "It's going to rain, I saw cows lying down on my way into town." I knew there had to be some substance to that prediction because I had heard the same from Mother.

Though they would not meet until both were living in the Detroit area a couple of decades later, Mother, nee Mary Evanish, came into the world about 50 miles from where Dad, Louis Cogut, was born. They came from what was then Austria, a place they always called "the Old Country."

Weather forecasting was a common pastime amongst the Caro-area farm population, and Mother was in the thick of it, being mainly interested in the long-range form of it. She told me that a ring around the moon was a reliable long-range rain predictor unless, of course, atmospheric currents changed. And then there was our dear friend, an Italian immigrant on the next farm, who knew drought or damaging storms were caused by the "wronga power." Dad never joined in folklore-type predictions. The only time he ever seriously consulted with me about a weather forecast occurred years later when I was home on leave after just having finished weather forecaster school, and we were trying to attach new shingles to the old farmhouse's roof in advance of a storm. If Mother was the long range forecaster, Dad was the somewhat non-committal short range predictor. Eyeing the alfalfa or June clover while glancing at the clouds, he'd say: "It's time to cut the hay." He knew there'd be enough dry weather to cut it, dry it, and stack it up in the hayloft in the new barn. Having lost his affluent suburban Detroit life style with the stock market crash of 1929 and the concurrent lay-off following 20 years with Henry Ford (unemployment and retirement payments were still not a part of the American scene), and being faced with making a living off farming at a time when farm product prices were in the cellar, he tried not to look at what the days would bring. No truer patriot ever lived on American soil. When economic events seemed overwhelming, instead of criticizing his government, he'd take the violin down from the hook on the living room wall and play a spirited Old Country tune. Our immediate cares would then drift away on the fading notes of the violin. Like so many Great Depression-era children would later say, "We were poor, but we didn't really know it."

If you would only let it soak in, the events of farm life could provide a great background for dedication to the task at hand and for discipline in general. Daisy, our cow, for example, always had to be milked somewhere between 5 and 6 a.m. — there could be no exceptions. That was my job when Dad went to Detroit to work in the war plant. I would get up at 0500, trim the built-up carbon from the wick of the kerosene lantern to keep the flame from smoking the glass, light the wick and trudge off with the lantern to the barn to milk Daisy before heading for classes at Caro High School. On the farm, procrastination was a word that had yet to be discovered. On the day in late March or April that Dad picked up a handful of soil and pronounced it just right by temperature and moisture content, you were sure to see him head for the cellar to carry out bags of seed potatoes. Retrieving the hand-held potato planter from the tool shed, he began to plant. If we didn't call him in for a lunch break, he would be planting those potatoes without stopping until daylight faded completely away. A man of great determination, when he saw the farm needed a barn, on the advice of his friend, Albert Abke, who also showed Dad, with a peach twig, where to successfully dig a well, and with whom he sometimes lapsed into one of his several Old Country languages, this time German, he bought rough-sawn hard maple lumber from a local saw mill. He selected the hard maple and most likely got a good price for it because there was no market for it. Every carpenter knew it would be too hard to drive a nail through hard maple. Ah, but *he* knew you *could* drive nails in it if you built your barn in a big hurry, before the green wood had a chance to dry. And that's what he did. This was in 1939. The barn still stands, the dried hard maple holding nails in a death grip. That kind of determination had a way of rubbing off on the children. We did things that I now find hard to believe we would even have attempted. When we were eleven years old my boyhood pal George Abke (Albert's son) and I tried to construct a deep swimming hole on Cogut Creek, the stream we named ourselves that ran behind the barn. With rocks, shovelsful of dirt and tree branches thrown in a frantic race against the rapidly increasing force of the hydraulic head, we built dam after dam that would hold the waters long enough for one or two of our dives and then give way in a great flood of dam debris. We gave up and decided to go swimming at Cass City, so we *walked to Cass City*. **Cass City was at least 15 miles away,** on State Highway M-81. Though paved on its westward stretch toward Saginaw, eastward to Cass City it was then gravel and not good for riding our bikes. M-24, the highway running by our farm, was also gravel. When the Army Air Force made me a weather forecaster, I tried to recall those days of dedication to task and the similar need for prompt response as pilots popped into the weather station for their flight briefings. If you hadn't taken the time to prepare yourself with the meteorological situation beforehand, providing, of course, that you were given the time for it, you'd be shooting dice with the forecast. And if the forecast was bad you didn't ever want to face those pilots again.

The town's and county's (Tuscola County) most impressive public building was the white marble courthouse with the huge window depicting Lewis Cass, in stained glass, "signing a treaty with the Indians." Cass had been Michigan's only Presidential candidate[1] until Tom Dewey came along. Dewey was from Owosso, which was not far from Durand, but he had left his Michigan roots and had moved away to New York.[2] The river that ran just in back of our farm was named the Cass River, after Lewis Cass. George Abke and I often rode our bikes down the sandy road to the dam on the Cass where we'd spend half a day fishing. Not long after I began fishing there I became convinced that successful fishing in the Cass correlated well with a day of light rain. Then after more fishing days had come and gone, I became equally convinced that a bad day of fishing also correlated well with a day of light rain. We were always hoping to catch a Northern Pike, but they were too smart for us. One sunny day in May, as I walked along clear, cool Cogut Creek, I saw a magnificent Northern, perfectly motionless, in a secluded curve of the stream where the water eddied back in a slight counter current. There he rested, free from having to fight the main current. Though the stream carried many schools of 20 to 25 suckers that swam upstream from the Cass in the springtime spawning season, when flooding made the creek as wide as some parts of the Cass, this was the first time I had seen a Northern in it. Of about 20 inches in length, he was a spectacular sight. I tried to entice him with a hook baited with an angle worm but he ignored it, preferring, no doubt, a minnow which I did not have. I was sort of glad he

---

[1] Lewis Cass was governor of the Michigan Territory from 1813 to 1831 and ran for President on the Democratic ticket in 1848. He also held the posts of U. S. Senator from Michigan, was Secretary of War, Secretary of State and Minister to France. He held important commands in the War of 1812.

[2] Readers will be quick to point out that Gerald Ford, who was from Grand Rapids, was a Michigan man who *did* become President, but of course Ford had been appointed to the job, not having run for it initially.

didn't take the bait and could remain in the creek. With the warm sun reflecting off his speckled side and the golden sandy creek bottom as a backdrop, he was a sight to warm a boy's eyes. I returned to the same spot on the creek bank the next day, but he was gone. Concluding that observation of the Northern would correlate with a sunny day in May, thereafter, on many succeeding days of that month and also in succeeding years when May rolled around again, I looked for the Northern at that spot. But he never returned. Many years later I would remember those conclusions when new meteorologists tried to use only raw statistics to the exclusion of current atmospheric dynamics in day-to-day weather forecasts.

Just behind the farm there were Indians who lived on a small reservation.[3] They most likely were descendants of the Indians who had been a party to the treaty signed by Cass. The Indians would often take the short cut across our farm as they walked to town. On their way to school, the Indian children would pass through our yard and often join me for the remainder of the walk to the one-room Connor School. To an outsider who was not familiar with their reticence, our walks during which nothing was said might have seemed strange. I wondered about their silent demeanor and thought that it might have been that they didn't really like me. That thought was quickly set aside one day when I was in sixth grade and five non-Indian boys, bigger than I, knocked me down into a snow bank on my way home from school and were about to rub snow in my face and meant to do much worse when Silas Fisher, a slight but strong Indian boy only a little bigger than any of the five, caught up to us. Without saying a word, he tossed the ruffians one by one into an even deeper snow bank from which they peered out, and seeing that Silas had no other business with them, slowly extricated themselves and ran away, crying. Silas and I then walked home. I wanted to thank him but silence was of course the appropriate manner of our association. So I said nothing. As we walked along with the only sound being the creaking of snow beneath our boots, I was sure he must have seen my proud grin. But we both knew it would have been quite improper of him to let me know he saw it.

I have been dwelling on the town of Caro and its surroundings to help set the backdrop for the town's connection to the military, for this is a book about the military almost as much as it is about meteorology. A number of the town's streets were named after Civil War generals, and of course it was mandatory that we children join our parents in a ride into town in the Model A Ford, bedecked with five US flags attached to its radiator cap, and then stand with our parents on the Caro sidewalks to watch with pride as the county's sole surviving Civil War veteran, in his black Grand Army of the Republic uniform, wearing the black General Grant-style hat, marched at the head of the Memorial Day and Fourth of July parades. But it was during

---

[3] Our township was named "Indianfields."

World War II that that small town in the heart of Michigan's thumb achieved its great claim to fame. Caro, Michigan was Maynard Smith's hometown.

In the Hall of Heroes at the Pentagon, one can view a plaque commemorating S/Sgt. Maynard Smith, the first enlisted man to be awarded the medal of honor in the Army Air Force. Known by Air Force buddies as "Snuffy" after Snuffy Smith of the Barney Google comic strip, and by Caro friends as "Hokey," he had denied the Luftwaffe their prize of shooting down his B-17. It was S/Sgt. Smith's first mission and his B-17 was badly crippled by antiaircraft shells and Focke Wulf 190 cannon fire.[4] He was the belly gunner, meaning he rode below the rest of the crew, shut off from them in the cramped ball turret. Two crew members were badly wounded, three others bailed out. A fuel tank erupted. Fire raged in the waist section and radio compartment. Stored ammunition exploded. The hydraulic mechanism that would open the hatch above Smith wouldn't work. He hand-cranked it open to climb up into the fuselage proper. As the Focke Wulfs approached from different directions, Smith alternately manned the other gun positions, making the German pilots think the B-17 with 229649 on its tail was still a fighting machine, and that they had no safe approach angle. In between manning the guns he applied first aid to the injured, fought the fire until the fire extinguishers ran out, threw ammunition overboard to keep it from exploding in the fire and to reduce the weight the crippled plane had to carry, and helped the pilot fly the plane back to England. After that 306th Bomb Group B-17 of the 423rd Squadron was examined by the ground crews, it was judged too badly shot up to be repaired and was scrapped.

When word of his exploits reached home, stories about Smith began to surface. Like a lot of people who lived in or near Caro, Hokey must have enjoyed jokes and doing things that might be thought of as somewhat less than sedate. By one account, before joining the Air Force he had once caused a bit of commotion on Caro's State Street when he rode his pony into the Rexall drug store. When Secretary of War Henry L. Stimson arrived at the airbase in England to present the Medal of Honor, the *Detroit News* reported that Smith had to be pulled off KP (kitchen police duty), a duty assigned for some infraction of discipline. And that, it was said by many on Caro's street corners, was completely true to his character. Andy Rooney, the well known TV personality and World War II vet who calls things as they are, wrote this about him: "The Air Force loved to give medals and they had good reason in Snuffy Smith's case."[5] Hokey's fame may have had something to do with the fact that a number of my friends from the Caro area would later join the Army Air Force, including, I suppose, myself.

---

[4] My brother Bill built an exact replica of Smith's B-17. His model had a six-and-a-half-foot wingspan, was flown without mishap and was written up in the *Tuscola County Advertiser*.
[5] Andrew A. Rooney, *Pieces of My Mind,"* Atheneum, New York, 1984, p.223.

Bill Cogut (left), Air Force veteran and Fort Shawnee, Ohio Councilman, and Joe Joseph, his son-in-law, with the model B-17 bomber Bill made and named *"The Spirit of Caro."* It is a one-sixteenth-scale model, an exact copy, including markings, of the B-17 in which S/Sgt. Maynard Smith of Caro, Michigan became the first enlisted man in the Army Air Force to win the Medal of Honor in World War II. In the upper photo, just like the original, Bill's B-17 circles its landing field.

Chapter 2

I Take Meteorology

"Private Cogut! Front and center!" barked the three-striper we called a "buck sergeant." Interrupting a moment of deep reflection, his command made me acutely aware of the fact that I had been totally immersed in juggling the pros and cons brought forth by Andy's question: "You're taking meteorology aren't you?"

I had been thinking how much our lives depend on pure chance. Like a flash from the airfield beacon on a low-visibility night, the plan for a course of action intrudes itself. It has to be implemented quickly or the entire thought may fade into nothingness, perhaps never to return. If acted upon, it can have enormous consequences. This could be the time when you can change your life significantly for the better or — that chance flash of insight may be highlighting an opportunity that you will grasp and later regret. In those few brief moments following Andy's question I had been dwelling on similar thoughts when I might better have been serving myself had I been focusing on the military occupations that could be open to me.

We had just finished basic training. On the previous day, Flight 678, our unit, was marched at the Army's specified rate of 120 30-inch steps per minute from out of our bivouac under majestic southern pines. Marching was our way of life. Footsore, we tromped on, leaving those little tents pitched on soft pine needles at the bivouac area to return to the airfield itself and our assigned homes of drab tarpaper-covered barracks. Built on posts, the barracks were not much more appealing than Great Depression-era hobo shacks. In that squadron area, we rout-stepped onto delightful walkways made of loose seashells held within 2X6 planks, the shells softly giving way with each step. At night, moon-glow would reflect off the shells in a most charming way, making one forget, if only for a little while, that basic training was not an appealing way to spend time. On this morning, we marched once more from the squadron area — this time to the hangar in which offices had been fashioned. We spent the day there, taking a battery of aptitude tests.

The buck sergeant consulted our test results as he assigned to us a certain military occupational specialty (MOS). Once placed in that specialty, our fates would be cast in stone. For those of us like Andy and I who were considered "lifers," the three-year Regular Army enlistees, the decision could have enormous consequences that might even spill over into our civilian lives once those three-year hitches were completed. For the others with shorter commitments, the 18-monthers, or our one and only two-year enlistee and, in particular, the draftees, the military classification mattered little. They were only marking time. Until the day they'd be freed from the Army, they were merely marching in place.

A quite noticeable rivalry surfaced between the troops of the different classes of commitment. "What made you enlist for three years when you could do 18 months standing on your head," the 18-monthers taunted. We three-year men tolerated the 18-monthers in good humor. Like us, they were, after all, volunteers. Unlike the draftees, no one told them they had to join the military. The draftees, we thought, were the most unlike us. They were with us through no choice of their own and had no need to be diplomatic. "Only an idiot would enlist for three years," was their common refrain. There was also a geographical difference that reinforced Flight 678's in-ranks rivalry. We volunteers were mostly from Michigan and Wisconsin and were proud of it, while most of the draftees were from places like Boston; East Orange, New Jersey; or Pawtucket, Rhode Island and were just as proud of their origins. Our draftees didn't seem to know or care that a three-year enlistment carried with it the promise of a choice of overseas theater and, most importantly, the choice of Army branch. Instead of taking the chance of being assigned to the ground combat arms — the Infantry, Artillery, or Armor — we three-year men made certain we were going to the Air Force. But then, after having made that decision, after signing up for three long years to make sure we'd be in the Air Force, we found ourselves serving with 18 monthers and even draftees who somehow managed to be in the Air Force just as we were! We observed that the draftees were older men, and in spite of our reacting as if *they* were the ones who were mentally deficient, we did silently respect the wisdom we believed should accompany advancing age. So their remarks served to reinforce a suspicion that perhaps we had zigged when we should have zagged. Just before lights out, after the card games, jokes and arguments had subsided in our tarpaper home, we three-year men contemplated our futures in the dim light of the bare 40-watt bulbs that hung from cross beams nailed to two-by-four rafters. And we tried to remind ourselves of why we had enlisted for the three-year term, reasons that had once seemed quite rational.

Nowadays, three years seems to be a terribly short period of time. Back at Keesler, where we spent most of our time attending GI parties (translation: scrubbed the blackish wooden floors until they actually began to turn white), marched at double time to the obstacle course, and worked 16-hour days of KP duty, three years seemed an absolute eternity. At 50 dollars a month on Army pay, we didn't have the money to have cars then, but now a person's car payments often exceeds three years, not to mention far exceeding 50 dollars, and we think little of it. Whether it was a wise decision or not, we had enlisted for three years and were determined to make the best of it. And while on KP scrubbing smelly garbage cans, we whistled past the graveyards of three-year commitments. We took comfort in imagining that, like the short termers, we too were only marching in place and were only there to gather material for a book. We would call it *Three Years Behind the Mess Hall,* a title we'd steal from Richard Henry Dana's novel: *Two Years Before the Mast.* And then we'd soberly note that his term of service was shorter.

"To what field do you want to be assigned?" The buck sergeant asked as he extracted my 201 file from the clutter of records on his desk, some of which had spilled onto the concrete floor. Expecting the worst, that my only option would be some non-technical field such as cooks and bakers, I was surprised by his question.

"What *are* the fields?" I asked.

"Anything the Army Air Force has," he replied, not wanting to be bothered with a complete answer to my question which, after all, a recruit shouldn't think he was entitled to in any case. But he did feel magnanimous enough to add: "Your test scores are high enough to let you into any of them." Well, this indeed was a bit of good luck. But now it would be nice to know exactly what fields this Army Air Force had to offer to one who was qualified for any one of them. I decided to sort of force the issue.

"I'd like to be a pilot," I said. He didn't respond, body language telegraphing that he had no time for jokesters. I clearly meant it as a joke. Pilots had previously been trained through acceptance into the aviation cadet program. But this was May of 1946. President Truman had terminated the World War II Period of National Emergency in March, the official date for the end of the war. The pilot training program consequently had either ended or had been severely cut back. In any event, married men were not accepted into the program, and I was married. A few months before enlisting, I had married the lovely Marie Nordstrom whom I met while we both were inspectors in the Detroit-area war plant, inspecting B-29 engine parts, some of which Dad had made. It was the same plant he had left the farm for back in 1942, and it was he who had suggested, upon my graduation from high school, that I might also gain employment there. In a sense, he had changed my life in two major ways —through making me more intimately acquainted with aircraft and by making it possible for me to find that most perfect girl.

Marie's father, Walter Nordstrom, a descendant of Swedish immigrants, also worked in a Detroit war plant, the Dodge Main plant, making Army trucks. Her mother, Martha Golden, and father met while they lived on farms only about a half mile apart near Hastings, Michigan. Martha's grandparents came from Wimbledon, England, had settled in Ohio, and then to the Michigan farm. The lure of high-paying auto-plant jobs brought Martha and Walter to Detroit where they settled in a northern suburb.

Other than the fact that they obviously had pilots and other commissioned personnel such as navigators and bombardiers, and enlisted men who were gunners on the bombers, when I enlisted I knew absolutely nothing about the jobs available in the Army Air Force. About a year before I had been told by the Navy that I had been accepted as a combat aircrewman in their service which meant that I'd be the pilot's radio operator when not firing the plane's machine guns. But with the war approaching its conclusion, that training program ended before the Navy could call me up. When I enlisted in the AAF, being a radio operator was about the only job I knew even a little about, but now that the job horizon lay wide open to me I was not prepared to settle for being a radio operator, and being a gunner also held little attraction.

In the short time since February, when I had enlisted in the Army Air Force upon the recommendation of a corporal in the recruiting office in Royal Oak, Michigan, I had only been stationed on two Army installations, here at Keesler and at Ft. Sheridan. Except for learning how uncomfortable it was to stand in inoculation lines that wound out from the dispensary around two or three other typical wooden World War I buildings where we were taunted by bemedaled veterans, who called us "Needle Bait" and "Jeeps" (the last being a term that I would never learn the significance of unless it was that, being their replacements, the veterans held us as close to their hearts as they did their beloved Willys Jeep), I had learned little about the Army at Sheridan, not even so much as to be introduced to KP since the mess hall jobs were all handled by German prisoners of war. The POWs, who wore woolen US Army shirts and trousers that were dyed black, were much too efficient to benefit from any help we GIs might provide. I didn't know it then, but that was going to be just about the cleanest mess hall I'd ever eat in during my entire military career. After scrubbing down the interior to a virtual sparkle, those German POWs then took hoses to wash down the *outside* of the building. I would never see that happening at any other military installation. Once, at breakfast, while extending my tray to one of the German mess attendants, a tiny bit of oatmeal from his ladle splashed onto my olive drab blouse. Grabbing a wash cloth, he dashed around the counter and began to clean away the offending spot, while I tried to protest to no avail that it was no problem and he shouldn't bother with it.

There were a couple of things that I learned at Sheridan that, should I have remembered them, could have proved useful in my future military career. It all began one day when the duty sergeant, looking in our barracks and seeing that I was the only one there (the others were out, trying to buy nickel candy bars at the PX), announced that I was to be fire guard that night. He explained that meant that not only was I to see that no fires erupted, but that I was also *the man in charge of the barracks.* Now, that was a responsibility I did not want to assume, but, after all, this was the Army and I was in the Army. I resolved to do my best in discharging that duty. Everything went well until the veterans began returning from their night of carousing in "Chi town" (our fort was just a few miles north of Chicago). I was surprised at how many of the veterans had gone to town. Some of them had class "A" passes which they only had to show to the MP at the gate to be allowed out of the compound, but even those who had no passes had gone to town, slipping out, as I learned later, under a hole in the fence.

At midnight, a somewhat inebriated technical sergeant (five striper) made his way back to our barracks. His battle jacket (also "Ike Jacket" after General Eisenhower who instituted the style) was decorated with four rows of "fruit salad" (medals), and a blue-background shoulder patch with white "A" in red circle of General Patton's Third Army. I noticed he did not have the purple heart. Seeing he had a willing listener, he lost no time in explaining that he had been with Patton throughout the drive across France and into Germany and made it very clear that he was justifiably proud of that service.

Somewhere around three o'clock in the morning the more completely inebriated soldiers returned. One could hardly have blamed them for having raised a few (?) toasts in commemoration of having survived the most all-invasive war in history. In a short time, however, the confusion brought on by John Barleycorn caused a fight to break out on the second floor between two who couldn't agree as to which bunk they belonged. The tech sergeant and I then approached them and, while trying to separate the combatants, the sergeant lost his balance and tumbled head-over-heels down the stairs. It was then up to me to get him to the base hospital. At the hospital, I learned that the good sergeant, after having survived unscathed throughout Patton's dash across Europe, had broken his arm in the Ft. Sheridan barracks that was then under my jurisdiction and a full report of what had happened was subsequently required. With that episode, I could see it would be advisable in the future to avoid similar "man in charge" kinds of duty. There was a lesson, too, in the behavior of my fellow recruits who were not to be found when the duty sergeant came around looking for help. On the day after that incident, I learned I was going to Keesler and took the news with a sense of great relief. And now, to return to that May day at Keesler...

"Which jobs require the highest test scores?" I asked the buck sergeant. I thought the jobs requiring the highest test scores would be deemed critical and that a critical job could keep you off KP. I had learned more than I ever wanted to know about KP at Keesler. It was a job in which I wished to acquire no additional experience.

"Cryptography and meteorology," the sergeant replied, adding, "Go back out in the hall, take a seat and think it over. But don't take too long. I go off duty in 15 minutes." This choice that would determine my occupation for at least the next three years, for the period of this enlistment and perhaps longer, was not going to be the result of long, and careful consideration. I had to choose one of the many career fields the Army had to offer and choose it fast. I desperately needed 15 minutes of clear, lucid, logical thought. Taking the sergeant's advice, I sat down on the long bench in the hallway along with other equally confused recruits and tried to imagine what the two career fields would be like. I knew that the guys working in cryptography were assigned to the G-2 Section of the Army, the intelligence section, and that seemed glamorous. Ian Fleming had not yet invented his James Bond character, but the picture of myself that I conjured up was every bit as dashing as Ian's future double-oh-seven.

At about the 13th minute of the allotted 15 I thought the choice had to be cryptography. But would the job really be interesting? Being mostly secret work, you wouldn't be able to tell anyone about it. You wouldn't be able to explain, in your modest way of course, how you had, in a single-handed manner, enabled the capture of Joseph Stalin's most sinister spy. The required secrecy did appear to be something of a downside. At that moment, Andy again interrupted my thoughts.

"Did you take meteorology?" he asked.

"I haven't decided. What about cryptography? And what would you do in meteorology?" As strange as it will seem to everyone reading this, I was completely unaware of that mysterious field called "meteorology." One must remember that, while Nikola Tesla had foreseen television in the early days of the 20th century and television was demonstrated at the New York World's Fair as early as 1939, it was not available as a broadcast service until well after World War II. Most American families would not have television sets with which to view weather programs until about 1950. Weather information might have been available by commercial radio broadcasts but radio was not available to us for some time because the power lines had not yet been extended to our farm by the REA (Rural Electrification Authority). Then, sometime in 1938, Dad went to town and bought a battery-powered Zenith radio complete with short wave capability and with a far-ahead-of-its-time wind charger that was mounted on that two-story farmhouse's roof. The wind charger, with a propeller shaped similar to that on an aircraft, would charge the radio's 6-volt car-type battery. For awhile, we then were able to receive weather forecasts, but these were terminated with the US entry into World War II. All weather data were then classified secret by the federal government in order to deny the information to Germany and Japan. Even though we had heard weather forecasts for that short period before the government curtailed them, I had not the faintest idea of how the weather information was gathered, not to mention how the forecasts were made.

"Come on. Take meteorology," Andy urged. "It's a good job. You'll be looking at clouds, drawing lines on maps and even coloring the maps." And that clinched it. On the farm, I frequently found myself looking at clouds to try to learn what the farmers saw in them that led them to decide if it would rain; and when I was in first grade, I made a free-hand drawing of Santa Claus that must have been quite well done because it drew a suggestion of admiration from Dad. This was more important to me than my subsequent good fortune of being able to skip three grades of elementary school.[1]

Dad's approval was important because, to his children, he was a real life hero. He became the head of the family at the age of 10 when his father, at the age of 45, was murdered by the local aristocrat.[2] Grandfather had tried to collect payment for wheels he had made and delivered to the aristocrat, but that criminal then decided the simplest way of avoiding payment was to have Grandfather killed, which he did and with not so much as a word of disapproval from the government in that land ruled by the Emperor of Austria, Franz Joseph! After having established himself in Detroit with Henry Ford, Dad brought his younger siblings, Aunt Frances and Uncle Peter, to this promised land and tried to bring his mother and older sister as well. Grandmother Elizabeth, however, was turned back to the Old Country for a supposed eye problem (she probably only needed glasses!) and Aunt Katherine stayed behind with her. Thereafter, Dad sent money to them until the Great Depression made that a complete impossibility. Despite those tragic circumstances, Dad could see the humor in many events considered by others to be serious and had many light-hearted moments that he shared with us, but he was mainly a practical-minded man who didn't cotton to projects of small worth which a drawing of Santa Claus might be thought to be. Examining that work with apparent seriousness, he said: "Maybe some day you'll be an artist." Well, I could see that drawing lines on and coloring weather maps might require a certain artistic aptitude. It seemed that I might have a qualification for that job that would stand me in good stead.

So I took meteorology.

---

[1] I loved to tell my three daughters who didn't have the advantage of attending one-room country schools as I had, where conditions were more conducive to quick learning, that my education had been less sufficient than theirs because I never attended the second, fifth, and eighth grades, having been double promoted past each of them by my teachers in their rare moments of weakness.

[2] Alf Landon, 1936 Republican presidential nominee, once said on a TV interview that the reason European immigrants came to America was that the nobility wouldn't let them fish in the streams. Since I liked fishing, a similar reason would have been sufficient for me and might also have been adequate for Dad, but his reason was much more serious. He hated the Old Country and wouldn't allow his children to learn its language.

From 1942 to March 1946, weathermen in the Army served in what became known as the AAF Weather Service. They would have worn a shoulder patch like the one above except that the arc above the winged star would have read: AAF Weather Service. In March 1946 the weathermen were assigned to the Air Transport Command and their organization was renamed Air Weather Service. The winged star shoulder patch shown above would have been worn by other members of the Air Transport Command but even though they were under ATC, Air Weather Service personnel wore the patch without the ATC arc. *Patch courtesy of Robert Nordstrom.*

Chapter 3

Chanute, the Home of the Weather Schools

There are some incidents in life that stand out in memory in contradiction to all expectations. For me, one of these were certain moments in June at Chanute Field, Illinois, the Army Air Force's AAF Technical Training Command base. We began our Weather Observer Course there on May 20, 1946. At evening time, I liked to step out onto the upper bay balcony of our World War II-type barracks and take in the fragrance of new-mown hay rising from nearby fields. At a time like that on that base on the prairie, it would be serenely quiet until some mechanic revved up the aircraft engines on the hard stand, for Chanute was also a school for aircraft mechanics. The deafening roar sounded as if it was April of 1942 and all of Doolittle's B-25's were at full throttle at the same time, on the carrier deck, pushing for engine speed that would enable them to lift off for Tokyo.

For a few moments on that upper bay balcony you could forget the Monday morning incident in the classroom when one of the assistant instructors, a corporal or buck sergeant, roamed the aisles looking for students who had nodded off so that he could rap that offender's desk with a pointer stick. That was bad enough, but it was even worse to have the attention of the class riveted on *you* as the instructor shouted, his voice amply amplified by the microphone: **Wake that man!** And then, of course, you'd have the assistant instructor making a bee line for *your* desk to slam his stick on it with all the force at his command even if it was evident that the instructor had already awakened you. We could see there was a certain enjoyment that flowed from the stick-rapping procedure, an opportunity for a second or two of localized fame that the assistant instructor would not miss if he could help it. We suspected that one or two of them harbored an ambition for a career on the drums.

Just two months prior to my joining that elite group, the military weathermen, the name of their organization had been changed from AAF Weather Service to Air Weather Service. At that time, at Chanute, four different weather courses were taught. The Weather Observer Course was the basic course for all prospective weathermen. If he volunteered for additional training, however, the weather observer graduate might be directly enrolled in the Radiosonde Course or in the Weather Equipment Teletype Technician Course. When he completed the observer course or the follow-on course, the student would be sent to an operating location to practice his craft. After some years of experience in the weather stations, he might be selected for a return to Chanute to an adjacent classroom. This time for the Weather

Forecaster Course, the top level of meteorological training that would provide an education equivalent to and, many would say, superior to that obtained with a bachelor's degree in meteorology. Unlike the fine hangar-classroom of the observing and forecasting classes, the radiosonde classes were conducted in a wooden building not far from the flight line where balloon releases and hydrogen inflation-gas production operations could more easily be conducted. Except for its teletype technician phase that was taught at Scott Field, Illinois, the Weather Equipment Teletype Technician Course was taught in the hangar next to the observing classroom. Out of concrete and above the outer entrance to the hangar, some artisan had fashioned a large and attractive set of pilot' wings. Chanute impressed one as being a topnotch training base.

The projected attrition of those whose terms of service would soon be coming to an end, and they were in the vast majority, dictated that the size of the observing classes be large — if memory serves there were about a hundred in my class, and another class was conducted at the same time.

Dozing off in the weather observing class happened most often on Mondays after the students had returned from their weekend trips home, but it also occurred on other days. It was not that the subject was dull, in fact, I found it fascinating, but the instructor's microphone seemed to demodulate his voice, taking out the high and the low notes, producing a dull droning that filled all corners of the classroom. Sent to the school immediately following basic training where we had only been permitted to be off base on one day and for only a few hours at that, we had been away from home three or four months. For many of us, that was the first time away from the homefires. So, in the interest of keeping the students close to the learning task and not tired out from excessive travel, the school commandant instituted a travel limit. You could only remove yourself from the airfield a certain number of miles, but weathermen are a resourceful lot and it was the rare weather student who would admit the mileage restriction had placed a damper on his travel plans.

Not one of us had a car, and bus or train travel was out of the question, not only because of the cost of the fare but also because, in the pre-computerized finance offices of that time, the $50 per month pay that was due you would not usually be known at your new base for a number of weeks after your arrival. At that new base, it was not unusual to stand in the pay line for an hour before finally arriving at the paying officer's desk where you snapped to attention, saluted, stated your name, rank and serial number and only then learned that the paying officer couldn't find your name in the pay book. And, if he found your name, prior to having your money counted out to you, you'd be "red lined" if, when you signed for your pay, any slightest part of your signature happened to either overflow into the line above or the line below. Being red lined meant that a ruler would be laid on the page of the pay book on which your name had been typed and, as you watched, the first sergeant picked up a nibbed pen (this was before ball point pens became common) dipped it into a bottle of red ink, and very carefully drew a red line along the ruler's edge, a line that ran directly through the approximate center of your name. All this meant you'd not be paid until the next pay period rolled around — 30 days later. This was a particular hazard in my case since the "g" in my last name would naturally extend down into the line below unless I consciously forced it upward, thus confining the "g" to its authorized area and resulting in a signature that made me look as if I had been AWOL when penmanship class was convened in elementary school.

For our weekend dashes home, in my case to Michigan, we quite naturally resorted to hitchhiking. This could be about as safe as taking the bus. You and the driver of the car didn't have to worry about being held up or mugged. The warm feeling worked up for servicemen during the war still abounded. And of course, while you were trying to get the ride you'd be wearing your uniform because civilian clothes were not then allowed to be worn. Before you climbed into the car that had stopped for you, you soon learned to ask where the driver'd be letting you off. Even though you hated to let any ride slip by, you wanted to be left off where the prospect of getting the next ride was fairly good, at a crossroads of two highways or, even better, at the far end of a town. It wasn't good to be left on the edge of the road in a rural area far from any intersecting highway, particularly if it was late at night, for you could be waiting for hours before getting your next ride.

The great thing about hitchhiking at that time was the opportunity you had to engage in conversation with a variety of people. Most of the time, you'd be picked up by farmers who would only be going to town to buy chicken feed or some such thing, and so the ride would only take you to the next town. I much preferred getting a ride with a traveling salesman who could be going all the way to the suburb of Detroit where my wife, Marie, was staying at her parents' home, but listening to a farmer as he explained what the weather was doing to his crops made the short hops some of the most enjoyable rides. When the farmer learned you were a student of meteorology he'd usually take an even greater liking to you and in one instance the farmer, wanting to learn more about what we were studying at Chanute, took me 15 miles out of his way to the next town down the road.

I got a ride with a traveling salesman one day and it became an experience I did not want to repeat. Shortly after I entered his car, I saw that he obviously liked to drink intoxicating beverages and also liked to drive his car at its maximum possible speed, a combination that alarmed me. I extricated myself from that situation by energetically agreeing, even encouraging, that it would be good to stop at a bar. As a driver, he was a disaster waiting to happen, but he was also a great supporter of the troops. "Soldier, I'll buy you a drink," he said as he settled onto a barstool somewhere on the outskirts of Toledo. "And then I'll take you all the rest of the way to your home in Michigan."

It would have been nice to have shown my appreciation by joining him in a drink, though it probably would have made me sick as I had yet to have had so much as a beer unless you count the time when George Abke and I were eleven. George found a box of beer on the road while riding his bike to our farm, and we took it down to Cogut Creek where we cooled it in the fresh-flowing stream and then probably tasted it, though memory is vague on that point. Seeing that I could not get away from this likable though inebriated driver unless I used subterfuge, I excused myself, saying I was going to the bathroom, and with at least a tinge of guilt, slipped out a side door and onto the highway where I was lucky enough to get a ride before my companion began looking for me.

That ride took me to downtown Detroit where I began searching for a streetcar stop and the streetcar ride north to the end of the line at Eight Mile Road from which I'd take a bus home to Royal Oak Township, the city's northernmost suburb. It was 4 o'clock in the morning. The city was in deep sleep. Not so much as a police car's siren disturbed the quiet of the city's main thoroughfare, wide Woodward Avenue. In the daytime, the sidewalks and the avenue would be bustling. You couldn't walk along Woodward's sidewalks at such times without giving way to other pedestrians. At 4 o'clock in the morning, in the center of what was then the nation's fourth largest city, in that more tranquil age of our social development, you felt, and were, perfectly safe. Lit by streetlights, the silent metropolis seemed spooky but at the same time definitely pleasant, even enchanting. Woodward Avenue was the bottom of a canyon bordered by tall city buildings. I walked along, acutely aware that, for any number of city blocks that surrounded me, I was the only human being to be seen. My footsteps faintly echoed off the brick walls and large display windows of the J.L. Hudson department store, famed for its upscale merchandise, and I recalled with amusement how, in what might be thought of as a precursor to my present wanderings, I had alarmed my mother when she had let go of my hand for an instant at the bottom of Hudson's up escalator, and I then proceeded all the way up to the store's top floor, the 12th floor, where a frantic mother and store staff finally found me admiring a jackknife display. I was two years old. Many years later, with the advent of television broadcasting, that store would become well known across the country through the televised spectacular Thanksgiving Day parades it sponsored. Sadly, in the late 1990s the building was demolished, a victim of the movement of businesses to the suburban shopping malls.

Hitchhiking between Chanute and that northern suburb of Detroit was an interesting experience, but one of those trips convinced me that I'd better not do it any more. On this occasion, I left home earlier than usual, in plenty of time to make it back to Chanute, and feeling confident with the extra time that I had allotted myself, not too mention the confidence that I had acquired in what seemed like an uncanny ability to catch rides, I decided to accept a ride taking me west instead of down my normal route south to Toledo and then southwest down US 24 through Fort Wayne, Indiana. The way west would take me through Chicago. I had never been to Chicago, Carl Sandburg's "City of the Big Shoulders," and this would be a chance to see it even if there wouldn't be time for much sight-seeing.

At about 2100 hours (nine p.m.) I found myself somewhere in Chicago, on a thoroughfare called "Halsted Street." Being deposited in the middle of a big city with the hope of getting a ride, where most of the drivers weren't leaving the urban area, turned out about as bad as being left between towns on a rural road, but I was still optimistic. After trying to get a ride for an hour or so, I decided I'd have to get myself to the outskirts of the city, to the southside (Chanute was about a hundred miles south of Chicago), and after waiting for and eventually riding an elevated train and a bus, I found my way to the southern suburbs where my efforts to catch a ride went on for hour after hour. Finally, just when it seemed hopeless, a new Buick pulled up. "I'm going to St. Louis," he said, "how far are you going?"

Now this was certainly a lucky break, even if a little late. To get to St. Louis, the driver'd be going down US Highway 45, directly past Chanute Field, and when he learned of my predicament he floored the accelerator pedal, assuring me he'd get me to the base on time. I learned that he had not served in the war which I assumed was due to a physical impairment and a classification of 4-F. Unlike my experience with the drunk near Toledo, I had no problems with this fellow's penchant for speed. But in spite of all his efforts to coax the maximum out of his new car on that highway that then had no speed limit, by the time he dropped me off at Chanute's north gate it was about 0730 hours. Roll call, I knew, would already have been taken and my name would have been entered into the orderly room's morning attendance roster as an AWOL. The weather troops would be marching from morning breakfast at the mess hall to the classroom at the hangar. It was plain to see that my military career was cut off before it hardly got started.

Not knowing what else to do, I trudged along from the gate toward the squadron area and the orderly room where I intended to turn myself in, a most repentant AWOL. But as I approached the squadron area, I saw masses of troops milling around on the nearby parade field. What was this all about? One of my barracks friends spotted me. "Where you been, Ted?" he asked. "We thought you were AWOL." "I *am* AWOL," I said, "but what's going on here?" This was, he explained, a muster roll. A muster roll was an apparent holdover from the days when the commander on a given day needed to know just how many troops he could field. To find out, all the troops would be turned out and in this case, at least, their dog tags checked to verify identification. They were all surprised that morning, my friend said, when they were told they wouldn't fall out in formation in front of the barracks for the regular 5 a.m. roll call. Instead, all the troops from all the student squadrons were to gather at the parade field where their names were to be checked off a clip board, and with the confusion of the hundreds of milling bodies, our flight was yet to be checked. It was the only time in my military career that I was to participate in a muster roll. Saved by fate when I didn't deserve to be saved, I resolved then and there to stop hitchhiking.

I called Marie via long distance to tell her my hitchhiking days ought to end. To talk to her on the phone, I would call the people across the street from her Dad and Mom's house, and they would run over to bring her to the phone (not many homes had phones at that time). Marie and I decided that she would come to Chanute, or more precisely, to the town of Rantoul whose southern extremity adjoined Chanute's north gate. When she arrived, we obtained an upstairs room in a neat little white house owned by an Army officer's widow. She had two upstairs rooms to rent, the other was rented to a second lieutenant and his wife. I suspect the widow lowered the rent when she saw the arms of my battle jacket were conspicuously lacking the stripes by which the Army identified the rank of its soldiers. Even so, we were pushing it financially. I thought we could survive on my pay but Marie knew better. Without my knowledge, she found a job as waitress in Rantoul. I wished she hadn't, but did not object very much. Without doubt, we needed the money.

Marie's coming to Rantoul proved very helpful to me in my studies. Now, instead of spending my weekends rushing home and suffering not only from lack of study on those weekends but also from the lack of full attention in class brought on by the tiring aftereffects of all-night hitchhiking, we would sit under the giant elms in Rantoul memorizing such things as the call letters and international index numbers of the vast number of weather stations found on the weather maps. These had to be memorized as there wouldn't be time enough to look up their locations when one began to plot weather data on the maps. The weather station or airport call letters that are so familiar to present-day air travelers from the destination identifier tags on their luggage, were then only known to weathermen and the more experienced pilots. Some examples: "DTW" was the identifier for the airport at Detroit's western suburb and stood for "Detroit Wayne Major." "ORD" stood for "Orchard Place," one of Chicago's airports. "STL" for St. Louis, was easily apparent.

Most of the students were being introduced to subjects that they hardly knew had existed. There were: air masses * cloud names * 27 standard cloud types and the altitudes they were found at * equinoxes * solstices * weather codes * fronts * troposphere * tropopause * stratosphere * cloud ceilings and the ways of determining them * ceiling lights and the new ceilometer with mercury-vapor million-candle-power lamp * cyclones * weather symbols and abbreviations * anticyclones * synoptic charts * hygrothermographs * dry and wet bulb thermometers * 10-gram ceiling-measuring balloons * psychrometers * Robinson 3-cup anemometers * instrument shelters * maximum and minimum thermometers * facsimile machines * mercury barometers * microbarographs * altimeter settings * visual flight rules (VFR) * instrument flight rules (IFR) * eight-inch rain gauges * tipping bucket rain gauges * teletypes and their weather circuits * the taking and transmitting of weather observations * pressure gradient force * Coriolis effect * Buys Ballot's law * land and sea breezes * mountain and valley winds * PIBALS (for "pilot balloons": white, black and red balloons in 30- and 100-gram sizes) * theodolites and balloon tracking * plotting balloon angles to produce winds aloft * measuring cloud heights with clinometers * radiosondes * SCR-658's for tracking radiosondes * making hydrogen for balloon inflation * adiabatic lapse rates * plotting adiabatic charts * convection * advection * isobars * isotherms * isohyets * isotachs * the world's ocean currents, their names and directions of flow * etc. We had what seemed like a world of information to master and that in a relatively short time. A few minutes' lapse during which you had been wafted into slumberland by the instructor's steady-toned microphone could put you so far behind you'd be in great danger of not passing the end-of-subject test.

And so the time went by. We tried our best to absorb the material. During the 10-minute class breaks we lolled around on the grass. Those "who had 'em, smoked." I had never picked up that habit, and in fact, when the third week of the month rolled around I'd frequently be lending what little money I had to the smokers so they could buy their cigarettes. We talked about what our futures would be like. The draftees and the 18-monthers said we three-year men had found a home in the Army, implying that we never had one before. At one of those times, Andy replied, "One thing's certain, none of us are going to re-enlist," and casting a glance at me: "I know for sure Ted won't," and then, fixing me in his sight, "will you?" Of course, I had thought about it. I could see reasons why it might be a good idea. There were no good paying jobs on the outside that I'd likely qualify for. Industry was converting back to peacetime production. Instead of making B-29s, they'd be making Fords. Before joining the AAF, with Dad and brother-in-law Bob Nordstrom, I had applied for a job at the Ford Willow Run bomber plant where Kaiser-Fraser began building autos. But the returning veterans, as they should have, were having first choice at jobs and, excepting for brother-in-law Bob, we were never called up. By the time they called for him, Bob had given up on them. He was on active duty in Japan with the AAF's Air Transport Command. I knew I'd never again be hoeing corn, a job I had volunteered for as a boy to be of some help to Dad when things were tight on the farm. The sun was too hot and the rows too long. "Don't be too sure," I said in reply to Andy's question. "I just might be a 20-year man."

"I'll bet you twenty dollars you'll never re-enlist," Andy said. And we shook on it. A twenty dollar bet seems inconsequential these days, but it was a lot of money to someone drawing $50 per month. After graduating from weather observing school on July 26, 1946, Andy and I were sent to different bases, but we'd remember the bet as a bona fide obligation.

One weekend while at that school, Marie and I were witness to an unforgettable, even historic, event. On that day, we joined hundreds of residents from the town and surrounding farms to attend base open-house activities. While standing on the runway apron we watched in amazement as a flight of P-80 Shooting Stars silently dropped down for refueling. Nattily decked out with checkered scarves, the pilots were enjoying their moment of glory. They knew that none of us had ever seen a jet aircraft before. When the only aircraft you'd ever seen was the kind driven by airscrews, the P-80 was a sight to behold. It would be two to three years before jets would be so common in the Air Force that the only thing that would catch your attention when they were taking off or landing was the smell of their fuel which invariably made me think of the kerosene lamps and the kerosene kitchen stove back on the farm.

The Lockheed P-80, America's first production-line jet fighter, was built by Lockheed in 1944 and put into general service late in 1945. This one is on display at Holloman Air Force Base, Alamogordo, New Mexico.

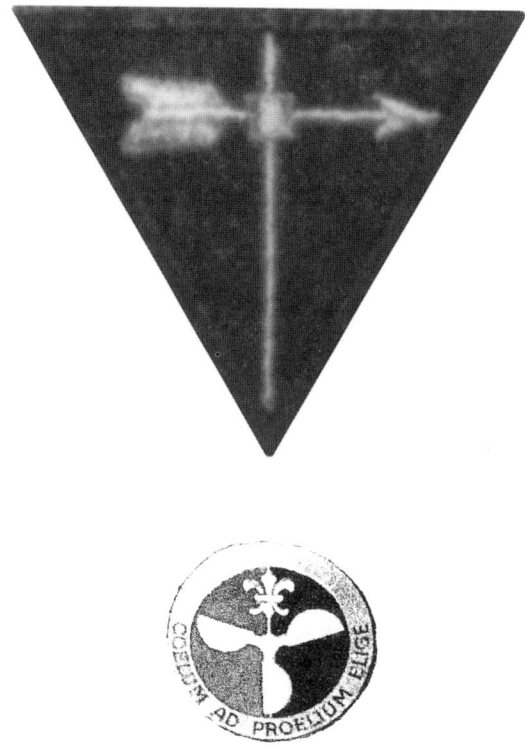

Shown approximately actual size are the AAF weather insignia authorized for enlisted men. The triangular wind-vane emblem, a cloth patch, was worn on the lower right sleeve. The circular pin, approved on 8 September 1942,[1] was worn on epaulets and the overseas cap. A *fleur-de-lis* on the pin, above the representation of the Robinson three-cup anemometer, commemorates participation of the US Army Weather Service in France during World War I. That was the first combat employment of a US Army weather unit.[2] In the anemometer's background, the left hemisphere is blue, the right is black, for the day and night operations of weathermen. *COELUM AD PROELIUM ELIGE* translates from the Latin as: "Choose the Weather for Action."

---

[1] William E. Nawyn and Rita M. Markus, *Air Weather Service: a Brief History, 1937-1991*, p.42.

[2] Lillian Nolan, Air Force Weather Association Historian, in article "Did You Know," *Observer*, Oct/Nov '99, p. 24.

As a private in the AAF and new graduate of the Weather Observer Course, while at home on his first furlough the author poses in May 1946 in the khaki summer uniform and then regulation brown shoes. Note the weather pin on his cap, the AAF shoulder patch, and the weather emblem on the sleeve. Is that serious look a reflection of the gravity of the meteorological instruction he has just absorbed?

Chapter 4

WETT

During one of those lighter moments in the classroom while we were getting hands-on training in transmitting weather observations over a teletype, one of the assistant instructors transmitted to our slave teletype "The Weatherman's Song," the words of which nearly every weatherman of that era soon memorized. I kept that yellow sheet of teletype paper. Sung to the tune of "McNamara's Band," these were the words:

THE WEATHERMAN'S SONG

Teletypes, synoptic charts
Anemometers spinning around
Our pressure lines are intertwined
Our fronts are underground,

The winds that blow
From high to low
Have blown me off the track
I'll have to throw my books away
And use the almanac.

Chorus:

We are the **men**
The **weat**hermen
We may be wrong
Oh now and then
But when you see
Our planes on **high**
Just remember we're
The ones who let them fly...

Second stanza:

I'll never forget
The day was wet
The general wanted to fly
He said, "My boy is it O.K.
For me to go on high?"

When I said, "No, it's going to snow"
You should have seen him frown
Say I'm the only one
Who ever kept the general down.

Repeat chorus

Two years later, when I acquired the Army's MOS 787 (military occupational specialty) of weather forecaster with the responsibility for briefing pilots and signing their flight clearances, I was cautioned to remember that I did *not* have the clout to keep a general on the ground. When briefing a pilot where hazardous flight weather was predicted, we could recommend that the flight not be attempted but could not prevent the takeoff.

But in the meantime, now that we had obtained the weather observer MOS of 784, we had our next assignments to ponder. I had a great desire to acquire all of the technical training that the Army Air Force might feel willing to grant me. There was at that time, two other courses that the new weather observer could volunteer for: the Radiosonde Course and the Weather Equipment Teletype Technician Course, abbreviated "WETT." There was also a course called "SFERICS" in which lightning discharges (atmospherics) were located by triangulation and their locations plotted on the weather maps, but that course was only occasionally convened and was not then available. I volunteered for the Radiosonde Course and was promptly told a radiosonde class had just graduated, and it was not known when the next one would convene. But if I wanted to, I could take the Weather Equipment Teletype Technician Course which was just then starting. I volunteered and was told I'd be in the next class which would convene in about two months.

While waiting for that class to begin, we were in what the Army Air Force called "casual status," which to the uninformed might mean that we only had to lay around sunbathing on the parade ground. Not so. To be casual meant that you were among the first to be tapped for those 16-hour days of KP duty. Altogether I spent 22 of those days scrubbing down the mess hall all the while thinking, "where were the German POWs when you *really* needed them?" It was an onerous duty, mainly because the mess sergeant could not make up his mind to let you go until it was past lights out at the barracks. You'd scrub the floor till it was squeaky clean and when you asked if that meant you could leave, the mess sergeant, not having anything else left to be done, would tell you to scrub the floor again, and again, and again. I desperately wished the trays that we ate off at meal times were as clean as that floor.

Eventually, on 23 September, we began the weather equipment class. That phase lasted only two weeks. We would learn such things as how the facsimile machine worked and that it was wise to not make contact with its 1,000-volt electrical circuit. The FAX machine is found everywhere these days and many probably think it is a fairly new invention. Actually, it has been around for well over a half century though its makeup has undergone a number of significant changes.

It was not new even then in 1946, but its makeup was considerably different from the current FAX. At that time, the image produced depended upon a slight burning or charring produced by strong electrical impulses that were directed onto a special gray-colored metallicized paper by a stylus that was mounted so that it was close to but not touching the paper. The stylus was stationary but the paper revolved on a drum that moved along a threaded bar. The on-off character of the electrical discharge at the stylus's point would produce a tiny dark dot on the paper (and a tiny bit of smoke) whenever the transmitting FAX scanned a dark spot on the image being sent. In that way, weather maps from a transmitting FAX were sent from the National Meteorological Center to weather stations across the country. The intensity of the signal would vary with the intensity (blackness) of the image being scanned, a feature not required for the duplication of weather maps but one used by the civilian wire services to transmit photographs. These "FAX maps" were entirely satisfactory but did have a few minor drawbacks. The charring produced by the stylus left a black residue on the maps that came off on the observer's and forecaster's hands, and the machine had to be phased by the observer in a hands-on procedure at the precise time that an audible alert signal sounded. If the observer were busy cutting and separating for display the teletype paper on which was printed weather observations from various weather circuits or if he were busy plotting maps or taking and transmitting weather observations over the teletype, he might miss the phasing signal and the west and east coasts of the US would not appear at the left and right edges of the FAX map but somewhere in between. The map would then have to be cut at that point and the two pieces taped together, returning the geography of the US to its more recognizable form.

We learned how to care for the various instruments in the weather stations, including the microbarograph, the hygrothermograph whose humidity sensor, we were told, consisted of strands of hair from blond Scandinavian women, and the balloon-tracking theodolite whose most critical adjustment was the precise setting of the line of collimation (its line of sight). Then, when the two weeks expired, we climbed onto a bus that took us about one hundred and fifty miles south to Scott Field, Illinois. Scott Field was the Training Command's center for teletype maintenance instruction.

But before leaving for Scott, I had a 15-day delay enroute (leave) between the two courses, and with the uncertainty of the impending move in the offing, Marie would return to her parents home in Michigan. We left by train for Michigan. Boarding that train was a bit of a memorable experience. We bought a train ticket in Champaign and then boarded a bus that would take us, we were told, to the train station at a place 7 miles south called "Tolono."

That "station" was just the beginning of an unusual travel situation, for there was no building where the station was said to be. Instead, we were dropped off the bus beside railroad tracks that literally had been laid in the middle of a corn field. With Pfc. (private first class) Fetherolf, a fellow weather observer graduate, who was from Detroit, we waited. After about an hour we three spotted a locomotive's smoke rising from above the corn tassels and somewhat to our surprise, for we had begun to think we had been taken on somebody's idea of a different kind of snipe hunt, the train screeched to a stop alongside. When we boarded we found every seat occupied by passengers who had boarded in St. Louis. The only thing to do was to sit in the aisle on our suitcases, which was what we did for that entire trip to the Michigan Central Station at Detroit, a rocking ride in which we did our best to maintain our balance, leaning our bodies from side to side to counteract the car's repetitive side-lurches over that distance of well over 300 miles. An elderly gentleman offered his seat to Marie, but as was typical of her, she would not deprive anyone else of his comfort. Along with Fetherolf and I, she endured that ride on the suitcases without complaining. Previously, I knew of Fetherolf only in passing, but we got to know each other quite well on that trip. In a little more than a year, Pfc. Fetherolf would die in the crash of a four-engine transport plane, a C-54, while on takeoff from Goose Bay, Labrador. He had been sent to the 8th Weather Squadron's weather observing site at Ft. Chimo in cold northeastern Canada. Fetherolf, and a few other weather observers were returning from the northern outposts where they had been on isolated duty. This was about a week before Christmas. It is my recollection that they had been released from their observing duties slightly before their 12-month tours were completed so that they could be home for the holidays.

We had a wonderful furlough — lengthy visits with both of our families. In their home near the state fairgrounds in Detroit, my mother had placed a blue star in the window which was the symbol that someone from that family was in the armed forces, a holdover from the practice followed during the war when many such windows were so decorated. (No one wanted to see a gold star so displayed. A gold star was displayed in the window of a home of a serviceman who had been killed in the war.) We had lived there during my last year in high school, the John J. Pershing High School. While waiting on State Fair Avenue for the city bus to take me to school, I often observed a lone soldier in his long olive drab overcoat with a rifle over his shoulder, patrolling the fairground's cyclone fence line. For that was then the POW impoundment for Italian prisoners of war who had been captured in the fighting in North Africa. On Sundays, some of the local patriots were scandalized by the Detroit girls of Italian descent who would set up their portable 45 rpm phonographs on the sidewalk immediately outside the fence and play Glen Miller tunes for the Italian prisoners. I had no problem with the girls' attempts to make life a little more pleasant for the prisoners, for it seemed clear to almost everyone that few Italians would have entered the war if they had had anything to say about Mussolini's dumb decision to join the Axis Powers as Hitler's partner.

A few weeks after the teletype technician phase began at Scott, Marie would rejoin me. In the meantime, while waiting for the class to begin, I lived in a World War I type barracks. We had a lot of idle time on our hands which the duty sergeant would try to fill with squadron details. This consisted mainly of feeding coal to the furnaces that kept the barracks warm. Scott, at that time, burned soft coal in its furnaces. Soft coal was the cheaper fuel, but it was particularly nasty on those cold fall and winter mornings for it would produce sooty flakes that were held within the morning's temperature inversion and then softly settle down, all over the squadron area, producing a black dandruff on the shoulders of your light-beige colored cotton khaki uniforms. That fall, it turned cold early, and we would have been much more comfortable wearing our winter olive drab uniforms that were made of wool. But the Army had a set policy. The changeover from summer khakis to winter OD's had to be made on a date set in stone by some obscure policy-maker, who was probably quite comfortably officed at some headquarters. There was also a set date for starting the furnace fires which did not then coincide with the changeover in uniforms. While still in our summer khakis and the furnaces not yet fired, we attempted to keep warm wearing our green field jackets even while inside our barracks home.

Remembering the lesson of the recruits at Ft. Sheridan who had been nowhere in sight when the duty sergeant came around looking for someone to fill his duty roster, I would leave the barracks and wander about Scott Field. On a few of these wanderings, I visited a recruit whom I had known back home in Michigan. He was very glad to see me and seemed somewhat perplexed that I was then still considering the possibility of a career in the military. The Army was not going to be the life for this fellow. Along with all the others in his barracks he was awaiting his general discharge, a new form of release from the military that had been instituted for those who were deemed to be unfit for the Army but not so unfit as to be a candidate for a dishonorable discharge. He and his associates, I surmised, were the principal reason those of us in casual status waiting for our class to begin were not tapped for KP duty. They obviously were pulling the KP shifts that otherwise would have been ours.

Meanwhile, we prospective students had exhausted our funds. With time on our hands, we thought it would be an opportune time to take in the sights of nearby St. Louis. We could take the base bus to St. Louis. That would be free of charge. But once there, without money we'd have nothing to do. The move to this base, just as had happened when we moved from

Keesler to Chanute, had apparently not been made known to the finance office. We had money that was due us, but we were not yet on the base payroll. Then, one of our number, Pfc. Fagan from East Orange, New Jersey, suggested that we go to St. Louis anyway. His plan: we could appear before the ladies running the USO in the Kiel Auditorium. They would have tickets to certain functions that they would give us for free. Four of us decided to accompany him.

The kind lady at the USO desk explained that the only free tickets she had were for a Quiz show called the Sinclair Quiz Club at Radio Station KMOX. Being a little early for the show, we killed some time in the auditorium, cutting a 45-rpm record in a recording booth. The four of us sang and recorded the weatherman's song. We were terrible. (I kept that record as a historic and sentimental treasure, not daring to play it but hoping that over time our voices would mellow out. About 20 years later, I did work up the courage to play it. Then, for the good of posterity, I tossed it.)

Somewhat dubious that this would be entertainment, but having nowhere else to go, we left, nevertheless, for station KMOX. At the doorway, our tickets were placed in a hopper along with those of all the others in the room which, surprising to me, was filled almost to capacity. This was in the era of the nationally aired quiz program called the "64 Dollar Question," and quiz programs had sort of caught the public's fancy. (Not the "64 *Thousand* Dollar Question," that expansion of the original was aired somewhat later.) As a preliminary to the actual broadcast, the emcee explained that five tickets out of the hundred or so in the hopper would be pulled and four of the five, with the fifth as an alternate, would be the contestants. For each question answered correctly, the contestant would be given a five dollar bill. A cleanup question given at the end, if answered correctly, would negate any previous incorrect answer so that the contestant could walk away with a maximum of 25 dollars. My name was drawn and I was led onto the stage as one of four contestants. With each correct answer that I gave, the emcee hit my open hand with a crisp new five dollar bill, the sound of which seemed almost as loud as Dad's ace hitting the table in a pinochle game. The crack of the five-dollar bill was immediately followed by the whoops and hollers of my four associates in the audience. I was doing well even with questions that seemed somewhat esoteric. But I stumbled at the fifth question which seemed to most to be the easiest — it was to give the price of the current air mail stamp. Having been sending all my letters to Marie by regular mail which did not then cost the soldier anything — he merely wrote "free" on the envelope — I was totally unfamiliar with the postal rates. So I missed that question. The answer: five cents. But the cleanup question — to give the name of the general who was known as the "General of the Allied Armies" — was easy. The Detroit high school that I graduated from was named after that general — John J. Pershing.

So we took our 25 dollars and had a great time in St. Louis and that would have been the end of this little story except that, on the very next weekend, being once again broke, our friend from East Orange suggested that we try for a repeat performance. So we all reported again to the USO, got tickets to the quiz program and, in a truth is stranger than fiction scenario, during the random drawing prior to the broadcast, out of that studio filled with people, *my name once again came up.* But the emcee, probably thinking that his listeners would believe the contest was rigged or perhaps simply wishing to avoid boring the studio audience and his radio listeners with this particular contestant, relegated my participation to sitting on the stage as an alternate with no chance at the five-dollar prizes. In the studio audience, heavy gloom descended upon the four wearing olive drab. As a consolation prize, the emcee presented me a Reynolds ball point pen, a not insignificant gift in 1946 as the ball point pen was then very new and greatly prized as a new-fangled substitute for the common fountain pen.

One beautiful day I met Marie at the Union Station in St. Louis, a fine old building with a black and white tiled floor and oak woodwork. (The station is now a shopping mall.) We bought a St. Louis newspaper and a sack lunch and walked over to a bench in a small park across from the Kiel Auditorium. While enjoying our lunch and looking through the want ads for a room to rent, an enterprising reporter from the Scott Field newspaper snapped our picture. He printed that photo, in which our faces were hidden behind the paper, on the front page of the next issue of his paper under the headline: *Behind the Headlines.*

We found a room in a large turn-of-the-century house in Belleville, Illinois, the nearest town. Another room had been rented to a very nice middle-age couple from Boston. He was an executive in a shoe manufacturing company in St. Louis. Just before my transfer to Scott, I had been promoted to Pfc. The promotion brought with it an increase in pay to $55 per month, a welcome increase but still not enough to put us in a position to buy even the most ancient automobile. The middle-age couple, however, had a fairly late model Chevy. On weekends, they took us to St. Louis. We viewed the attractive river-cruise ship, the *Admiral,* at anchor in the Mississippi, watched the swans at Forest Park, and reflected on how lucky we were to be where we were, to have such friends and to be looking forward to a future in the AAF's Air Weather Service and, taking the broader view, to be in the up and coming meteorological field which had been so greatly expanded during the war.

The course work at Scott was conducted in quite a different way from Chanute. After a week or two of elementary electronics in a normal class room setting where we learned the relationships flowing from Ohm's and Kirchoff's laws, we settled into hands-on training in a room full of teletypes. The machines we worked on were capable of 60 words per minute. A year or two later, in the weather station I was then assigned to, the machines were upgraded to 100 words per minute.

We dismantled and reassembled the machines and marveled at the genius who, in 1904, had invented that immensely complicated electro-mechanical device. Though probably not true, we were told the inventor had gone to the funny farm after inventing it. In addition to other components, we learned to identify and disassemble the code bars, the bell cranks, the latch, the pawl and the latch pawl latch. Actually, I think it possible that I misheard the instructor when he named that last part. It may not have been called the latch pawl latch. But even if fictitious, I like the name. Even now, like a tune that plays in your head at the sound of a word, I automatically think of the latch pawl latch when someone mentions "teletype."

Our tests consisted of returning to working order a machine that the instructor had maladjusted with a few turns of a screw driver or by cleverly hiding a small piece of paper between electrical contacts. Searching for the screwed-up components in those machines made us think the inventor really had gone bonkers after inventing that mind-boggling device. And then, on 23 November 1946, when it seemed we had learned more than anyone wanted to know about teletype machines, we graduated and were awarded the additional MOS of 790, weather equipment teletype technician. Marie and I spent Christmas at home, and then I was off again, without her, this time to report to the Continental Weather Wing at Tinker Field, Oklahoma.

On one beautiful fall day in 1946, Ted met Marie at the Union Station in St. Louis. They bought a paper and walked to the park across from the Kiel Auditorium. While they searched ads for an apartment, a reporter for the Scott Field paper snapped this picture which he ran on his front page under the title: "Behind the Headlines." Note that the newest auto in the background appears to be a 1939 or 1940 model. All auto production ceased in 1942 and did not restart until 1946. *Official photo USAAF, AAF Training Command, 3305 Base Unit AAF, Scott Field, Illinois, October 1946.*

A roving sidewalk photographer snapped this picture of Ted and Marie on the main street of Belleville, Illinois in November 1946.

Chapter 5

From Tinker to St. Louis to Grand Central[1]

"Lose the service records of any of these men and the Army will hang you!" said the staff sergeant in charge of the Continental Weather Wing orderly room. The staff sergeant, he with the sleeves adorned with three chevrons above and a rocker below, and considered at that time in the development of the U. S. military as ranking quite close to God, was speaking directly to me. The staff sergeant was in charge of troop departures. Nowadays, when someone says: "Do this, or don't do this or they'll hang you," we laugh because we know it is just a silly attempt at idle conversation. But at that time in our national history, with the desperate measures born of World War II still occupying the mind, a young person would take that remark to mean they *would* hang you if they felt like doing it.

In addition to the service records of 10 privates and four Pfc.s (including myself), the staff sergeant made it quite clear that I had been made responsible for their bodily presence. My mission: to deliver the 13 soldiers and their service records to 8th Weather Group Headquarters at Ft. Totten, New York. Ft. Totten was 1500 miles away. Trying to find the bright side of that situation, I thought it was good that the staff sergeant had not said I'd be hanged if I lost any of the 13 men, that would happen only if I lost one of the 13 service records (14, counting my own). But then, he probably just forgot that part of my instructions.

You may wonder, as I did, "Why me?" Why was I the lucky (!) one to be put in charge? The answer was, really, quite simple. The man in charge would be one of the four Pfc.s. I was one of those. The other Pfc.s were: Ralph F. Eberle, Anthony Rendulic Jr., and Glen S. Wood. Granted that rank dictated the man in charge had to be one of the Pfc.s, I thought of asking "why *this* Pfc.?" but immediately thought better of it. It all went back to the time-honored Army practice of seeing things in alphabetical order. Alphabetically, Theodore L. Cogut, *me,* would outrank the other Pfc.s. Selection by alphabet also was the reason I had pulled more than my share of KP and other disagreeable squadron details while in basic training and while in casual status at Chanute. I often thought of how much easier life could have been if the civil servant who checked Dad through Ellis Island would have spelled his last name with a K, as it was spelled in the Old Country, and not with a C as that official apparently thought the Americanized version

---

[1] If you see a resemblance in this title to a famous baseball double play combination, be assured there is a connection but only in the similarity of the difficulty of executing that play to the difficulty of accomplishing the task assigned the author as described in this chapter.

should be. I figured that "Cogut" spelled with an initial letter K would have gotten me out of at least 30 days of KP.

This was January of 1947 and Tinker Field was not only a temporary staging area for weather people but also for those veterans of the war in the Pacific, the sleek B-29 bombers. Aside from an MP detachment (military police) we relatively few weathermen were the only ones there. Tinker Field, abandoned by units that had once been there, looked like a ghost town.

Acres of B-29s, carefully parked wing tip to wing tip, stood in silent rows. As I observed those still proud, shiny birds, the largest remaining collection of the 3,965 that had been built during the war, my mind drifted to the time, only two years before, when Marie and I had worked in the Detroit war plant as an inspector of B-29 motor parts. Detroit was then known throughout the world as the "Arsenal of Democracy." Its main automobile plants and the smaller plants that supplied them with certain components used in auto production had all converted to producing war materials. At adjacent Willow Run, Henry Ford built the world's largest manufacturing plant where he turned out B-24 bombers. My future father-in-law, Walter Nordstrom, worked at the Dodge plant making Army trucks. Detroit was a city that was intent on, and knew it could, defeat the Axis powers. When I graduated from high school, Dad, who had left the farm in 1942 to work as a diamond bore machinist in one of those war plants, suggested that I might want to try to get hired at the same place.

The tenor of the times dictated that you do something in support of the war effort. It was a fervently held belief, and it swept the country. If you couldn't serve in the armed forces, you could work in a war plant. I told Dad I'd try for the job. He knew I had been wanting to join either the Army's or Navy's air services, and I like to think he was relieved that I wouldn't be offering myself up to be killed. But I had misgivings, why would they want to hire a fresh high school graduate for so important a job? Nevertheless, I borrowed Dad's '40 Ford and drove to the war plant, being careful not to exceed the nation-wide war-time 35-mile-per-hour speed limit which had been imposed to conserve critically needed gasoline and tires. On this trip, which was within the city, keeping within that speed limit was easy. But if you were driving out in the country in that car that you knew could do 100 miles per hour, because on one day when you had lost your good sense you *did* push it to a hundred, 35 miles per hour was a killer. But you did it. It was, after all, only a small inconvenience. And besides, like the war time poster "A Slip of the Lip Can Sink a Ship," you didn't want to be the weak link that could be responsible for losing the war.

Upon arriving at the plant, the security guard gave me a temporary pass and directed me toward the employment office, saying I'd get a photo badge if I were hired. High on a flagpole above the gate Old Glory reached out for the breezes and beneath it flew the Army-Navy E-flag for efficiency, a banner that was awarded to war plants that were demonstrating great efficiency in work critical to the war effort. At the employment office, I was given an application which I promptly filled out; and as an indication of how desperate were the times, I was hired as soon as I handed in that application.

And now, here at Tinker, I wondered how many of the B-29s carried the motor parts that Dad had made and Marie and I had inspected? I knew it would not be long before those magnificent machines would be reduced to scrap. That wonderful example of American ingenuity would be known only in memory or perhaps seen by just a few as they walked through a museum.

One museum at which the B-29 figured prominently was the Air and Space Museum in Washington, DC A few years prior to this writing, rabble-rousers who either were too young to have had to face the life or death consequences of World War II or had managed to avoid them, staged protests regarding that B-29, the *Enola Gay,* which had been piloted by Brigadier General Paul Tibbets when he dropped the atomic bomb on Hiroshima on August 6, 1945. To think that anyone would even entertain the thought of an apology for dropping the atomic bomb, considering that it was the Japanese who had initiated the conflict that had caused the deaths of thousands of Americans as well as peoples of other nationalities and of the Japanese themselves, is beyond comprehension. I say this while thinking of the Japanese friends I have known since the war. More polite and friendly persons could hardly be found. The bomb, terrible as it was, actually saved the lives of additional thousands of Americans and Japanese, lives that would have been lost in an invasion of the Japanese home islands.

There is only one B-29 that flew in World War II that remains as it was when it ran on routine bombing runs of Japan. It is *K-40*, on prominent display at the Pima Air & Space Museum in Tucson, Arizona. Named by its crew *Sentimental Journey* and also *Quaker City,* it flew 32 bombing missions against Japan and shot down one Japanese Zero fighter. On one of these missions, it lost seven feet of its vertical stabilizer that became lodged in the bomb bay doors of a B-29 flying directly above it. With the only directional control being the ability to alternately change power in the right- and left-hand engines, the crew miraculously brought it back safely to an American base. In March, 1999, we were fortunate enough to have been in the museum audience when George Litzenberg, whose uncle was one of that war-time crew, gave an inspiring talk documenting the history of B-29 *K-40.* We enjoyed it so much we returned to hear his presentation two additional times. George wrote a poem regarding *K-40,* copies of which at this writing are available from him at the museum. With heartfelt thanks to George Litzenberg, we are repeating his poem:

Where did all the years go?

We left as boys and came back as men,
Let's hope no one has to do that again!
We took a plane brand spanking new,
Started as strangers and came back a crew.
We were young and slim, our backs were straight,
Our eyes now dim, we know our fate.
We had to go, a job to do,
Our friends all went, we had to too.
Now years have past and soon we'll rest,
The whole world knows we did our best.
Above the clouds our spirits will soar,
When life is o'er, it's through the next door.
We'll join up again on another plane,
Take off for the heavens, a crew again.*

George Litzenberg
February 19, 1998

For a long time now, Tinker Field (redesignated Tinker Air Force Base in 1947) has been surrounded by the Oklahoma City metropolitan area. But in January of 1947 it was in an agricultural setting, a few miles outside "Ok City's" city limits. It was a pleasant base made even nicer for us by the fact that the guard house contained a number of involuntary residents who were tapped to perform KP duty, thus exempting us from that onerous chore.

But now we were leaving Tinker. We were eager to finally be assigned to a base where we could practice the craft of weatherman. We proudly wore the triangular blue and orange weather patch on the right hand sleeves of our battle jackets and the weather service pin on our epaulets and overseas caps. What was lacking in our lives were the "teletypes, synoptic charts, anemometers spinning around...." You get the picture.

---

* The crew of *Sentimental Journey:* 1st Lt. Lester E. Gilbert, Aircraft Commander; F/O Lawrence C. Bennett, Pilot; 2nd Lt. Kenneth D. Hurley, Navigator; 2nd Lt. Orley W. Van Dyke, Bombardier; 2nd Lt. Jonas Carpenter, Radar-Bombardier; M/Sgt. William T. Prim, Flight Engineer; M/Sgt. Theodore A. Lewis, Central Fire Control; S/Sgt. Joseph W. Le Bon, Radio Operator; S/Sgt. Wencel J. Bohr, Left Gunner; S/Sgt. Norman A. Whipple, Right Gunner; Sgt. Robert W. Washam, Tail Gunner; T/Sgt. Harry L. Temple, Crew Chief; Sgt. James B. Keith, Assistant Crew Chief.

We gathered at the rear of the deuce and a half, the 2 1/2 ton truck, and were about to climb aboard when one of the Pfc.s said: "Private Zee (not his real name) is missing." I'm not sure but memory tells me Tony Rendulic, who had entered the AAF from Pennsylvania, was the one who brought me that bit of unpleasant news. I think it was Tony because he was the one who could be relied on to know what was happening. Like me, he was a three-year man. Cool, steady, a soldier with a clear view of where he was going, I was impressed with his conviction that he would reenlist at the end of his hitch. He told me once that he was going to be a 20-year man, a comment that none of the rest of us could yet bring ourselves to make or even envision seriously. Of course, I had thought about it, but I could see many downsides to the prospect. Tony and I would lose track of each other after this trip, but I would meet him again about 10 years later when I was stationed at Selfridge Air Force Base in Michigan as a weather forecaster and by then on my way toward being a 20-year man. He had remained in the maintenance field and was well thought of at Selfridge.

We stood there for about 10 minutes, waiting for Private Zee to make his appearance, when the truck driver walked up to me saying: "You and your men are going to miss the train if we don't get out of here *now*." So, in my best command voice I announced: "Let's move out!" and we jumped on the deuce and a half. Half of my mission, that to deliver *all 13 men* to Ft. Totten, New York, was already in a shambles. I patted the breast of my battle jacket to reassure myself that the 9x12 manila envelope I had tucked inside it, the envelope containing the service records, was still there. It was. At least, I was still safe from an appointment with the hangman's noose.

With a sigh of relief I also noticed that the white legal-size envelope containing the train tickets and the meal tickets was also safely tucked within the battle jacket. I pulled that envelope open and examined it for the first time. The outside was marked as follows:

TRAFFIC SECTION

WAR DEPARTMENT

ARMY AIR FORCES

    Name of Traveler Pfc <u>Theodore L. Cogut & THIRTEEN OTHERS</u>

                                  From <u>OKLAHOMA CITY, OKLAHOMA</u>

                                  To <u>BAY SIDE LONG ISLAND, NEW YORK</u>

                                  Via <u>SLSF: NYC:LIRR</u>

"SLSF" stood for the St. Louis and San Francisco Railroad, "NYC" was the New York Central Railroad and "LIRR" was the Long Island Railroad.

My only other experience as part of a troop shipment by railroad had occurred almost a year earlier on that troop train that had taken us from Ft. Sheridan to Keesler Field. This time, we'd be going by commercial rail, free from the military restrictions that had been imposed on us on that previous trip. The white envelope contained tickets for five meals, each ticket indicating that the meals were for myself and 13 men. And this train had a Pullman car. The train tickets, titled: "THE UNITED STATES OF AMERICA, WAR DEPARTMENT — TRANSPORTATION CORPS, indicated they were for berth assignments in car 151 out of Oklahoma City. There was no doubt that the accommodations would be infinitely better this time. Unlike the smooth hardwood seats of the troop train, we'd have upholstered seats. Nor would we have to sit on suitcases in the aisle as Marie and I had done on that trip from Tolono, Illinois to Detroit. Instead of box lunches such as we had on the troop train, the meals on this trip would be taken in a dining car at tables with white table cloths. There would be waiters who would see to it that we were happy diners. After enjoying the meal, we'd sip coffee; and through spotless picture windows, like a tycoon on a trip to inspect his far-flung properties, we'd admire the scenery that floated by. And then, when it was time to hit the sack, a kindly porter would come by to make our beds, and with car 151's gentle rocking, we'd be lulled into sleep while visions of home fires flitted through a succession of velvety dreams.

As soon as we had all clambered aboard the truck, we signaled to the truck driver that we were ready to go and he rolled us out of the barracks area with impressive speed, heading for the Ok City railroad station. We slid around a few corners of the barracks streets, bounced a bit and then someone yelled "Stop the truck, Zee's coming." And so he was, running with the speed of a lion hunter who has missed with his last rifle shot, Zee caught up to the truck, climbed aboard. I had been prepared to soundly chew him out in the unlikely event that I should see him again, but I was so relieved to know that I'd be able to complete the mission in what would appear to be fine style that I merely smiled at him, acting altogether as if I was happy to see him, which, of course, I was.

At the Oklahoma City train station we hopped out of the deuce and a half, slung our barracks and duffel bags over our shoulders and rout-stepped down the station platform. I brought the men to a halt as we arrived alongside car 151. "This is going to be a long trip," I said. "On our way to Totten we'll be changing trains in several stations. It'll be easy to get separated in the crowds. So stick with the group. If I lose you, you'll be on your own — without transportation and without meal tickets."

I could see they were listening attentively to my little speech, and that was indeed heart-warming. Though still not much more than recruits, it was quite evident these men were real soldiers. But as I stepped up into car 151 behind the last of my group, there, behind me, running along the platform to join us, was Private Zee. He had dropped out of the group when we jumped from the truck and while the rest of us were making our way to car 151, he had wandered into the station, taking in the sights. Of course, he had missed my entire presentation about not getting lost. "Don't over-react," I said to myself, drawing upon my farm-bred sense of optimism, "things can still go well."

And they did — for awhile. In fact, as the train eased away from Ok City, we felt as if we had fallen asleep and awoke in fantasy-land. The train had barely reached its normal speed when a helpful porter sought me out to brief me on the meal ticket routine. He ushered us into the dining car. We actually had a choice of menu items. That was something we were not accustomed to. When the food arrived, we found it to be excellent and the service was superb and this even though the waiter knew that young soldiers wouldn't have money for tipping. Through the windows a distant ranch house lay nestled near the thin, silver reflection of a winding creek. Smoke lazed from its chimney and the rancher's cattle speckled the sides of the rolling, beige-colored hills.

After dinner we walked back through the cars to our familiar car 151 where we waited in the aisle as a porter prepared our berths. Someone was actually making our beds! You had to be a young GI to appreciate *that.* Now there were stars over the prairie and an occasional moving light below — a car moving along a section-line road. And then, off to dreamland.

Then, in the morning, breakfast with the same superb quality of food and more viewing of the countryside where the prairie had given way to woodlands, mainly oaks and walnut trees. We passed through a number of small towns, and then the hitches connecting the cars clanged from the first car to the second, to the third, clanged from car to car on down the line to our car and then to the cars following behind us. An application of brakes, a sudden wrenching that almost toppled those standing in the aisles. It was not a wreck, just the deceleration as we entered the rail yard at the Union Station in St. Louis.

As the train jockeyed for its parking position, the conductor pulled me aside. "We're going to have a four-hour layover," he said. "If you want to, you can let your men go into the city."

I thought that would be nice. I thanked the conductor and told the men they could go for a little sight-seeing if they so chose. Except for the Pfc.s who were graduates of the Weather Equipment Teletype Technician Course at Scott, most had never been to St. Louis before. The others, the privates, were graduates only of Chanute Field's Weather Observer Course and would not have had as much opportunity to have visited that city. I knew they would enjoy the break from our travel routine.

But before dismissing my charges, I gathered them together just as I had before we had boarded car 151 in Oklahoma City. "The train will leave at 1400 hours," I said. "Be back here no later than 1330. Now go out there and enjoy St. Louis." They all affirmed their understanding of the departure time and assured me they'd be back on schedule. All of them, that is, except Private Zee who had left on the run as soon as the words saying they could leave were out of my mouth. Once again, he had missed my pronouncements.

I ambled off into the familiar surroundings of St. Louis's picturesque Union Station. This was where I had met Marie just a little more than three months earlier on the day she came in on the train from Detroit and where we had caught the train back to Detroit to be home for Christmas. With the uncertainty of the location of my next assignment, she had stayed home while I was at Tinker.

A lone soldier in a railway station seemed an easy mark to a zoot-suited pimp. Eyeing me from under a maroon-colored hat with a brim about half a foot wide, the zoot-suiter appeared to be in danger of tripping on his ridiculously long gold watch chain that dangled down toward sharply pegged trousers. He wore a jacket with extra-wide lapels and about a half foot of padding at each shoulder. In length, the jacket resembled an overcoat, reaching down below his knees. Pausing to give me adequate time to take in that grand appearance, he eased pointed-toed shoes toward me and said he had, "Girls, lots of girls, all kinds, all colors of girls." With the courage born of the boxing successes I had in high school where I had learned there was advantage to be gained with the aggressive stance, I told him to leave while keeping an eye out for a switch blade which we believed zoot-suiters carried. If he had a switch blade, he did not show it and chose to discontinue the meeting. (Although the zoot-suit fad was then about to end, about three years earlier it was at a peak. Despised by many servicemen probably because they had, for whatever reason, escaped the draft, the zoot suiters were then engaged in a riot by about 2500 soldiers and sailors in Los Angeles.)

I settled down on one of the oak benches and was joined by some of the troops who, wanting to be sure not to miss the train, were returning early. At 1300 hours we began a leisurely walk along the station platform toward car 151. Along the way, others of our group caught up with us.

Back in our seats in car 151, I began to tick off the names of the troops that had returned. Five were still out, but there was plenty of time. Four of the five joined us at 1350 hours. They regaled us with stories of the great time they had in St. Louis. The missing man, of course, was Private Zee, and I held out little hope that he'd make it back in time. At 1359 I presented my missing man problem to the conductor. "Don't worry about it, sergeant," he said, "we can hold the train for a little while to give him a chance to get back."

"Thanks," I said and, pointing at my one stripe: "for the promotion. I'm only a Pfc."

"To me," the conductor replied, " you're a sergeant. Anyone who'd take on the responsibility for this many men for as far as you've got to go, must be a sergeant."

Well, of course, the kindly conductor, a man of about 55, didn't know the Army as we did. He didn't know the Army had put me in charge because my last name began with a "C." But to think that, just to keep me from losing a man, he'd hold up the entire train! Trains were known for keeping to schedules at all costs. They usually got to their destinations at the time they were scheduled to arrive. That was the "law" of the railroads. I had to remind myself that the "big one," the most destructive war in history, had ended only a little more than a year earlier. That could account for the willingness to hold up a train for just one soldier, but it was still almost unbelievable.

As we waited, trying to make small talk, the conductor said, "One of your men said your next stop is Grand Central Station. Have you ever been there?" I had to say that I had not and waited to see if the conductor would enlighten me. All that he said was this: "It's a mighty crowded place."

At 1410 the conductor reached into his vest pocket, extracted his gold railroad watch, flipped open its hinged face cover, studied the time, then said, "We'll wait a little longer."

At that moment, I wanted nothing more than to have Zee face the consequences of being AWOL. At the next moment, I wished he could make it back so I wouldn't have to face the commander at Ft. Totten to report him missing. I didn't want to have to explain that, somehow, in this non-combat situation I had managed to lose a man.

At 1415 the conductor again examined his watch. "We'll wait five more minutes," he said. "Five more minutes." And looking directly at me: "If he doesn't get here. The Army will know you did all you could." Well, I wasn't too sure about that.

At 20 minutes past the scheduled departure time, the very moment to which the conductor had made his last extension of the train's departure time, as if he knew how much extra time he'd be granted, Zee bounced into car 151. He slipped into his seat, made himself comfortable and looked around with a somewhat disgusting expression, as if to imply this was some train — it didn't even run on time.

I leaned over his seat. "Lose your watch?" I asked.

"No," Zee replied. He looked as innocent as a new-born babe. Then seeing that some elaboration seemed to be called for, added: "I had a mother in St. Louis." I wasn't sure if this meant he had visited his mother in St. Louis or if he merely once had a mother who lived in St. Louis. To think that he even had a mother who would claim him would have come as a surprise to most of the passengers of car 151. If he did have a mother, I was convinced she could not then be with the living. He would surely have driven her to an early grave.

With that little crisis now past, I walked through the car, chatting with the men, letting them know how much farther we'd be going before stopping again. Then, settling in my seat, I dozed fitfully, insecure in the knowledge that I would surely face a test of leadership in New York City.

Grand Central Station — "The crossroads of a million private lives." I thought of that introductory line of a popular radio program as we walked through the crowds. Reconciled to the inevitable, I was sure I would lose Zee here and, this time, never find him again. And now I thought that wouldn't be so bad. There was no way I could keep him in tow through the milling throng, and I would no longer spend any thought on the consequences.

But it didn't happen. And that, I realized, might have been predicted because Zee would not be doing the thing that others thought he might be doing. We gathered in a group, stationing ourselves at one spot within that magnificent building, and waited for transportation to Penn Station. Zee sort of kept himself in sight, about a few dozen paces from our group, disappearing from our view for a minute or two at a time as the dense crowd, men and women and a few children held close by nervous parents, hurried past.

I began to relax, thinking that his experience at St. Louis could have taught Zee a lesson. I was thinking: "This is going to be O.K. after all," when two white service caps, the instantly recognizable headgear of the military police came into view above the mass of people. "Is that one of your men?" they asked, nodding at Zee. I had been thinking of Zee's unusually good behavior and was smiling. The MP's, all business, were not smiling.

"Yes...why...?" I couldn't see what he could have done that they had spotted and I hadn't. For the first time on this trip, even though he had given me trouble enough, I dared to think that if he continued his present behavior I could, as unbelievable as it might seem, actually be pleased to have the rascal in the group.

"Get him straightened out," one of the MP's said. Zee's infraction was so obvious they knew they didn't need to say more. But what was it? For about the first time during this trip he seemed to be doing what he was supposed to do, and this was at the crossroads of a million private lives where one might expect him to be trying to be a showoff. At that moment, in his best anti-military manner, Zee adjusted the angle of his cap so that it rested nonsquarely upon his head. He could see out from under the cap with only one eye, vision from the other eye being blocked by the rakish angle of the cap's shiny visor. He obviously thought he then exuded an overpowering charm which could be employed to attract the young ladies who hurried past with the crowd. That was it. The cap. Zee was out of uniform.

  The rest of us wore the overseas cap, the one you could fold and loop under your belt when indoors. The overseas cap was then legal headgear. The service cap, the one with the visor, which Zee was wearing, was not (although it would be legal in a year or two). Most of us actually preferred the service cap which could be purchased off base at Army and Navy surplus stores. We thought it just looked better and would give some protection in the rain. Some of the recruits would purchase the service cap and would toss it on the floor and stomp on it until it lost its flat top and crisp edge, making it resemble the pilot's "fifty-mission crush" cap. While home on furlough or pass where the MP's would not likely be, the recruit would strut about under that cap and all who saw him, no matter what they thought of him in civilian life, knew that the soldier had made his mark in the Army Air Force.

  I turned to the MP's. "Thanks for pointing that out," I said. I'll take care of it." And I did. In language whose meaning could not be mistaken, I told Zee to shove that hot rock head gear back in his bag and keep it there. He did.

Top: A B-29 weather recon ship at Dhahran Airfield, Saudi Arabia in 1951. *Bottom:* B-29 *Sentimental Journey* on display in its hangar at the Pima Air Museum in 1999. From left to right: Howard "Henry" Klier (he was stationed with the author in 1947 at Westover Field, Mass.), Anne Casaday (the author's sister), Ardis Klier, Marie Cogut, the author.

Chapter 6

8th Weather

We were waiting again. This time at Penn Station where we would connect with a Long Island Railroad train that would take us to Bay Side, Long Island and Ft. Totten. Penn Station, particularly for a Michigan farm boy, was a sight to see. Covering more than seven acres, its architectural lines, it was said, had been modeled on the baths of the Roman Emperor Caracalla.

It was there, within Penn Station's stately structural designs, amidst the rushing throng, that one of the privates, who had been carrying a silent guitar throughout the trip, finally became inspired to strum a few tunes. I thought he could have been waiting for a properly upscale audience. Though he might eventually have become quite exceptional, the objective music critic would probably then have considered him some steps removed from world class; but as he picked at that guitar beside our pile of duffel and barracks bags, we knew he was nothing short of great. In that little corner of Penn Station we joined him in song and were able to produce enough good notes to keep the bad ones from being overly obvious. Our singing was pleasing enough to stop passersby, at least those who were not in exceptional hurries. They gathered around us, applauded after each performance of the old standbys — *Home on the Range, Red River Valley, When It's Apple Blossom Time Along the Wabash, Deep in the Heart of Texas*. We thought we got better with each rendition.

Then we warmed to the military songs — *Lilly Marlene, Coming in on a Wing and a Prayer* and the artillery song: *The Caissons Go Rolling Along*. If I had known then that one day I'd have an intimate rendezvous with the artillery, the artillery song might have seemed more than just a good military tune. In fact, it was so good that the Army itself would later take it from the artillery and make it the official tune of the entire Army.

It was possible that our faithful Penn Station audience might have believed they had lucked upon a spontaneous performance of an official Army choral group who were taking the opportunity to rehearse their program on the way to some important New York City musical event. Feeling a little bolder, and because it seemed appropriate to the situation, we even tried *Sentimental Journey,* but that one was a little beyond us and our voices trailed to an embarrassed, laughing stop. The crowd, now quite sure they were listening to amateurs, applauded nonetheless. Encouraged, we fell into the last stanza of the "Weatherman's Song" as an encore, much better this time than that attempt four of us had made at recording it at the USO in St. Louis.

I'll Never Forget
The day Was Wet
The General Wanted to Fly...

When I said "No, It's Going to Snow....etc. [1]

Of course, we were frauds. None of us had ever had anything to do with operational meteorology. And our only connection with planes was to view them from the sidelines. Or, as at Scott Field, from the upper bay balcony of our barracks where we would run to when we heard a P-80 making its takeoff or landing. (The jet plane was still unusual enough to capture that kind of attention.) We had been having a much closer association with trains than with planes; and now it was time to climb aboard another one, the Long Island Railroad train. Next stop: Ft. Totten. Planes on high, here we come. Well, not exactly.

Ft. Totten, like Tinker Field, was the location of just another weather headquarters — the 8th Weather Group. There was no one at Totten who took weather observations or made weather forecasts, just a bunch of pencil pushers. Totten was an old moss-covered coast artillery fort. We would spend a week there and during that week, we never saw a plane. We waited there for the orders that we hoped would finally take us to our operational weather assignments and whiled away some time back on the Long Island Railroad, taking trains into New York City, playing tourist at Times Square.

"Show me where it says in your service records that you're to be assigned to the European Theater of Operations," the staff sergeant at 8th Weather Group Headquarters said as he handed back to me the stiff, manila-colored form known as my 201 file. I looked at the form. Sure enough, the letters "ETO" were nowhere to be seen on it. But as I passed it back to him, holding it parallel to the floor, I could see the indentations where the letters

---

[1] See Chapter 4 for the complete song.

"ETO" had been made and then obviously erased. When I had enlisted at the Army recruiting station in Royal Oak, Michigan, in addition to the usual promise to military recruits of lifetime medical care for me and my dependents should I stay in the Army to retirement, in exchange for a commitment of three years I had been promised two things: my choice of Army branch and of overseas theater. I selected the Army Air Force and the ETO.

I recalled how the corporal at the recruiting office had immediately promised me my choice of Army branch, but did not promise me choice of overseas theater until after I began to waver in my enlistment decision. Giving him the benefit of the doubt, he might have thought he had the authority to promise choice of overseas theater and had later been overruled. But it was also possible he had a quota to fill and had deliberately lied. That possibility I quickly put out of my mind. At that time, I was still a mere neophyte in regard to Army ways and just didn't want to believe a corporal could be capable of such shenanigans.

A few months later, another explanation came to mind when the person in charge of travel pay disbursements at the Continental Weather Wing, who, as I recall, was not a weatherman, was court-martialed and presumably sent to prison for stealing the travel pay of the soldiers passing through the Wing. This was money that was to reimburse the soldiers for travel expense incurred in reporting to Tinker. There were several of us, including me, who became his victims, with losses of more than a third of a normal month's pay. The proceedings of his court martial, in which his transgressions were outlined, were sent to those of us whose travel pay had been stolen. Knowing that he had done even worse, I thought it likely he also could have erased my preference for the ETO in order to help ensure I'd be sent to an isolated station where there'd be little opportunity to consult with the Army's legal offices about that missing travel pay. If he was willing to risk everything to get the travel pay of young soldiers passing through his organization, he would not have thought twice about altering a service record. I finally concluded he was perhaps more of a suspect in the case of the missing ETO notation than was the recruiting office corporal.

I might have been suspicious about an entry made in pencil, but I never dreamed anyone would alter that record. Even if I had, I would never have thought the promise could be broken. At that young age, it was easy to feel a sense of betrayal. My concern was not directed at the Army Air Force, the service I loved, but at the culprit who had somehow made his way into that service. I supposed I could make a fuss, could try through the Army's legal department to right the wrong and, if carried to the extreme, might even achieve the cancellation of my enlistment. But I didn't want that. I wanted to fulfill the expectation of becoming a real, practicing, weatherman. And so, like my lost travel pay, I resolved to put the service-record tampering out of my mind and to make something of an enlistment that I believed held out the promise of good things to come.

I could see that the good things were probably not going to arrive in the immediate future. The duty stations of the 8th Weather Group were predominantly those isolated weather observing sites in the frozen North. I could picture myself shoveling a path to the weather-instrument shelter at some place like Padloping Island or Indian House Lake.

At the end of that week at Ft. Totten, we were gathered again at a train station — our destination: the 8th Weather Squadron Headquarters at Westover Field, near Springfield, Massachusetts. The town closest to the base was Chicopee Falls. For this train trip, no one was put in charge. Though we were traveling together, we were on our own. As we were about to board the train, someone noticed that Zee was not with our group. He had been scheduled to travel with us. But we were on our own, so it was nobody's worry. There was some hesitation, one or two of our group acting as if perhaps we were obliged to find him. They looked at me. It wasn't my place to say it, but I did. "Let's go," I said, and we were on our way to Springfield.

As the train slowed down for a short stop at Hartford, Connecticut, through the windows we saw Zee running alongside. In a flash, he was with us again. No one asked him how he had managed to overtake the train, but he told us anyway. When he saw he had missed the train he began hitchhiking. Catching a ride in a fast car, he apparently convinced the driver it would be a patriotic plus should he manage to overtake the train so that one dedicated soldier could arrive at his destination on time. In a devil-may-care dash down the New England roads, the challenge was accepted and met. Having completed his tale, Zee waited for our amazed responses with a Cheshire-cat grin spread wide across his face. But of course we had all learned not to be surprised by anything Zee might do. We ho-hummed and went on with reading magazines, working cross-word puzzles, staring out at the passing New England landscape.

That was my last association with Private Zee. An ending to this episode, that would comport with the way such stories usually end, would have Zee rising to great distinction in the service. And that might even have been predictable. He had a flair for resourcefulness, a favorable military attribute. And I am sure that some of you who served in the military will say you must surely have met him and that he did indeed rise to high rank. Perhaps fearing that might be true, as we scattered to our far-flung operational assignments I never tried to discover where his subsequent travels took him.

We arrived at Westover Field well after midnight and were thrust into the care of a staff sergeant — the charge of quarters (CQ). (When you were a private or Pfc., your immediate decision-maker was more likely to be a staff sergeant than, say, a master sergeant or captain.) This was a cold January night. Not knowing quite what to do with us, it appeared, the CQ sent us walking toward a few lights that twinkled in the frigid air far across a field. It may have been a parade ground, but it seemed too far across for that. It was hard to be sure in the dark. "Out there," he said, "where the lights are. They'll take care of you there."

"Which lights?" I asked. There were three little islands of lights in the far distance.

"Those," he said, waving at all three. We moved out over the icy field into the face of what seemed to be gale force winds. With eyes watering in the freezing blast, I tried to decide which cluster of lights would provide warm shelter for the night. The decision was easily made, we chose the light that seemed the closest. With the straps of our barracks bags and ropes of our duffel bags cutting into our shoulders, we trudged on toward that light. It was late, and we were tired even before we started. In the moonless pitch darkness we stumbled along, bucking a persistent head wind of about 30 miles per hour.

Eventually, we arrived at the light. It was a barracks — the Army standard, a frame building. This one was completely deserted. A latrine with a shower room, a stairway leading to the upper bay, a bulletin board with no messages. We walked into an echo-filled chamber with floor boards that creaked in the cold. Whoever had been assigned to this barracks was long gone as was the soldier whose duty it once was to run the heating system. But except for small breezes that filtered through cracks in the window frames, the building did provide shelter from the gale that continued to rage outside.

With no heat in it for who knew how many days, every board of which that barracks had been made had become cold-soaked. We were inside a wooden refrigerator. There was some talk of going back out to look for a warmer building, but we were dog tired and the distance to the nearest of the other two lights in that continuing gale was a challenge no one wanted to face. The thought of moving quickly cooled.

The barracks was furnished with steel GI cots. A rolled mattress had been placed at the foot of each cot. There were no sheets, no blankets, only mattresses. It was as if that deserted building had been put on a moth-balling list. I laid my poncho on the bare springs of the nearest cot and flopped, fully

53

clothed to include that heavy olive drab overcoat that we called a horse blanket. I was still wearing my combat boots and the leather government-issue gloves with woolen inserts. To retain maximum body heat, I kept my battle jacket and overcoat buttoned all the way. Then I unrolled the mattress, letting it spring down to cover me as if it were a blanket. My companions began to fit out their cots in the same way. But the mattress was a poor substitute for a blanket. Being stiff, it wouldn't droop down to completely cover your sides where you remained exposed to the frigid air; and its heavy weight pressed you into the springs, keeping you from adjusting to any new sleeping position. But it helped to conserve just enough body heat to keep you from freezing which I then believed you otherwise could have. We had no thermometers but my subsequent years of focusing on weather and temperatures tells me the actual temperature that night in that barracks was something on the order of 10 to 20 degrees Fahrenheit.

As I lay there lapsing into intermittent sleep, I dreamt of how this first day at Westover Field, my first day at an operational weather base, was like my first day at Ft. Sheridan which was my very first day in the Army. Ft. Sheridan was on the Lake Michigan shoreline. It was the 6th of February of 1946 and, like this day, we had also arrived at night, at 2300 hours. About 15 of us had just come in on a bus from Michigan. We stepped off the bus at Ft. Sheridan in the teeth of a determined wind that was blowing at us from off the frozen waters of the lake and were left in the care of a CQ who opened the door of his orderly room just long enough to see how cold it was. And it was bitterly cold. In about a half hour, the CQ peeped outdoors again, this time to tell us to remain standing there, outside, and he would try to find out what to do with us. Our arrival had apparently been a complete surprise. He began working at that task of finding out what to do with us from within the warm walls of the orderly room. It must have been a particularly tough job for we waited and waited and waited. This being our first contact with Army life, we were, of course, wearing civilian clothes. I was wearing a royal blue overcoat. The good thing about it was that it was a stylishly cut, one might even say, dashing, garment. At the Richman Brothers' clothing store in Detroit I had been attracted to it by its luxurious, soft nap. The bad thing about it was that it was pathetically thin, a shortcoming I had not noticed when I was civilian who could limit his exposure to inclement weather, going indoors at times like this or into a heated automobile. But it failed to meet the test of being measured for effectiveness against that Lake Michigan gale. The wind bore through it with energetic vengeance and with it came a relentless progressive penetration of the cold. I could hear my neighbor's teeth chattering above my own. The CQ did not return again for at least an hour. He was about to tell us where we might find a warm barracks when I awoke to the swirl of snowflakes that the New England gale blew through the cracks in the window frame. It was morning. I pushed up the mattress and brushed off a miniature snow drift from it that had blown in through the window cracks during the night. We walked out in the sun.

From a passerby we learned where 8th Weather Squadron had established its headquarters and the location of the mess hall. After breakfast we trekked over to the headquarters. It was located in one of Westover Field's hangars. Gone and almost forgotten was the misery of the night before. This was the morning of the day of our first assignment in operational meteorology! This was *it!*

Captain Laverne Beneke briefed us on our assignments, matching names with the operating locations within the squadron that were short of weather observers. Those operating locations of 8th Weather were strewn along a wide and varied stretch of geography and over vastly different climate zones. It was possible, in that outfit, to be sent to record the presence of cottony cumulus over the balmy beaches of the British Crown Colony of Bermuda; or, in the other extreme, to wait in the snow for parachute re-supply drops at outposts near the Arctic Circle. Even now, the names of some of those duty stations evoke a sense of adventure. Indian House Lake, Labrador; Padloping Island, Baffin Island; Simuitak, Greenland; Mecatina, Quebec; Cape Harrison, Labrador; Ikateq, Greenland; Grondal, Greenland; Lagens, Azores (the spelling of this Portugese air field would later be changed to "Lajes"); Kindley Field, Bermuda (where we all wanted to be stationed but knew our chances were slim to none); Meeks Field, Iceland; Sondrestromfjord, Greenland; Upper Frobisher, Baffin Island; Fort Chimo, Quebec; Goose Bay, Labrador; Harmon Field, Newfoundland; Presque Isle, Maine; and the base weather station at Westover Field itself.

The headquarters personnel of 8th Weather were the kind of men you felt proud to be associated with. Like the commanding officer, Colonel Frederick J. Cole, they were thoughtful and considerate of the men in their charge. In the midst of laying out the operational weather assignments, Captain Beneke asked if any of us might have a reason for not being sent overseas. One of our group spoke up for me saying that my wife, who was staying at her parents' home in Michigan, was expecting a child. (As that might mean someone would have to go in my place, I did not think it proper to bring it up myself.) There was no discussion, and not another of the group had anything to offer. That almost offhand suggestion by one of my friends was all that was needed to have Captain (later Lieutenant Colonel) Beneke assign me to duty at Westover's Base Weather Station, the squadron's only stateside base other than Presque Isle, Maine. At Westover, it would not be overly difficult to obtain transportation for visits home. This demonstration of compassion for an enlisted man who lived at just about the very bottom of the rank ladder, was as surprising as it was heart warming.

The author, a brand new member of the 8th Weather Squadron, poses at the squadron's day room where the barracks-dwellers could relax when not on shift. Inside were a pool table, some books and a few overstuffed chairs.
Photo taken 23 March 1947 by Howard "Henry" Klier.

Chapter 7

A Weather Job at Last — Sort of

The rest of our group were scattered to the many duty stations within the 8th Weather Squadron's domain. A good friend from Chicago, William J. O'Dwyer, who had been at Tinker at the same time that I was there but who left for 8th Weather before me, and with whom I whiled away some of the off-duty time playing pool in 8th Weather Squadron's day room, was sent to Lagens, Azores as a radiosonde operator (MOS 942). (For the first time in my life, while playing 8 ball with O'Dwyer, I ran the table. Perhaps thinking that my performance was clearly beyond what anyone would expect of me and therefore made the game one that did not depend on skill and consequently was hardly worth bothering with, O'Dwyer, who was a good pool player, said he'd never play that game again and as far as I know he kept his promise.) Like Andy Anderson, William O'Dwyer was someone you thought you'd always keep in touch with. He wrote me a letter from the Azores to which I responded and then we never heard from one another again.

Westover's Base Weather, the station with the call letters DOJ that would later be changed to CEE, was an impressive assignment. Here, flight forecasts were made for flying crews who took off on the long hauls. When taking off for a flight over the Atlantic, they usually flew in the C-54. The pilots of these piston-engine planes would routinely choose 17,000 feet as their cruising altitude (about one-half the altitude of today's jet airliners). Some of them would be touching down in Europe over half a day later. To assure adequate separation, they would return to Westover at an altitude of 16,000 feet. Sometimes, because of wind conditions, they would drop down, usually cruising at 9,000 feet eastbound and 8,000 feet westbound. One of the most frequent destinations was Lagens, Azores. This was 2200 miles away or about 9 to 11 hours flying time.

Westover Field was an aeronautical Grand Central Station. It was the Military Air Transport Service (MATS) center for military air travel to and from Greenland, Iceland, Europe, North Africa, the Middle East and the island bases of Lagens, Azores and Kindley Field, Bermuda. For Army personnel and especially for we weathermen, it was a crossroads where we met old friends from tech school days and made new friends of those who were passing through for reassignment or separation from service.

When a weather troop came through Westover he usually would stop in at Base Weather to see if there was anyone there he knew or just simply to see what our operation was like. One of the many who passed through was a recently re-enlisted corporal who had been in the AAF Weather Service during the war. I asked him why he had re-enlisted. "It was just to get back into weather," he said. "You'll understand when you've been in it awhile. Army weather gets in your blood."

Some seven years later, when I passed up a chance for a second re-enlistment and was a civilian working on a highway survey crew, I would think of the corporal's remark as the traffic zipped by on Telegraph Road, a race-track-like highway, near Detroit. As I tried to steady a level rod with cars and trucks speeding past alongside, in both directions, their slipstreams tugging at the rod and my clothing, and thinking all the while, why am I here? I would truly understand what the corporal had been trying to tell me.

The corporal also said this: "The weather service is the only place in the Army where you'll find a major doing the same job as a corporal." He meant this as a favorable comment on the employment of weather forecasters. I thought he had been carried away with exaggeration, as that seemed to be more democracy than any military organization could absorb. In future years, I would find his remark to be quite true. I never saw a corporal on forecasting shift but I myself would later be forecasting as a buck sergeant and would be relieving a major just as he would relieve me, depending simply on the rotation of the shift schedule. A buck sergeant was just one rank above corporal. In any event, I never saw that the scheduling of forecasters of quite different ranks caused any problems, a result that I attribute to the fact that all forecasters I knew were gentlemen regardless of rank. Partly, I believe, this was an inevitable effect of being so intimately involved with meteorology. Dealing with the intricacies of weather forecasting brought about a certain humbling, which had to favorably affect character.

At that time, the comment made by the corporal was of little interest to me. I had a quite different area of responsibility to contend with. I was a graduate weather observer and a graduate weather equipment teletype technician but was given a job that required none of those skills. Put to work in the weather message section, I edited various meteorological observations which were radio-transmitted to the far-flung 8th Weather Squadron outposts. This information was coded. There were separate codes for the several kinds of weather observations. Usually, these were strings of numbers in five-digit groups. The weatherman at the receiving end, at one of the Far North outposts, would determine if the code sent was a PIBAL (for pilot balloon), a radiosonde message, a three-hour synoptic code of surface weather observations, a six-hour synoptic code of surface weather, etc.

The PIBAL message consisted of wind directions and wind speeds at one thousand foot height increments usually to somewhere above 10,000 feet or to as high as the observer could track the balloon with the theodolite. Radiosonde messages also provided winds aloft, the balloon in this case being tracked with an electronic device, at that time, the SCR-658. It had an all-weather capability, which of course the theodolite did not have. The SCR-658 would track a radiosonde tied to a balloon to as high as the 200-millibar level and above or to an altitude of seven or more miles. The radiosonde transmitted temperature, humidity and pressure information. The three- and six-hour surface weather observations were comprehensive reports that contained more information than was provided on the other surface weather hourly observations. The six-hour observation, for example, would contain the depth of snow on the ground. Surface weather observations were taken every hour on the hour. Normally, an observer at the forecasting stations would plot only the three- or six-hour observation on the surface weather map to be analyzed by the forecaster.

We were, in effect, a weather message redistribution center. On a typical shift, I would walk over to the chattering teletype machines, rip the latest weather reports off a machine's yellow roll of paper, reports that were sent to us by weather stations that might be thousands of miles away, and would separate out the ones we wanted to send to the outposts. With a sweep of a grease pencil, I obliterated all information that we thought would be superfluous to the receiving weather personnel. An economy of information was vital as there was a lot of information to be transmitted, all sent by tapping out the data in communication code on a manual tapping key such as would be done by the telegraph operator in a Wild West movie. A sergeant of the Airways and Air Communications Service (AACS or "double A C S" as we referred to it) tapped out page after page of the five-number groups. Over the eight-hour shift, the sergeant was virtually continuously tapping that key, and he did so with such speed that I sometimes had to hurry to keep him supplied with the numbers.

Now if you were the double A C S sergeant, you had to *believe* there were better ways to be spending your life. Once in a while, when I thought the glaze covering his eyes seemed particularly impenetrable, I would vary the routine. Instead of marking the copy with that usual black grease pencil, I'd hand him a sheet marked by a red grease pencil or even with a green grease pencil. Having an ambition to do cartoons, on an occasional weather sheet I thought I would draw the ubiquitous cartoon face of Kilroy, with his large rounded nose hanging over the top of a wall above the words "Kilroy was here," but always stopped short of it as that would have constituted a degree of levity that would not seem quite proper in a proper military person as I imagined myself to be.

My weather work was not entirely confined to editing. Occasionally, I'd be called on to substitute for the weather observer as he left for the chow hall or was simply not there because the weather team was playing a game in the base fast-pitch softball league and that observer was on the team.

One day when the weather team was in desperate need of another player, I was asked to play and was substituted into the game in the ninth inning. I tagged the first pitch for a clean single (I later described it to friends as a *smash* single). The game ended in that ninth inning. I was not called on again and was saved from having to display an ability to also strike out in grand style. For the rest of the season, strangers looking at base bulletin boards marveled at that amazing athlete who carried a 1000 point batting average. This was like the forecaster who makes a forecast known for its great difficulty that is later acknowledged for being "on the money," gaining for him a reputation for accuracy beyond what he knows he deserves.

With all the practice, becoming proficient as a weather message editor was unavoidable. At different times, even when not at work, as a result of my newly gained proficiency, I would find myself translating numbers into the teletype's alphabetical equivalents. The teletype printed numbers in its upper case mode and letters in the lower case mode. If the teletypist forgot to switch cases, the machine would print letters when numbers was wanted and vice versa. The skilled weather observer soon learned to ungarble these messages just as I did. To this day, I find myself making curious sense of groups of numbers such as: 6-9-7 2-3-4-3 1-7-8-5-3 5-4-8-5-3. If you are interested in the translation, you can look up the corresponding letters from the key below.

1 : Q
2 : W
3 : E
4 : R
5 : T
6 : Y
7 : U
8 : I
9 : O
0 : P

There are times when you know things are going about as smoothly as anyone can expect. It was that way as the spring of 1947 rolled onto the calendar. I had become as skilled a weather editor as anyone would want to be. On March 20th, the Army Air Force bestowed upon me the rank of corporal. My name was being carried on the bulletin board as a batter with a thousand batting average. As I walked to the chow hall for breakfast under that Massachusetts sky, there was more blue in the stratus overcast than almost anyone else could see; in the afternoons, when the air was crisp with passage of a cold front, cumulus that popped up was whiter than the purest snow. Adding to the headiness was the exhilaration in knowing you were in a field where you might help seek that world of knowledge yet to be discovered.

And 8th Weather was a good outfit. As a tribute to its weather personnel at the many operating locations in the Far North, some gifted individual had designed the best squadron patch possible. It pictured a snowman in top hat carrying an umbrella. He stood behind an eight ball that was balanced at the top of the globe. There was snow underfoot and of course there were clouds in the sky. We wore the patch on a pocket of our field jackets. It was also used as letterhead for stationery and a large version, about five feet in diameter, was painted on the outside wall of the 8th Weather day room. Another talented person, I believe this was Staff Sergeant Tom McClung, used the motif in setting up the squadron newsletter. Its title: *Behind the Eight Ball.*

Tom McClung worked in 8th Weather Group Headquarters that had moved up from Ft. Totten and, like 8th Weather Squadron, had located its office in a hangar. This gave us two headquarters at Westover with the 8th designation. Tom was given the responsibility of writing the group's Manual of Operations, which we called "MANOPS." 8th Weather Group might be thought of as being top heavy in rank, so much so that a staff sergeant, such as Tom was, who in the usual military office might be thought to be exempt from that duty, would be the one to sweep out the office at the end of the day, a chore that was performed each day at 1645 hours. So Tom wrote into the MANOPS the requirement to report any changes in operations at precisely 1645 hours and by teletype. Since he was the one who knew the requirement, it was natural that Tom was the one to walk over to Base Weather, where the teletypes were, and send out the report. While Tom was sending out that daily message, which typically consisted of "no change," back at Group someone higher in rank was sweeping out the office. No one ever said weathermen were not a resourceful lot.

About fifty years afterward, I would meet Tom and another good friend of that time, Henry Klier. Tom and Henry had both served their three-year hitch and left the service to go on to successful civilian careers. Henry recounted how we had hiked up a trout stream to its beginnings and how, long before email was even dreamed of, we had developed a code for communicating with each other by teletype as he left for reassignment to Wold-Chamberlain Field at Minneapolis. At Wold-Chamberlain, Henry (who is now known as "Howard"), worked in weather observing, became an impressive weather-map plotter and substituted for the forecaster on occasion. All the information, and there was a great amount of it, that went into the weather station plot on the surface weather map had to fit under a dime. This could only be done by an accomplished map plotter using a fountain pen with a very fine point which had to be replaced from time to time as the plotter literally wore out the nib. A fast and accurate map plotter was the forecaster's best friend. Henry was one of those.

In spite of how well things were going in that spring of 1947, I was not content. Number 1, the lovely Marie was 500 air miles away; and number 2, I was not working in the weather specialties I had been trained for. As my assigned duty, I would much have preferred to be calculating the height of the base of the afternoon cumulus using the pseudo-adiabatic chart and then record that along with all other pertinent weather information on the weather observation form, the well-known "weeban." (Officially, the WBAN, which stood for Weather Bureau, Army and Navy. This form was used by all three weather services to record hourly weather and special weather observations.) I wanted to fill a 30-gram balloon with the precise amount of helium that would provide the required rate of rise of about 600 feet per minute. I wanted to track that balloon with a theodolite, plot its elevation and azimuth angles on a wind plotting board and calculate the winds aloft. If I couldn't do that, I wanted to use the other skill I had acquired in the Air Training Command. I wanted to tear into the teletype machines, extract the code bars, the code bar shift levers, the print hammer, the operating bail latch, the oscillating rail slide, the escapement lever, the rocker shaft cam plate, the carriage return function pawl, the bar bell crank, the shift solenoid, the platen idler spur gear, the main shaft and all the connecting parts. I wanted to lay them all out on my desk. Then I would reassemble them and see if I could make that 60-word-per-minute teletype type at 100 words per minute. At that time, many years before the advent of laser printers, we all knew 100 words per minute was *FAST*. (The teletype actually was later modified to print at 100 words per minute by the manufacturer.) Weather editing was a necessary function, that was clear; but I had stripped the protective paper coils from too many worn-down grease pencils. I wanted to get my hands on more tangible weather things.

Top: Weather observers at Westover Base Weather in 1947. The man on the right is plotting a weather map which will be analyzed by the forecaster in the adjoining room. Note map table — it is typical weather station furniture.
Bottom: Sgt. George Popadines holds a poster made by the author for Armed Forces Day at Selfridge AFB in 1950. The teletype machine was set up on the ramp for public viewing and was a great attraction.

Chapter 8

Hitchhiking with the AAF

When there was a lull in the incoming weather collectives and the chattering in the teletype room had ceased entirely, producing a disconcerting silence, or when I was simply off duty, I liked to saunter over to the forecaster's map table and watch him analyze the weather charts. One day, while off duty and beginning my three-day break, I strolled over to Base Weather to check out the map analysis. Staff Sergeant Paul Luty was the duty forecaster. As soon as he saw me enter the pilot's weather-briefing room, Sgt. Luty stopped me: "Run over to Base Ops and grab a parachute," he said, "I've got you a flight to Selfridge Field!"

Selfridge Field was located close to the northeastern city limits of Detroit. The lovely Marie, who was then expecting Leta, the first of our three beautiful daughters, was staying with her parents in the house that her father had built with his own hands, in Royal Oak Township. This was a northern suburb of Detroit, later incorporated as the city of Madison Heights. It was only 15 miles from Selfridge Field. Getting a hop to Selfridge was a fantastic bit of luck.

There were just three little concerns. First, unknown to Sgt. Luty, and even if he did know it, it would have been of little concern to him, was the fact that I had never been inside an airplane before, not even in one that was resting on the ground. Second, the aircraft was not the tried and true C-47, the one that I expected it to be and the one in which most domestic flights from Westover were made. It was an AT-11. I had never seen an AT-11 before and wondered a little about its reliability. Third, the pilot and copilot were reservists, first lieutenants from Windsor Locks, Connecticut. They were what we called "weekend warriors." Obviously, they wouldn't have had a lot of recent flying experience. But they were going to fly the AT-11 to Selfridge! What a break!

I met the lieutenants as they and the flight engineer, a staff sergeant, were walking out of Base Ops. The lieutenants were wearing their parachutes and had already checked one out for me. The pilot was carrying my chute. After explaining how it was done, he helped me strap the chute on and checked the fit of the straps. Satisfied that the chute was adequately snugged to my back, he then told me how to pull the D-ring attached to the rip cord and cautioned not to do it until I had counted out "one thousand one, one thousand two," etc. I knew I would never jump out of that AT-11 but nevertheless appreciated the care he was taking in briefing me. We climbed into the AT-11, the pilots up front, the staff sergeant in back with me.

A twin-engined aircraft, the AT-11 was a modification of the more familiar C-45. It was designed to be a bombing trainer. In outward appearance, the most noticeable difference between the AT-11 and the C-45 was the AT-11's plastic nose which was used by the bombardier trainer as his aiming station on practice bomb runs.

The AT-11 also had bomb-bay doors, of course. I settled down next to those doors on the canvas bench-like seat, listened to the 11's rolling wheels and, when the rolling ceased, watched the Berkshires shrink below us. We were headed west toward Selfridge. So this was what it was like.

Though it was only about 20 miles from where I had lived when I finished high school in Detroit and only 70 miles from where I had spent my boyhood on the farm, I had never set foot on Selfridge Field. During the war, there had been controversy regarding the base when French fighter pilots who were training there buzzed boaters on adjacent Lake St. Clair.

I recalled that, except for those unpredictable elements that intervene and change the course of your life, I might myself be piloting a machine such as this AT-11 on a flight to Selfridge — though I would rather that it be a Thunderbolt or Mustang. I had been a post-*Spirit of St. Louis* child and had read of Charles Lindbergh's famous flight in my third grade reading book. Later, when in high school, my best-liked subject was a special developed-for-wartime course titled "Aeronautics." The school, like many others of that wartime era, had geared itself to support the war to the extent that a high school could. Another course in that vein was titled "Fundamentals of Machines." I found its artillery-trajectory problems to be quite fascinating. Our physical education classes were almost entirely devoted to boxing in which I was the coach's favorite if not always a willing contender, and the football field had been converted to a replica of a military obstacle course.

When we were about an hour out, the staff sergeant flight engineer went up into the cockpit area and stayed there for quite a long time. "We're having trouble with one of the engines," he said, after returning to his seat.

"Will we have to turn back?" I asked. I certainly did not want to turn back. We had conquered the takeoff, one of the most hazardous intervals of a flight, and, as long as the coming down was slow and sort of controlled, I would rather that it be at Selfridge. I was already visualizing one of those home-cooked meals my mother-in-law was famous for and could almost conjure up the enticing fragrance of her home-baked rolls and pumpkin pie.

No doubt sensing my desire to continue the flight to Selfridge, the flight engineer gave me the answer I wanted to hear. "I'm going back to talk to the pilots," he said. "This plane is new to them and they'll probably want to return to Westover. But I know this ship. We should be able to make it to Selfridge." From Connecticut and with a last name that ended in "ski," he was also a weekend warrior but the matter-of-fact way in which he continued the discussion of the engine problem bespoke a confidence born from having, it was clear, much more experience in keeping an AT-11 aloft than the pilots had in flying one. He returned to the pilot's compartment as I watched through the circular windows for any indication of a turning back. As much as could be learned from looking down at the conical points of evergreen trees that poked up from a ground surface uniformly covered with snow, our heading hadn't changed. We were over Canada, on the shortest track to Selfridge, and both propellers were then turning in what I took to be a normal sort of way. Through the slots where the bomb bay doors imperfectly met the aircraft body, I could see clouds drifting past below us. A chilling wind blew up from those cracks, which now appeared to be alarmingly wide, and brushed past my face. This was the month of March and the free air temperature at our altitude would have had to be close to zero on the Fahrenheit scale. There was no cabin heat, or at least none that you could sense. I was wearing the standard winter class-A uniform, woolen olive drab trousers and battle jacket with khaki shirt and tie. Over that, I had on the standard green field jacket zipped up as close to my chin as possible. If I had thought of how cold it was going to be, I might have dressed warmer. I might even have worn that horse blanket, the heavy, wool olive drab overcoat. You certainly were no fashion plate when you wore that GI overcoat, and so we consistently avoided it, but that garment was the best possible match for old man winter, even on his meanest days.

"The pilots are going to keep going," the engineer said after another trip forward. "I think I've convinced them the engine problem is not serious enough to turn back. In a little while, anyway, we'll be past the point of no return." If I hadn't wanted so much to get to Selfridge, that comment might not have seemed so reassuring.

As you look at the ground from a plane that is at cruising altitude and cruising speed, in this case at 8,000 feet and about 150 miles per hour (we normally would cruise at about 160 mph but had an effective head wind of about 10 mph), you get the false impression that you are not going anywhere very fast. That thought seemed to compound an increasing physical misery — I was growing colder and colder. The only part of my body that was not chilled was my back, the strapped-on parachute was providing insulation. Parachutes, I could see, could be nice things to have, even, or especially, when unopened. I tugged on the straps to make the fit even tighter, hoping that would help keep more body heat from escaping. I thought I was practically frozen everywhere but where the parachute provided that protection from the wind rushing up through the bomb bay doors. In the midst of wondering how to keep from freezing, the engineer matter-of-factly said the problem engine was still giving some trouble, but, he added, it was not of any consequence as we were now committed. We were past the point of no return.

I ran my fingers over the D-ring and tried to imagine, if and when I fell, whether there would be sufficient coordination between mind and body to enable my brain to signal my arms and hands, telling them they should pull the rip cord. And if that signal should get through, I wondered if my arms and hands would be too cold to function, and if they were not too cold in a temperature sense, would they obey the signal or would they simply become frozen in whatever position they were then in, victims of a case of sky fright. In any event, there wouldn't be much to worry about, because there'd be no option. The engineer said they would simply open the bomb bay doors and drop me through that opening if a real emergency developed. And then, I mused, I would know if I could pull the rip cord.

But there would be no pulling of the rip cord. "We're on the approach to Selfridge," the engineer said, and we looked down to see ice boats skimming along on Lake St. Clair. The base bordered on the lake. The pilot called in an emergency to Selfridge tower, skipped the circling approach, headed for a straight-in landing. As we touched down and began rolling down the runway, crash trucks and ambulances appeared on both sides, racing along with us, the drivers apparently expecting to see the AT-11 careen off in a fiery crash. It was as smooth a landing as anyone could have wanted. Even with that bad engine, we rolled down the runway as gently as a fair-weather cumulus cloud would skim the tip of the Berkshires. Weekend warriors did not deserve the jokes that were often told about them.

As the pilot came back to open the door, I let go of the D-ring, saying: "Nice landing, sir." And as I stooped through the doorway, I took the time to pat the smooth skin of that AT-11. It was a beautiful day. Visibility fifteen miles plus and only an altocumulus deck in the blue sky.

The pilot and copilot intended to leave the next day for the return flight to Westover, so the four of us arranged a meeting for the next morning. We would meet at Base Weather, immediately next to Base Operations ("Base Ops"), in the south end of the brick Base Headquarters building.

Obviously laid out with efficiency in mind, at all bases that I knew of, that was the operational facility arrangement. Base Weather was located immediately adjacent to Base Ops. (Hanging on one of the walls of Selfridge's Base Ops, the pilot would see framed photos of Seversky P-35s flying above Lake St. Clair, reminders of the first-line aircraft that were probably based at Selfridge in pre-war days.) This allowed the pilot to draw up his flight plan and get his clearance form in Base Ops, then he'd merely walk the few steps to Base Weather, the next office, for his briefing on takeoff, inflight, and destination weather. After the weather forecaster signed his clearance form, the pilot returned to Base Ops to file his flight plan with the operations officer. He was then ready to go out to the ramp and run up his aircraft for takeoff.

I left Selfridge immediately for the reunion with Marie and my in-laws. My arrival was a complete surprise. It was a Saturday afternoon and Marie's father, Walter Nordstrom, was home. He worked at the Dodge plant in Detroit which, for the past year or so had been back to producing civilian vehicles following its wartime production of Army trucks. He was proud of my new corporal's stripes and dragged down from the attic his World War I Army artillery uniform. To show he was still in good shape, I suppose, he tried on that tight-collared olive drab jacket. It was a fairly good fit and that was as predictable as was Walter's generous character. He was a very active man who not only built, with his own hands, the home they lived in, but also had excavated a basement after the house had been built, a daunting task for the faint of heart, but no problem at all for him. There never seemed to be a day when he did not extend himself to help others who needed help before it would occur to them that they should ask for help. When it comes to evaluating his daughter's husband, the typical father has reservations. If Walter had any concerns about me, he never expressed them; and now, as we compared our two Army uniforms, we established as much of a bond as men of his and my generation would ever let anyone perceive.

The next morning, after a difficult farewell, I went to my folks home in Detroit, about three miles away, for a brief visit and to have Dad drive me back to Selfridge. I wanted him to see the planes of which he had a natural interest arising from having machined B-29 motor parts during the war. There'd be no B-29s at Selfridge, a base that was devoted mostly to fighters, but I had another motive — I wanted him to see a weather station. He had a no-nonsense way of looking at things and, though he would have been proud of the fact that someone in the family was wearing an American uniform, I always suspected that he might have viewed the meteorological profession as a refuge for those who didn't want to face up to doing something worthwhile with their lives. That thought might have been reinforced through a few less than helpful weather broadcasts we had received by radio back on the farm where accurate anticipation of the coming weather was critical. At Base Weather, the pilot was already poring over the surface synoptic chart that had just been analyzed by the duty forecaster. He had been briefed by the forecaster and relayed that briefing to us. A moderately strong low pressure system and its attendant warm and cold fronts were located southwest of Selfridge and were moving northeastward. In the standard weather map analysis procedure, the warm and cold fronts were drawn on the chart in red and blue pencil, respectively. A wide area of rain and drizzle that enveloped the low center in the shape of a huge comma, extended into the Selfridge area. It had been shaded in green by the map analyst. This was a colorful map, some would even say artistic, which told a bad flying-day story. The pilot explained the map situation to Dad who listened with intense interest. (Like almost everyone else at that time, excepting meteorologists and pilots, Dad had never seen a weather map before). After noting that flying conditions were too poor to return to Westover, the pilot said we would wait until late in the day before attempting the flight. This would give the weather system time to head up into northern Canada. To a neophyte airplane passenger like myself, the pilot's cautious approach to the weather was something to be welcomed. In the future, after I had become a weather forecaster, while briefing the occasional hell-bent pilot, I would remember the professional approach of that weekend warrior and wish all other pilots were like him. He was actually waiting for the ceiling and visibility conditions to improve so that he could fly under visual flight rules, VFR, which meant the ceiling had to be at least as high as 1,000 feet and the visibility at least three miles. If those conditions were not met, the pilot would have to fly under instrument flight rules, IFR.

I was still concerned about the AT-11's bad engine. I brought up the subject with the staff sergeant flight engineer. He had spent the remainder of the previous day working on the engine and the problem was corrected. Since he knew no one in the Selfridge area and would have to spend the day reading magazines or walking across the street to the PX annex, the cafeteria, to drink coffee alone, I asked him to join me in a visit to my in-laws, which he did. He especially enjoyed my mother-in-law Martha's cooking which was never anything but superb, and he had a chance to compare his uniform with Walter's which came down out of the attic again for only about the second time in 29 years. After the meal and the uniform ritual, Walter pushed in a spring-loaded secret panel he had built into the upper kitchen cupboard and brought out a prized specimen of aged whiskey, which the menfolk dutifully sampled, and then it was back to Selfridge for the uneventful return flight.

It had been a good trip. Hitchhiking with the AAF, I could see, was going to be much better than bumming rides by car over those Mid-western roads had been. This trip had ended with the bonus of an extra day at home. Weather was being good to me. I knew it would be.

With that success, it was inevitable that I should try for repeat performances. Thereafter, on my breaks between shift strings, I was frequenting seen near the forecaster's desk, trying for another flight to Selfridge. But that would only happen once more. In the meantime, I took hops to other Mid-western bases.

I developed a sort of routine. I'd take the first C-47 headed for Wright-Patterson Army Air Field at Dayton, Ohio. From there, it was a fairly easy bus trip home. If the weather was good, I'd hitchhike. On the way back, I took the train out of the Michigan Central Station in Detroit.

Once, after waiting through a number of shift breaks with no planes going to Wright-Pat, I grew a little desperate. I took a hop on a C-47 going to Greater Pittsburgh Airport at Coraopolis, Pennsylvania. After touching down at Greater Pitt, I found there was no bus or any other form of ground transportation to take me on my way. So I started out hitchhiking by ground again. This began as more hike than hitch. After a long walk up and down hilly Pennsylvania blacktops — which really was quite pleasant, it was spring and the scent of lilacs blooming in the yards beside the road filled the air — I managed to get a succession of rides that were all-too-short. No one seemed to be going more than a half dozen miles. Well, I couldn't blame them for not wanting to go very far. This area that was their home was beautiful country, but I could see that, timewise, this was not going to be a very pleasant trip for me. The C-47 had carried me about 475 miles, the distance still to be traveled was about 300 miles. When I finally arrived home, it was precisely the hour to turn around and head back for Westover. I never flew in to Greater Pitt again, but in spite of the disappointment of not having any time at home, I still look on that trip as one of the most pleasant.

All of those flights, with the exception of two, were in the old reliable C-47 that had been in existence since the middle 1930s. One day, I caught a hop in a C-46. From a distance it looked something like a C-47. It had two engines like the 47, but its rudder was somewhat rounded as opposed to the 47's sort of squared off rudder. And when you got up close you could see it was bigger than the 47, and the pilot's and copilot's seats were up much higher from the ground. This C-46 was going to Lockbourne Army Air Field near Columbus, Ohio. I had never heard of Lockbourne before, but being close to Columbus it would be even a little closer to home than Wright-Pat.

The flight was smooth. We arrived after dark, at about 2100 hours. A few bumps on landing, but O.K. As we taxied to a stop, through a window

I saw a second lieutenant with shiny gold bars ride up in the jeep with the FOLLOW ME sign. Ever since a day in basic training when a self-important 90-day wonder said, "stand closer to the mirror when you shave soldier," I tried to avoid second lieutenants, but this one, wearing pilot's wings and the AO armband which signified his duty was that of airdrome officer, was very friendly. He asked what I did, and I said I was a weatherman at Westover.

"You'll want to talk to our guys at Base Weather," he said, and he promptly took me there. Riding along in that jeep I reflected heavily on the thought that I had been in the Army well over a year and this was the first black officer I had ever seen. And it seemed strange to me that I had never before wondered about not seeing a black officer nor even a black enlisted man. And of course the troops I then met at Base Weather were black, and it suddenly became clear that the entire base must have been manned with black servicemen.

The technical sergeant forecaster and the two observers, who were corporals, were eager to talk shop. What weather station reports did we receive over our teletypes at Westover? Were those teletypes and the other weather equipment the same as theirs? Yes, they were perfectly identical, even to the standard weather station furniture with lighted map tables and swinging map display panels. Did we have much of a problem with IFR conditions? Was there much low stratus to contend with? They had been having a lot of low ceilings which had improved only lately.

We could have gone on comparing notes indefinitely, but sensing that I must have been eager to get on the road for home, they showed me where to get on the base bus which took me to the Greyhound station in Columbus. For the rest of that trip, I marveled at the thought that I had just visited a parallel universe. That was the way it was in that spring of 1947. During the very next year, President Truman signed legislation that would eliminate black-white separation in the military.

Although the C-46 seemed reliable enough, I had heard from pilots that the C-47, with its nose closer to the ground as it began its roll for takeoff, was easier to fly. I thought the C-47 was the most reliable plane in the AAF, but I got a hop in another plane one day that seemed even steadier.

It came about more than a year later when Captain Cruthirds, who was commander of Westover's Air-Sea Rescue Unit, looked me up at the weather station to tell me he was going to fly to Selfridge. I had only had that one other hop to Selfridge, in the At-11, which he knew about because he and his wife had previously rented a room, as Marie and I had, in the same upper story of a frame house in Willimansett, Massachusetts and our wives became good friends. Marie, at that time, was again back in Michigan, awaiting the arrival of our daughter Willa. This flight was to be just the opposite of that trip in the somewhat unsteady AT-11. We flew to Selfridge in the captain's B-17. It was a plane such as this that Maynard "Snuffy" Smith from my hometown of Caro, Michigan had flown in on the day he fought the onboard fire and the Focke-Wulfs and became the first enlisted man to win the Medal of Honor in the AAF. I thought it would have been some coincidence if this were the same plane, the one with tail number 229649, but that, of course, was not possible as his 17 had been so badly damaged it had to be scrapped. This particular former war bird was now doing additional duty in air-sea rescue work. It had a lifeboat strapped to its belly that could be dropped to the ocean or other body of water should that need arise. The bomber droned along, steady, over Massachusetts, across New York state, over Canada, the sound of her four engines merging into a single unchanging hum that was loud but imparted the feeling that all was well and would stay that way. We circled down over Lake St. Clair. The big plane's wheels kissed the Selfridge runway. Through the window, the base seemed quiet, serene. No crash trucks. No ambulances racing alongside.

In the succeeding years in the service and for many years afterward, I saw many thousands of aircraft and looked for another AT-11 but never saw one. The AT-11, I began to think, never existed. The wind whistling through those bomb bay doors, and the chilling prospect of actually using a parachute for which it was designed must all have been a figment of my imagination. Then, one day about a half dozen years ago, I visited the air museum in Macon, Georgia. And there it was. There, safely on the ground where I thought it looked best, was the aircraft made by Beech and called, as I would learn, the "Kansan." It could have been the very same one that took me to Selfridge in 1947. And more recently, I was not surprised to find another at the Pima Air & Space Museum in Tucson. I was not surprised, for that museum, in my opinion and in the opinion of many others, is the best military aircraft museum in the world.

Top: A Beech Aircraft Corporation AT-11 "Kansan's" Plexiglas nose in front of a North American AT-6 "Texan." Both aircraft saw service as trainers in the Army Air Force during World War II. The At-11 was used in bombardier and gunnery training. Photo taken in the USAF Museum at Wright-Patterson AFB, Ohio.

Bottom: This is an R4D, the Navy and Marines designation for the Army Air Corps and Air Force C-47. Manufactured by Douglas as a DC-3, it was the typical civilian airliner of the 1930s. This aircraft probably has no end life. As "Puff the Magic Dragon," it was used as a mini-gun platform in Vietnam and is still employed in many ways today. Photographed at the Pima Air and Space Museum, Tucson, Arizona.

The author's father-in-law, Walter Nordstrom, in his Army Artillery uniform. Photo taken in 1917 at a studio near his base, Camp Custer, Michigan. A man who kept fit all his life, thirty years after this picture was taken he could wear that jacket just as he had at age 22, his age in the photo.

Made by Curtiss, the C-46 Commando was the largest twin-engine cargo plane of World War II. Because of its size and tail-wheel-style landing gear, the pilots sat far up off the ground and found it more difficult to taxi, for example, than the C-47 with similar landing gear. This one is in a hangar of Tucson's Pima Air and Space Museum.

Chapter 9

Family Life in Massachusetts and a Christmas Surprise

On an anniversary of D-day, June 6, the first of our three lovely daughters, Leta, was born. Marie and I felt we could not then afford the price of rent in the Westover area, so she remained with her parents in Michigan. In the fall, with my promotion to sergeant, she would join me. A three-stripe sergeant (buck sergeant) did not draw the kind of salary that would completely free us from what had been an unending stretch of financial worry, but we were determined that we would somehow make a home for our little family in Massachusetts. I took a short leave, got a hop to Wright-Pat, took a bus to Michigan.

That hop came up so suddenly that I had no time to secure an apartment. I was about to skip that free ride, and pay for public transportation home which I could hardly afford, when S/Sgt. Luty and others at the station insisted that I should take the hop and they would, they assured me, have an apartment ready for us when I returned.

We packed everything we thought we would need in a large steamer trunk and my father-in-law, Walter, drove us in his 1940 De Soto to Detroit's Michigan Central Station where we boarded a coach. The train would take us to Springfield, about five miles from Westover.

Somewhere near Buffalo our utterly charming and heretofore perfectly behaved daughter began to tell us in embarrassingly loud tones that she was not happy.[1] It was her mealtime and Marie had discovered only minutes before that the bottle with baby's formula had been jostled during a station stop and was broken. There was no more formula. What to do? A friendly porter came to our assistance. He took us past the white-tablecloth-covered tables of the dining car laid out with silver eating utensils and decorated with fragrant flowers and on into the car called the "galley" where he explained our predicament to the chef who was obviously quite occupied with preparing the evening's meal. The chef produced a bottle of milk from his refrigerator, saying, as he handed it to Marie, "You can warm it on this stove," pointing to the blackened cooking range fitted with nickel side rails that kept things from sliding off as the car made its occasional sidewise, parallel-to-the-track or vertical lurch.

---

[1] If my other daughters, Willa and Pamela, should by chance be reading this, I want them to know that they were just as charming and equally as well-behaved.

We were then traveling over a particularly rough set of tracks. The train was rocking to the extent that it was difficult to keep one's footing let alone try to operate an unfamiliar stove. So the kindly chef, who probably was not often interfered with while preparing the meal that would, I judged, be instantly recognized throughout the dining car as nothing less than exquisite, took time out to warm the baby's milk. And then, back in our coach, everything was again serene.

When we arrived at Westover, the weather troops took us to the apartment. It was at Chicopee Falls, the community about three miles to the south-southwest of Westover. The teletype identification of "CEE" for Westover was an adaptation from Chicopee Falls.[2] We rode there in Jim Ryals' 1939 or '40 Oldsmobile. Jim, a corporal and observer at that time, was one of the few observers who had a car. As would often be the case with friendships we made in the service, I would lose track of Jim after my tours at Westover, but we would briefly meet again years later when we were both stationed in Washington, DC, and he had advanced to Chief Master Sergeant. As we drove up the tree-lined street, you could see Jim and the others were truly proud of what they had accomplished. Finding an apartment for someone else clearly was not the kind of assignment most people would volunteer for. What if, after all your searching, and apartments were not easy to find, your efforts were not appreciated? As we turned into the drive of an attractive brick home that rose slightly up from the street behind a perfectly manicured lawn, I could see them glancing at Marie and I to see if we approved of their selection.

The apartment had been built into the second floor of the home. The landlord welcomed us inside, led us up the stairway on newly stained oak steps. The trim boards and the walls looked as fresh as if the house had just been built and we were to be the first occupants. It was much nicer than what we might have hoped for. In the exchange of pleasantries, the landlord learned that I was a weatherman. At that, he led us immediately out to the driveway where we could have a good view of his chimney. "Look at the top of the chimney," he urged. "Those three missing bricks were knocked off in the hurricane of '38." I was familiar with that hurricane. The news of that devastating storm had reached us even way back on the farm in Michigan. It had killed 600 people and caused property damage of about 300 million dollars. Given that the rest of the house and yard were in immaculate condition, it was apparent to me that the landlord had not had those bricks replaced in the nine years since the storm because he considered their absence to have some sort of historical significance, to be a kind of weather badge of honor. A man who approached meteorology in that way was one I could

---

[2] In addition to Chicopee Falls, many of the service families lived in the nearby communities of Chicopee and Willimansett.

enjoy associating with. Everything seemed absolutely perfect. This would be a wonderful place to live. But there was one little problem.

When they rented that fine apartment for me, my weather friends might have thought I was independently wealthy. Or, perhaps, in their enthusiasm at having been so successful in finding so nice an apartment, they just didn't remember how little a buck sergeant was paid. I didn't want my friends to think I didn't appreciate their efforts, but we couldn't take the apartment. The weather guys, disappointed, understood. I then explained to the landlord that, while I was sure we would feel very much at home in a house that had been touched by the hurricane of '38, my salary just wouldn't allow us to stay there. We would, of course, be helped in our budgeting by buying our food at the base commissary. Coffee, for example, was sold in light-brown, one-pound bags for 19 cents. This was what we called "GI coffee," it was coffee prepared expressly for the military with no brand name. Bread that had been baked at the base bakery was also available at low cost. These items and others that similarly gave the military family a break, made the commissary of that day a much more attractive place to shop than the commissaries of today, but we still could not afford to rent that fine apartment in Chicopee Falls. In fact, if food stamps were then available, we would surely have qualified for them.

We found a place we could afford in a small upstairs room in nearby Willimansett, about two miles away. The house, a frame structure somewhat in need of paint, was nothing like the apartment in Chicopee Falls. It was old and had settled over the years. It had been a long time since the windows had rested tightly in their frames. As winter approached, the landlady, who lived downstairs, stuffed rags in the cracks around the windows to subdue somewhat the wind that whistled through the rooms. On Mondays, Marie would open one of those windows, braving the cold, and hang the clothes out to dry on a rope loop that ran from a pulley attached to the outside of our room to another pulley attached to the house next door. The rope was pulled each time an additional article of clothing was pinned to it, thus moving the garment away from our house. When dry, the garments were retrieved in reverse fashion, and the rags that had been removed from around the window when it was opened were stuffed back into place to retain as much heat as possible during the cold days.

One night as I was returning from working the evening shift, I stepped off the bus in Willimansett, walked up to the door at the stairway leading to our upstairs room, and discovered I had forgotten the key. It was about 0200 hours. There were no lights in the house. Marie was asleep and so was everyone else, even the residents of the neighboring houses. Though I had always been proud of baby Leta's perfect deportment, this was one time when I wished she would awaken crying for her bottle. Weren't babies supposed to squall at about that time for the 2 a.m. feeding? Well, not this baby, at least not this 2 a.m.

Willimansett was normally a quiet community and that night it was absolutely still. The only sound was that of my footsteps on the walk. I knocked on the stairway door. No response. I wasn't surprised that the knocking couldn't be heard at our room on the second floor as there was another door at the top of the stairs that would tend to block the sound. The time honored trick of tossing pebbles at the window near which your beloved is sleeping was an idea that came to mind. With no streetlights and only a sliver of a moon, it was hard to find pebbles. Nevertheless, I found a few, tossed them. No response. It was obvious that pebbles of a size that would break the window were needed for that procedure to work. I might have been able to wake the landlady by knocking on her door which was on the ground floor, but I didn't want her to think she had rented the room to an incompetent and perhaps generate in her mind second thoughts about having rented to us. If she removed us there probably would be no other apartments available and my little family would have to return to Michigan.

I didn't want to call attention to my predicament. It was a simple error. I had merely forgotten my keys. I thought I was being reasonably quiet, but I must have made enough noise to have been noticed. Sometime close to 0300 hours as I dozed where I sat on the outer steps I was caught in the twin beams of two long D-cell flashlights. From the safety of their squad car, two of Willimansett's finest called out: "What you doing there? Who are you? Step up closer to the car."

Few of us, after having worked a night shift, are at our perky, bright-eyed best. I knew I didn't look like the sharp soldier that I might have been the afternoon before when that shift began. I was wearing my class-A khaki uniform, cotton trousers and shirt. The bottom of the khaki-colored tie was no longer where it was supposed to be, tucked inside the shirt between the second and third buttons. The starch in my uniform had lost its effectiveness a half shift ago. There were only wrinkles where creases in my trousers used to be. It had been about 20 hours since I shaved. In the eyes of the police officers, I knew I had to look suspicious. As I arose from the steps, they got out of their car, hands on holsters. A perfectly normal reaction, I thought.

"I live there," I said, pointing at the second story.

"Is that so? If you live there, why are you lurking around out here?" I began to formulate an explanation that might be convincing, knowing full well it wouldn't be, when they told me to remain standing there. They returned to the car, got on their radio, checking, no doubt, to see if any burglar matching my description had been reported.

The noise generated by their radio awoke my beloved spouse. She came down the stairs, unlocked the door, rescued me only minutes, I thought, before I was to be taken to the station house for booking and the interesting experience of lodgment in the crowbar hotel.

During that fall, on my days off, we liked to visit the Springfield Art Museum. We'd bundle Leta up in her snowsuit and we'd ride a bus to Springfield. With the passing of the years, all but one of the paintings have faded from memory. The one that remains was a canvas with dimensions the size of an ordinary door. Faithfully depicted on it was a gray farm house door upon which hung a hunter's bag of the day — ducks, rabbits, etc. You felt as if you could step forward, brush the game aside, reach out and open the door, so realistic was the work. It was probably the rural image, an evoking of my boyhood on the farm when I'd go hunting for pheasant and rabbit with my single shot 22 or full choke 16 gauge that made me remember that particular painting. If there had been a painting of a weather map I'd remember that too. (And, in those days, unlike the computer-produced weather maps of today, many of the charts plotted by the observer and drawn by the forecaster could easily, in my admittedly biased opinion, be considered works of art.)

At about that time, Base Weather received word from 8th Weather Squadron that the squadron would be sending students to a weather forecasting class that would convene at Chanute. Our Station Chief, cool, competent Master Sergeant Karl Abrahamson, who was also a forecaster and who later was appointed a warrant officer, asked if I wanted to apply. I then held the two Army occupational specialty codes (MOS) of 784, weather observer, and 790, weather equipment teletype technician. I had resolved to take as many technical courses as would be available to me, reasoning that the person who properly prepares himself just might be sitting in the lucky chair should opportunity make its appearance. I wanted to add the MOS of 787, weather forecaster, to my Form 201, the personnel record, but I felt it was something that would have to wait. It seemed to me it would be a bit too ambitious to aspire to the 787 MOS at that time. The challenges of the weather forecaster's job, not to mention the challenge of the weather forecaster's school itself, would be impressive. Perhaps the timing would be right some half dozen or dozen years in the future.

If anyone else would have asked, I am quite sure I would have responded in the negative, but I had immense respect for Sergeant Abrahamson. On several occasions while at Westover, before Marie joined me there, the Abrahamson's invited me to their home as a dinner guest. Abe would drive by the barracks in his 1940 Packard and take me to their home. We sat beneath a reflecting chandelier at a table setting elegantly done by Mrs. Abrahamson, complete with crystal glasses and wine. With their well-behaved elementary school-aged boys doing their best to look interested in the dinner conversation, which, of course, frequently turned to technical aspects of the weather, I drank in the warmth of that brightly lit family scene, so utterly unlike the barracks to which I would soon return. The barracks scene was dominated by two rows of bunks, one on each side of a center aisle bordered with occasional 4X4 posts. The foot-end of the bunks were aligned against the posts and were made up with olive drab blankets. The soldier's footlocker stood immediately in front of his bunk. Beneath the bunk were his low-quarter GI shoes and his combat boots. At the routine Saturday morning inspections, some soldiers would put the boots on display in their prescribed location beneath the bunk and hide their low quarters in duffel bags because the boots were not expected to be, and could not be, shined, whereas an insufficiently shined pair of low quarters could earn one a gig. (My old friend from those barracks days, Henry Klier, recently reminded me that we had been gigged one Saturday for having "indiscriminate bird droppings on the balcony." Life in the barracks was a precarious existence.) The boots could only be described as ugly. Made mostly of a rough gray leather, they were laced to a normal shoe height and topped with an orange-brown leather band. Two buckles attached to each band secured the upper boot to the leg. In the evenings, a few bare light bulbs hanging by wire from rafter supports cast criss-crossing shadows across these furnishings.

The Weather Forecaster Course had acquired the reputation of being a tough school indeed. It was said that any number of students that entered the school had washed out for academic reasons. For that reason, I might have avoided applying for it; on the other hand, if Sergeant Abrahamson thought I was qualified, it could be reasonable to think I could handle it. So without thinking much about my chances, I applied. In any event, I was sure there was little chance of my being accepted. The weather forecasters I knew were considerably older than I. It seemed obvious that men with more experience would be preferred. And so, as the weeks went by, somewhat secure in the knowledge that I wouldn't see orders sending me to Chanute, with my grease pencils I continued to strike out irrelevant data groups in the weather collectives, aligned my combat boots beneath my bunk for the Saturday inspections, and forgot about the application.

One day, we were called in, one by one, to see the Station Weather Officer, Major Frank Angus. He asked us if we wanted to volunteer for some other branch of the Army. There were a number of choices: the Signal Corps, Medical Corps, Chemical Corps, Transportation, or the combat arms: Infantry, Artillery, Armor (battle tanks). This had come about because the old AAF, the Army Air Force, was to be no more.

We had known something of the sort was in the offing, but the need to make that kind of career decision still came as a shock. The triggering event occurred on 18 September 1947 when the War Department was abolished and succeeded by the Department of Defense. At the same time, the Department of the Air Force was created. The old tried and true Army Air Force, the organization having its origins in the earliest days of aviation when, as the air section of the Signal Corps, canvas and sticks held together with glue and wire made up the War Department's front-line aircraft. That organization became known as the Air Service in 1917, the Army Air Corps in 1926, and the Army Air Force in 1942. By that time, the aircraft inventory would include such famous planes as the P-47 Thunderbolt, P-38 Lightning, the B-17 Flying Fortress and others. Major Angus advised that we did not have to transfer to an Army branch if we wanted to stay in the organization we knew so well. We would then simply be transferred to the new Air Force.

I believe we all thought opportunity was knocking, that we were getting in on the ground floor, for everyone of us at Westover's Base Weather, and everyone else that I knew of then, chose to go with the new Air Force. Years later, in the 1960s, after I had returned to the Army as a warrant officer in the Artillery Branch, I met a lieutenant colonel at a fire base in Vietnam who had, back then in 1947, actually elected to transfer out of the proposed Air Force. The difference in our rank appeared to indicate that some of us who thought we were getting in on the ground floor, with that new independent branch of the military, had stayed closer to the floor than would otherwise have been the case.

So we all came out on orders transferring us from the Army and into the new United States Air Force. There was no need for changing duty stations, we simply went on with our weather work in the same place and in the same way as we had when we were in the Army.

In December of that year, Marie and I and baby Leta took a furlough to be in Michigan over Christmas. This was our second Christmas while in the service and we felt fortunate indeed that we were able to spend both of them at home. It was a great furlough. We visited family and friends, opened presents at the two family Christmas trees, and just relaxed in the thought that we were going to enjoy 30 days with absolutely no care in the world. Then, with only half the furlough gone, a telegram: I was to report immediately to Chanute Field to attend the Weather Forecaster Course.[3]

---

[3] More than 50 years later, at a reunion of Air Weather Service members in St. Louis, my friend Tom McClung, from the old 8th Weather Squadron headquarters, told me that he had been slated to attend that class but had declined. It then seemed evident that I had been an alternate selectee, taking Tom's place.

Top: The author's brother-in-law, Robert Nordstrom, a mechanic in the AAF, poses beside a P-51 at his airbase in Japan. The letter partly showing in his pocket announces the birth of his niece and the author's daughter, Leta. *Bottom:* This typical World War II-era barracks is preserved at the Pima Air and Space Museum in Tucson, Arizona. The author lived in buildings identical to this at Chanute Field, Scott Field and Westover Field.

Chapter 10

Air Training Command's Weather Forecaster Course

Over the years, the weather forecaster classes varied in length. At the time I attended, it was 26 weeks (six months) long. Not long after I graduated, it was lengthened to nine months and later, I believe, was made even longer and the subject matter was broken up into two phases taught in separate courses.

Weather Forecaster Class 01148 to which I was assigned contained, perhaps, the most interesting student body that had ever gone through the forecasting course. Admittedly, there were a few uninteresting chaps such as myself who had little service experience, but the great majority of students in that class had resumes that were indeed captivating. These were the veterans of World War II, the "big one." Now in the service as sergeants, most had been flying officers who had separated from the service when hostilities ended. Probably disappointed with civilian job prospects and finding they could not regain their commissions, they entered the Air Force as enlisted volunteers.

In the course of our 26 week school, I would get to know the ex-officers very well. They were an easy-going lot, not overly impressed with rank and possessed of a genuine concern for others. They had been there and back and did not need to be troubled by the then current saying: "Watch how you treat others on your way up; you'll meet them again on your way down." It was obvious they had deserved their commissions. When the Korean War broke out a couple of years later, I was happy to learn that most of them, if not all, regained their commissions.

They had been lieutenants or captains, but were now serving in what we called the "first three grades," that sort of elite grouping of the top enlisted ranks — in ascending rank order: staff sergeant, technical (or "tech") sergeant, master sergeant. I was told that their grade determinations were based on how long they had stayed out of the service. Most of them actually managed to return to service as master sergeants. Some had to settle for tech sergeant. Only one or two, in that class, were made to fall all the way back to staff sergeant. I didn't know any who had to accept the next lowest grade of sergeant (with no prefix), the rank I held.

During the war, most of the ex-officers had served in the commissioned aircrew positions. We had former pilots, navigators and bombardiers. On combat missions, most likely in B-17s, B-24s or B-29s, they had undergone experiences no one would want to repeat. I felt it was no more than right that they should have been made first three graders, but the fact that they held those positions caused all promotions to those grades to be frozen. No buck sergeant, such as myself, could then hope to be promoted to staff sergeant.

On two occasions, while attending the forecasting course, the value of a promotion to staff sergeant became abundantly clear. The first was the day I was called out of class to be told I had to work KP. That was quite a shocker. Aside from the fact that any of us who had ever pulled KP detested that chore with a vengeance was the fact that, after all, I was then a sergeant. In all the organizations that I had been connected with, sergeants did not pull KP. And even if you granted that they had too few lower grade airmen to fulfill the KP requirement, one would think that the mere fact that I was then a student in one of the most valued courses the Air Force had would have insulated me from that onerous duty. And it was no small thing to miss a class day. The weather forecaster's school at Chanute provided training in meteorology that was probably better than the training that could be obtained in fulfilling university requirements for a bachelor's degree in that science. In order to accomplish that level of instruction in a mere six months, the course work had to be heavily concentrated and tremendously accelerated. But mine was not to reason why. I worked KP that day seething in the knowledge that the rest of the class would be receiving instruction that I would know nothing about.

The second disappointment deriving from being a mere buck sergeant was much more important. Just as we were about to graduate, warrant officer appointments were opened. This selection process was typically conducted on only certain years and over a short time interval. As an example, this opening occurred in 1948. To my knowledge, that would only happen one additional time — as I remember it, in 1952 or '53. In a decision that seemed strange to many of us, not long afterward the Air Force would altogether abandon that rank. The warrant officers who were then on board would be permitted to remain as warrant officers but no new warrants would be appointed in the Air Force. In the 1948 selections made in the weather field, the candidate had to take a test in his subject area. Since the test questions would reflect the course material taught in the weather forecaster class and probably were written by that courses' instructors, anyone, such as I, who had just completed or was about to complete that course would be a shoo-in for passing the test. But, as was usually the case when the "good deals" came on the scene, there was one little wrinkle — the applicant had to be serving in the first three grades. Buck sergeants such as I were not qualified. When that next window of opportunity for a warrant officer appointment arose, in the 1950s, I would be a tech sergeant. But in opposition to that old Army Air Force policy where anyone at or above the rank of staff sergeant would be eligible to apply, the Air Force tightened its selection procedure so that only those serving in the top enlisted rank, only master sergeants, could apply — with incredibly bad luck, I found myself once again just one stripe shy of the qualifying rank.

Weather map analysis was one of the most enjoyable classes in the weather forecaster course. Every day during that block of instruction, each student was given a map of the United States and Canada that was preprinted with weather observations. In analyzing this "surface weather map," the student drew contours with an ordinary pencil conforming to pressure values adjusted to sea level that were plotted on the map. He also located the positions of any fronts, troughs, or squall lines. Cold fronts were drawn with a blue pencil, red pencil was for warm fronts, purple pencil for occluded fronts, brown pencil for troughs, and green pencil for squall lines. All areas of precipitation were shaded in green, all areas of fog in yellow, and all areas of dust in brown (a small piece of window screen placed under the map while shading provided a pleasing stippling effect). The specific symbol for the kind of weather was marked on the map as a character about a 1/2 inch in size, in green except for the thunderstorm which was in red. Then, as a final touch, he would draw a red L at the center of each low pressure system and a blue H at the center of each high pressure cell. On the next day, the maps were returned with grades assigned for correct frontal delineation, accuracy of pressure analysis and the many other map attributes, not the least of which was the map's overall professional appearance.

Some of our students visited a classroom across the hall where a weather forecasting class was conducted for officer students. They assured us that a map for which we enlisted students would receive a grade of 75 would get the officer student something like 95. Whether accurate or merely an attempt at a joke, it was amusing for we knew that was as it should be, rank, after all, had its privileges.

To make an accurate scientific analysis of a weather map is something that every meteorologist with proper training can expect to accomplish; but to also portray all the wiggles in the pressure fields in pencil lines of unflagging smoothness and of uniform width is something that only a few analysts, those who are truly artistically gifted, can hope to achieve. Upon entering the classroom on the day after maps had been analyzed, the student body made a beeline for the bulletin board where a completely analyzed map, that had been plotted with the same set of surface weather observations from which the students had made their analyses, was on display. That map, the correct solution, would have been analyzed by Mr. Nobel, the civilian instructor who taught map analysis and a few other subjects. We would stare at his result and envy his master's touch. Not only were Mr. Nobel's fronts and isobars (lines of equal pressure) accurately located, but his isobars were absolutely unwavering in width. A number of us went to great lengths to duplicate what was called the "PA" (professional appearance) of Mr. Nobel's maps. I bought a number of mechanical pencils that I thought might hold the key to Nobel's absolutely uniform pencil lines. They did not.

In the succeeding years, I would draw thousands of weather maps and, always conscious of how Mr. Nobel's analysis would have looked, I would take pride in my smoothly curved pencil lines as well as in the appearance of the maps in general, but not once in all those years did I think I had equaled his level of eye-pleasing competence in map analysis. In addition to being scientifically accurate, his truly were works of art. I could easily imagine a collection of his maps hanging as a one-man show in some respected art museum in a major metropolitan area. "This must be an early work," one of the visitors would say. "Pessimism abounds in the preponderance of red L's as opposed to so few blue H's." Or, "This must be from his green period — look at all the rain in the Mid-west."

"Yes," another would agree, "and it has the mood of an El Greco, see the thunderstorm symbol over Toledo."

We studied meteorological theory from the recognized texts: *Weather Analysis and Forecasting* by Sverre Pettersen, *General Meteorology* by Horace Byers, *Dynamic Meteorology* By Bernhard Haurwitz, and others.

A separate block of instruction in surface map analysis was devoted entirely to tropical weather systems. It was taught by a tech sergeant who had served in the Caribbean and knew his subject. Unless you had a hurricane to contend with, the maps in those low latitudes could be immensely boring. There were no fronts and little change in pressure over comparatively large areas. "And," probably in reference to a then recently popular song, *Drinking Rum and Coca Cola,* the instructor said, "there was nothing to do there but drink rum." In contrast to the more obvious weather systems of the northern latitudes, when analyzing maps in that area one scrutinized the plotted data for signs of thunderstorms or lightning and the slight upward bend of isobars, any of which might indicate an easterly wave, usually a weather-maker, was traveling through the area, and one would look for the long east-west line of weather that marked the position of the intertropical convergence zone, or ITCZ, as we referred to it.

In a class called "Auxiliary Charts," we learned to analyze the earlier form of the thermodynamic diagram called the "Pseudo-Adiabatic Chart" which the Air Weather Service would replace, in the 1950s, with the "Skew T - Log P Diagram." These provided the forecaster a wealth of information such as atmospheric stability, including the tool with which to calculate the height at which cumulus clouds would form, the potential for thunderstorm development, and the altitudes at which icing would occur.

The basic data plotted on the thermodynamic diagrams came from radiosondes. Tied to balloons and with paper parachutes in their trains to more gently ease their descent to earth after the balloon had burst in the rarefied air high above, these instruments transmitted to their receiving stations continuous measurements of temperature, humidity and pressure. Tracked in those days by an operator seated at the SCR-658, the "bed springs" as we called it because that was what its antenna configuration resembled, the system also would produce winds aloft and do this, unlike the theodolite wind-calculating system, regardless of cloud cover and in all kinds of weather. By turning hand cranks that moved the antenna to adjust two green blips on an oscilloscope-like screen, the operator kept the SCR-658 aimed at the ascending radiosonde. Sitting outside during cold or wet weather was not a job that the operators enjoyed. During the Korean War period, the SCR-658 was superseded by the GMD-1, an electronic direction finder that eliminated the need for the outdoor operator.

All the training in the Weather Forecaster Course was aimed, of course, at developing in the student an ability to predict the weather. As a part of the course work, we were expected to make predictions of ceilings and visibilities, parameters critical to aircraft operations. But we would learn that we would not leave that course with the ability to make reliable weather predictions. That level of expertise would require more detailed study than any classroom could provide. Confidence in making reliable predictions would come to us after we graduated — at our weather station assignments where we would fine-tune the classroom concepts with experience gained in the field.

While trying to make those forecasts in the school, we employed all the documented forecasting procedures. One of these employed a mathematical formula, taken from one of our texts, which used the magnitude of pressure change over the preceding three hours to predict the future movement of a front. It was neat. You plugged the numbers into the formula and out came the future front position. Here was a positive, no fudging forecasting procedure. It seemed we had found a forecasting tool that allowed of no compromise, the hard, fast answer we were all searching for. But meteorology doesn't lend itself to the easy solution. It was beautiful but couldn't take into account the dynamic, changing pressure field which would, in turn, change the positive or negative accelerations of frontal movements. In back-calculating, it told you *precisely* where the front had been, but only *approximately* where it was going. This, perhaps, was our first introduction to the real world of meteorological forecasting. Like some other forecasting "rules" we would encounter, it was nice but not very practical.

About midway through the course, Marie and baby Leta joined me. They had stayed in Michigan until I could find a suitable room to rent. This was in one of the old Victorian-style houses in adjacent Rantoul. We made friends with a couple who rented one of the other rooms. He was an ex-officer, a master sergeant like so many of my classmates, who was just entering the next forecasting class. Not sure if he was doing the right thing, he expressed some doubt as to his choice of career, but his wife was gung-ho, not in the least doubtful, saying: "My husband was really fortunate in being assigned to this school, there's going to be a good future in meteorology." She had a vision of the field opening up in tomorrow's brave new world. And, I thought, like my luck in getting in at the beginning of the new Air Force, she thinks we are in on the ground floor of a science that would only increase the dramatic expansion it had experienced during the war. At that time, I was sure her assessment was not much more than idle chatter. But I would remember her remark when, in the 1970s, the environmental movement created more vacancies in meteorology than could easily be filled, and many a university or junior college, that never offered a meteorology course before, felt they could no longer ignore it. This would be quite a contrast to the situation when I had elected to enter the field. At that time, I was told there were only eight institutions of higher learning in the entire country that offered meteorology as a degree program.

On weekends, Master Sergeant Bud Hartel, who had a beautiful 1941 Lincoln, would take the three of us on outings to the university town of Champaign-Urbana. We would be accompanied by other close friends — Sgt. Collins and S/Sgt. Ablard. Collins was from the South and liked to sing well-known passages from operas. I lost touch with him after forecaster's school but believe he could have followed a singing career if he so chose. In a few years, with the outbreak of the Korean War, Bud Hartel would regain his commission and ultimately retire as a lieutenant colonel. Forty-four years later, I would meet him at one of the Air Weather Association's reunions. With respect to the Korean War, I assume that a number of classmates went there, but I never heard of any excepting a master sergeant from Ohio who, I was told, had his weather station overrun by the North Koreans or their Chinese allies.

That first half of 1948 was a pleasant period. And when the end of the course came, it seemed to arrive all too soon. Our class had started with about a hundred students and ended with little more than half that number. With graduation, orders came out that would return me to Westover Field where I would assume the duties of a fledgling aviation weather forecaster. I requested and received authority to take a delay enroute — we were going to spend some time at home.

We thought it was high time to become owners of an automobile. So we cashed in the war bonds that Marie had purchased when we worked in the war plant, all the money we had in the world, and bought a 1939 Lincoln Zephyr, a car that was nine years old. Built like a tank, but with luxury in mind, it had two glove compartments, one on the passenger's side and another on the driver's side, and two cigarette lighters, the additional one in the back seat passenger's compartment, though none would have been sufficient since we didn't smoke. It was a fine looking car, but it had one serious defect. The previous owner had not treated the Lincoln's 12 cylinders with the care that he might have. As we passed along those Indiana roads, so familiar from my days of hitchhiking while in weather observer's school, we were able to pass everything on the road except gas stations — it was not necessarily gas that we always needed, but oil for that worn-out engine. But we didn't let that bother us much. It was the Fourth of July and we were in a holiday mood.

As we slowed to enter the business district of Logansport, Indiana, all still seemed right with the world. Then, the gut-wrenching banging of a thrown connecting rod brought everything back into perspective. It *was* a holiday and there were no mechanics in town. There was nothing to do but abandon the non-operable 12 cylinders and take the Greyhound. We did that and ruefully concluded, on that bus ride to Michigan, that we had a lot to learn about buying automobiles.

Top: Baby Leta and the 1939 Lincoln Zephyr at Rantoul, Illinois.
Bottom: After the school had committed itself to our graduation and calculably would not rescind that decision, I drew a cartoon of the principal instructors as a rogues gallery. The numbers below the heads are the typical values, in feet, of the contour lines on a 700 millibar chart. The instructor on the left never told us what he meant by "the three hisses," his response to our unflattering reactions to the jokes he told. He would rise to high rank in the AWS. The cartoon was posted on the classroom bulletin board.

Chapter 11

Castles in the Air

During that July furlough, I took a day's trip with Dad back to the farm near Caro where he was to repair the roof shingles on the two-story farmhouse. He was then 59 years old and though still seemingly as spry as ever, I knew he could use some help. Just walking around the old place which had been unattended since Dad went to work in the Detroit war plant, brought forth a sense of nostalgia that was intensified by the fact that I had not been there since before I entered the service.

The farmhouse was a typical frame building that had been constructed at about the time of the Civil War. Up on its roof, we had a commanding view of Cogut Creek, the fields, the woods and the distant farms. I could still make out the old sled run, where I flew down the gully on my Flying Arrow and then onto the frozen creek. On one of those cold cold Michigan days, one of my boyhood buddies wanted to test the theory: that his tongue would stick to the sled's metal frame. Mother, our ever-dependable fixer-upper, rescued him but only after admonishing: "Don't do things like that!" He probably lied about his age when he joined the Navy in World War II. Some ship, I think it was the *Yorktown,* took him with it when it went down.

Only the huge back-yard maple with a trunk diameter of three to four feet, on which we had our tire swing, obstructed my view. Stretching to see beyond it, I looked far to the west where the wild strawberries grew and, beyond them, the blackberry bushes that stood at the edge of the popple woods. If you walked on past the popples you'd come upon the Indian reservation and the Cass River. One day, as I daydreamed lying on my back on the river bank, observing the drifting cumulus while chewing on a stalk of timothy grass, Helen Fisher, the aunt of my silent boyhood Indian friend Silas, noiselessly and completely unnoticed walked up to me. "You have a pike on your line," she said. Somewhat embarrassed at having been taken by surprise, I jumped up to see my bobber gone. I was sure it couldn't be a pike. It couldn't be a pike because, for week upon disappointing week, I had been hoping to catch a pike, but every time I pulled in my line with a fish on it, I'd see the much smaller rock bass. There were a few times when, by the fight, I was sure I had a pike on the line, but those were the ones, if indeed they were pike, that got away. "Grab the pole, throw the line high up over your head and onto the bank," Helen advised. That I did and there on the bank, flopping its speckled sides three to four feet into the air, was the must beautiful Northern Pike I had ever seen.

Mr. and Mrs. Rudolph Setla had the farm to the immediate north. On Sunday afternoons, after the polka program aired on the radio from the Hotel Winona in nearby Bay City, their three sons would put together their own musical program with Joe on the violin, Ted on the accordion and Walt on the drums. Theirs was a spirited brand of music whose notes traveled out to the farthest reaches of our farm. I marveled at the talent and determination of anyone who could take it on his own to learn a musical instrument as they had and then play with a verve that challenged the listener to detect a miscalculated note of which, as far as I could tell, there were none.

Dad was another who had accomplished the same. As a little boy in the Old Country, he happened to see a traveling musical group in the village and, while telling Grandfather about it, said he'd like to have a violin. "If you want one, you shall have one," Grandfather responded. And with that said, Grandfather began to collect the wood from which he made Dad his first violin. Dad's tunes, entirely constructed from the memory of those heard as a boy, was lively but in a subdued way. As I watched him pull the bow across the strings in the living room of this very farmhouse and concentrate with half-closed eyes to recall the coming changes in the melody, I imagined he was deliberately muting the tunes as if it would have been obscene for him to recall in a completely joyous vein anything at all about that faraway and deplorable land, even if it were only music.

To the east, directly across Highway M-24 on which our farm stood, was the Walter Hoag place, especially notable for its black walnut trees. Their son Thurlow and I used to crack black walnuts with Mr. Hoag's ball peen hammers on the concrete steps of their tool shed, the brown stain from the shells staining our hands for weeks to come. Mr. and Mrs. Hoag always seemed pleased to see me. Walter Hoag had been a truck driver on the Western Front during World War I. A man who loved to rib kids, he always greeted me with: "What do you know that you'll stick to" — a phrase that would come back to me many times in my future weather forecasting days.

The Shepherd's 20-acre farm abutted the Hoag place. Mr. and Mrs. Eugene Shepherd would suffer a terrible loss when their son, Chloral, who was a bit old to be drafted at the outset of our involvement in World War II, was called up in 1943 as the war's manpower demands grew more severe, left his late-model Pontiac in their care, and was killed on Kwajalein in the Marshall Islands in January or February of 1944.

Farther east and out of sight except for the large willow trees that grew in their yard on Riley Road, was the home of Mr. and Mrs. Ben Burrows. Ben Burrows was a great buddy of Dad's and the Burrows were frequent Sunday evening visitors. Ben had been a sergeant on the Western Front in World War I and had hundreds of wartime stories that he loved to tell. Dad, who had missed service in that war due to an eye injury suffered at the Ford factory, enjoyed Ben's stories. Ben said he had so impressed the commanding general during an inspection of their tented quarters at El Paso, that the general made him his personal driver with the primary duty of driving his daughter to wherever she wanted to go. Whether this duty led to any romantic encounter or not, Ben was too gentlemanly to say. After arriving at the trenches in France, Ben never wanted to see the German he was aiming at. Like Dad,[1] he had relatives on the German side, so, to avoid seeing a German who might be a relative, he would simply lay his Enfield rifle atop the sand bags and shoot with his eyes closed. Once, while separated from his unit and behind German lines, he was lured by the aroma of fresh-baked bread that wafted up from an underground bakery. Yelling "Hands up!" he took the German soldier-bakers by surprise. Thinking that a trade of bread for their lives would be a good bargain, they thrust armfuls of warm loaves upon him. He stuffed the loaves inside his shirt, hightailed for the deep woods, ate the bread, fell asleep while hidden under a pile of leaves, and awoke to the sound of American voices as US forces retook the area.

About a mile farther east on Riley Road, like the Shepherds, the Roketa family would suffer a terrible loss during World War II. Their son, Anthony, who had been a classmate of mine in the one-room Connor elementary school and was a couple of years older than I, was drafted late in the war and killed in the push toward Germany.

As I looked to the south, toward Eugene Bracci's place, the bachelor Italian farmer and another great family friend, I could see the vineyard where he grew grapes with which to make wine and where he

---

[1] With a German mother and Ukrainian father, Dad lived in a multi-cultural world. He had an intense interest in history and mastered English as the last of five or six languages in which he was fluent. He would easily shift from German to Ukrainian to English without losing his train of thought. Mother, who came to America as a baby and was raised in the Troy, New York area, had Ukrainian parents though her mother, who was raised by French nuns in an orphanage, favored Polish but sometimes spoke French when she didn't want others to know what she said.

would hide with his ancient double-barreled 12 gauge to seek retribution from the chicken hawk that made periodic raids on his chicken pen. As a boy, whenever I needed a haircut, Mother would send me to Eugene's where he clipped my hair with a hand-operated cutter. An immigrant of the 1910s or 1920s, he was proud of the fact that he came from Rome, the Eternal City, which he called "Roma." I was conscious of the fact that his trip to America began from a much more historically important place than Lemberg, the Austrian city that Mom had been carried away from as a baby, and infinitely more important than the small town of Sokol, about 50 miles from Lemberg, that Dad had left at age 17. When I knocked on Eugene's weather-beaten screen door, I would see him seated at a rough table, minus tablecloth, that somehow stood solidly on irregular but smooth floor boards worn down from years of his and some unknown previous owner's footfalls to the extent that knots in the boards, more resistant to wear, protruded upward a full half-inch. Dressed in a gray work shirt and gray trousers belted with a rope, he would look up from his Italian language newspaper on which his name had been printed as "Eugenio" and whose title, as I remember it, was *Voce di Populi,* and would cheerfully greet me with: "Ah, Teodoro!" Then, to my amazement, he would open a conversation as if he were talking to an adult and not a mere ten-year-old.

World War II was beginning, and Eugene bemoaned the fact that the Italians were on the wrong side. He often told me the world was in bad shape. The "wronga powers" in the form of Hitler and Mussolini, he solemnly explained, had gained the ascendancy. In all our visits, he would eventually repeat his great disappointment with world events. I tried to find ways to repay him for the free haircuts and one day, after telling me he was continuously missing the chicken hawk, no matter how carefully he aimed his shotgun, I said fighter pilots in the war were told that eating carrots would improve eyesight. About a week later, I met him in his carrot garden. He said he had eaten most of his carrots and had used up all of his shotgun shells, shooting at the hawk that I observed was even then circling overhead.

Unwilling to let Eugene live without curtains in that drab house sided with vertical boards that had never seen a coat of paint and had weathered to the point of matching the gray of his clothing, Mother made him curtains, ironed them with flat irons heated on our wood-burning kitchen stove, and enlisted Dad to install them on Eugene's old windows while he stood in wonderment, not knowing how to help and seeming altogether surprised that windows needed that kind of frippery. If not an exponent of *House Beautiful*, one had to admit he was an excellent farmer. He even drew praise from

Michigan agricultural officials who had stopped to view his magnificent stand of sweet corn, but he sold no foodstuffs, only raising enough field corn for his chickens and only enough sweet corn and other vegetables to provide for his personal food requirements, which did not seem excessive. Whenever I caught him at mealtime I'd see him with a bottle of home-made red wine that he drank while munching on chunks of hard rye bread bought in town. In that day before lotteries in the States, he talked about the Italian lottery and we assumed he lived off winnings he had won in his home city of Rome.

I could picture our cornfield. It used to lie between Eugene's house and ours, just beyond our orchard of a dozen or so apple trees and three or four pear trees. There were several kinds of apple in our orchard, some varieties that probably no longer exist. The Caro area climate was great for fruit growing, and in fact it has been written that "the largest apple ever picked weighed 3 pounds, 2 ounces, and it came from a tree in Caro, Michigan...." (from *Visit Detroit,* Detroit Metro Convention and Visitors Bureau, Detroit, Michigan, 2001.) A waving sea of golden rods now stood in the place of our corn stalks. I remembered the days when, as a boy of 10, I worked the corn field with Dad. He in the rear holding the handles of the five-toothed cultivator and I in front, leading Prince, the big black horse, by the bridle, keeping him on a straight path between the rows. Prince, we learned, was exceptionally smart but was possessed of a devious character. If I were not there to keep him from it, I know he would have taken great satisfaction, if not sheer delight, in running the cultivator right through the planted row, destroying the tender corn plants in the process. He could have done it in spite of all Dad's efforts to hold the cultivator between the rows. Yet Prince was generally a kind animal. On a number of occasions, as I led him around the turns at the ends of the rows, his huge hoof completely covered my tennis shoe, but knowing he had done that, he stepped down so lightly that I hardly felt it. And Prince would remind us that his IQ was not trivial. If there were human-like occupations among equines, I know that with Prince's IQ he could have been the community's weather horse. At precisely 5 p.m. of each cultivating day when we wanted to perhaps cultivate one or two more rows before quitting for the day, Prince would break for the barn and there was nothing we could do except to attempt to keep the flying cultivator from being damaged on a rock or a tree trunk. This was not a case of Prince making a single lucky guess of the time, for he would repeat that performance. He did it every day that we worked that field. I don't know how he knew it was exactly 5 o'clock unless, perhaps, he understood English and heard us commenting that it was almost time to quit. Many years later, in Vietnam, I would think fondly of Price as I wore the black-horse shoulder patch of the First Cavalry Division.

If you were raised in that part of rural Michigan, you early learned that it was bad form to advertise your accomplishments. In fact, it was unforgivable, and that even applied to conversations with your parents. Nevertheless, I managed somehow during that 100-mile trip to the farm to nonchalantly mention to Dad that I was now a graduate weather forecaster. So, it followed that, on that morning, as we were preparing to nail the first pieces of replacement shingles to that steeply sloping roof of the two-story farmhouse, Dad asked the obvious question: "Theodore," he asked, "will it rain before we finish the job?" There were few times in the succeeding years of forecasting that I felt the pressure of the profession more intensely. When he asked if it was going to rain, we were staring at bare roof boards whose protective shingles we had just removed. If it rained before we finished the job, the farmhouse that had served our family so well and was now in its advanced age, would suffer damage it could not easily withstand. But there was more. I wanted to show this man whose entire life had been dedicated to efficient, honest work, that I had learned something that could be put to practical use. Although he had been diplomatic enough to not say it directly, Dad didn't appear to share my enthusiasm for the science of meteorology. When he met his friends after I had graduated from observer's school, he introduced me this way: "This is my son, Theodore, he works in astronomy." Of course he could see the value in being able to foresee the changes in the weather — his days of farming during the Great Depression would have made that obvious. I thought he meant it as a joke, for he loved jokes, but I also took it to be a subtle way of saying it was better to have your eyes on the stars than to have your head in the clouds.

But to return to the question. It is a first rule in weather forecasting to know your initial conditions. Up there on the roof, the only weather data available to me were the wind and an admittedly spectacular view of the sky. It was extremely little upon which to base a forecast. So my reply should have been that I didn't have the meteorological charts that were necessary to make a forecast. But this would probably be the only chance I'd ever have to give my father a forecast and I wanted to do it. And much more than that, of course, it was very important to my credibility that the forecast verify. What about those initial conditions? The wind was light and from the southwest. Stretching to examine the horizon above the peak of the roof while being careful not to raise my center of gravity to the slipping point, I looked far to the southwest, past Cogut Creek and the popple woods. There, in the distant southwest, I saw a perfectly shaped single line of the somewhat rare cloud formation called "altocumulus castellatus" which we abbreviated in the hourly weather reports as "acc." No other clouds were in the visible sky — just that beautiful line of middle-level cloud shaped like towering castle turrets. Way back in weather observer school an instructor told us such clouds were indicative of a particularly unstable atmospheric layer. Being a middle-level cloud, that layer would be at least 7,000 feet above the ground. Below it, due to the good radiation of the previous night's cloudless sky, the air had cooled and stabilized. If there were enough heating to destabilize that ground-based layer, thunderstorms would be a reasonable prediction. And on this day, with the sky clear except for that one line of acc cloud, there would be ample opportunity for heating. So I said it. Pointing to the southwest, in my best sort of nonchalant conversational tone, I said: "Those clouds are predictors of thunderstorms that should develop in the afternoon."

After our lunch break, as we hurriedly tacked in the last of the new shingles under a rapidly developed dark overcast, thunder broke loose. The downrush from a cumulonimbus cloud swept past as we scurried down the ladder and threw our tools into the trunk of the '40 Ford in the midst of a violent downpour. As I drove the Ford on that 100-mile drive back to Detroit, I was quite sure Dad had briefly looked at me in a sort of wondering and perhaps even admiring way, probably thinking — "Could this boy really be a weather forecaster?"

The author, on furlough after graduation from Weather Forecaster Course, introduces one-year-old Leta to the horses at his mother-in-law's (Martha Nordstrom) family farm near Hastings, Michigan.

Top: The Army Air Forces silver badge authorized for weathermen with six months' service in the AAF and graduation from the Weather Observer Course or the Weather Forecaster Course.

Bottom: At the time the author attended weather observing and forecasting schools this cloud was called altocumulus castellatus. Its name was later changed to altocumulus castellanus. On the observer's official recording form, the WBAN 10A (called "weeban"), this cloud would have been abbreviated as "acc." WBAN was the abbreviation for Weather Bureau, Army and Navy. To conserve space on the form and to save time as the reports were transmitted over teletype, abbreviations were extensively used for clouds and weather types.

Chapter 12

The President's Plane is Running Out of Fuel

The prospect of duty at Westover not as a weather editor, as I had been during my previous stint, but principally that of briefing aircrews for their long over-water hops, mainly to the Azores and Europe, was indeed intimidating. Upon reporting to Westover Base Weather, while familiarizing me with my new forecasting duties and wishing to impress upon me the seriousness of my new job, I was told that the forecaster would have to pay for any plane that was lost due to his faulty weather forecast. In fact, I was told, a tech sergeant in Alaska had lost a B-25 due to unforecasted bad weather and was thereupon promoted to full colonel since he couldn't possibly pay for the plane on a tech sergeant's pay. I looked for any signs of insincerity on the part of my informants but could find none. I still think they really believed it, but of course, I did not. I had been away from the farm too long to fall for that one, for even a colonel's salary wouldn't go far in paying for a B-25. For anyone who might be interested, I should add that in my subsequent weather career, in all my years in the Army and Air Force weather services, I never seriously thought I knew a colonel who acquired his rank in that way, though I'd be less than honest if I did not admit I wondered about it a time or two.

I quickly learned there would be no time to dwell on such thoughts. About a month before my return to Westover, the Soviet Dictator Joseph Stalin, who was as feared by the democracies as Hitler had been and was even more murderous, having killed 20 million Ukrainians by simply starving them to death and having sent untold thousands to Siberian slave labor camps, had elected to impose a blockade on Berlin that began to impact heavily on our workload. The blockade, more precisely, was imposed on West Berlin, a Free World island occupied by the Western Powers that had been established in 1945 by the Potsdam Conference attended by Stalin, President Truman, and Britain's Prime Minister Churchill. Responsibility for West Berlin had been divided into three zones. The French occupied the northwest, the British and Americans occupied the southern portion.

As the primary east-coast base for MATS, the Military Air Transport Service, successor in 1948 to ATC, the Air Transport Command, and the organization to which the Air Weather Service was assigned, Westover was the jump-off point for Project Vittles, the name given to the stateside portion of the Berlin Airlift — the Western Allies' effort to supply West Berlin with life's necessities and keep that important city in the Free World orbit. For Stalin, who was experiencing heady success in gobbling up Poland, Czechoslovakia, Hungary, Rumania, Bulgaria, Yugoslavia, the Baltic states and East Germany, to secure all of Berlin, a capital of world-wide significance, would have been a crowning achievement that would predictably lead to even more acquisitions. Berlin's fall could not be tolerated.

Berlin's fall to the communist sphere would signal to the world that Stalin's way was the way of the future. In the eyes of the Western Allies, the city had to be defended at all costs. But could it really be supplied by air? If the effort failed, like the proverbial ripe plum, Berlin would fall into Stalin's waiting hands.

The Berlin Airlift was truly an ambitious undertaking that many believed could not succeed. The sheer magnitude of the requirement was mind-boggling. What would it take to keep the 2 million inhabitants of West Berlin supplied, even with only the basic requirements? Back during the war, in the winter of 1942-43, in another case where a large-scale air supply effort was attempted, Hitler failed miserably to provide the bare necessities needed by Field Marshal Friedrich Paulus' doomed Sixth Army at Stalingrad. In June of 1948, no doubt emboldened by that memory, Stalin gambled that the Western Powers either would not attempt air supply or that the effort would fail just as it had at Stalingrad.

But the Western Allies were determined that Project Vittles would succeed; and the British and Americans assigned their airpower capabilities to the task. They quickly began flying into Berlin the vital necessities — food, clothing, fuel. It became the greatest airlift in history. In his radio broadcast on July 5, 1948, the news commentator Lowell Thomas provided some of the early Berlin Airlift statistics: "American planes," he said, "alone have flown 1,115 cargo flights in nine days. The mileage comes to nearly seven hundred thousand, enough miles to circle the world twenty-eight times."[1]

The blockade and airlift would extend well into 1949. Day after day, the airlift continued. At the beginning, flying weather on the continent was good. Then, as fall approached, fog made the operation hazardous. Still, the flights continued.

Being a brand new forecaster, my duties were confined mostly to assisting the duty forecaster in putting together the many flight folders that the exploding number of fights to Europe then required. Within a few months, however, I was awarded the skilled military occupational specialty (MOS) of 787, weather forecaster, and was thrust into the thick of the fray, preparing forecasts for and briefing pilots on the winds and weather to Berlin,

---

[1] Lowell Thomas, *History As You Heard It,* Doubleday & Company, Garden City, New York, 1957, 347.

over 4,000 miles away. During many of these flights, fog or low stratus ceilings made the selection of an alternate landing field, a flight regulation requirement, usually Paris or London, a most difficult task. Most of the time, the planes would touch down for refueling at Lagens, Azores, a distance of 2,200 miles. Others made refueling stops at closer airfields such as Kindley Field, Bermuda; Goose Bay, Labrador; Gander and Harmon Field in Newfoundland.

Preparing forecasts and briefing aircrews kept me too busy to reflect on the thought that I must then have been one of the youngest men in the US Air Force to hold the skilled weather forecaster MOS. If this were part of an experiment to see if youngsters could handle the job, I hoped I wouldn't be the one to disappoint them.

It was at Westover that I first became associated with and would learn to admire that special kind of soldier — the warrant officer — who was both officer and a top specialist in a technical field, in this case the weather field. There were two warrant officer weather forecasters at Westover. Whether they were told to do so or merely wanted to set a youngster on the right path, the warrant officers in any case became my weather forecasting tutors. They were models of patience and invariably gave me the impression that they would take as much time as needed in demonstrating the forecaster's various tasks, not the least of which were the explanations of the scientific bases for the forecasts. And for this brand new weather forecaster it was especially gratifying to know they meant it when they said there were no dumb questions. I would learn a lot from them.

The preparation of flight-level forecast winds is one of the aviation weather forecaster's most important tasks. For the long hauls over the Atlantic, we broke these forecast winds down into successive zones of five-degree intervals of longitude. These zones usually were laid out along a track called the "rhumb line." On a rhumb line, the angular relationship of the plane's track to true north is always the same. This must make the navigator's job easier, and it was probably why that was the most popular track even though it was not the shortest track. The great circle was the shortest track. On our maps, the great circle was laid out by simply drawing a straight line. The rhumb line, however, was laid out as a gently curving line, south of the great circle everywhere except, of course, at the takeoff and landing points.

It is interesting to note that Charles Lindbergh chose the great circle for his solo 3,600-mile flight from New York to Paris on May 20-21, 1927. No doubt chosen because it was the shortest course, and fuel was his primary concern, its navigational requirements might have provided the additional advantage of helping to keep him awake during the 33 and one-half hours of that flight.

In addition to the rhumb line and great circle, the pilot could opt to "fly the pressure pattern." This procedure involved keeping the plane flying at the same atmospheric pressure rather than letting it pass through the normal constantly changing pressure values. In navigating this route, the pilot would be staying at a constant indicated altitude as measured on the pressure altimeter while also staying at a constant real altitude over the ocean. The flight then would be conducted along a certain contour of the many that gave shape to the high and low pressure fields. It would take a northward turn to run up to the apex of a pressure ridge and would swing southward as the pressure contour dipped into a trough. That kind of track had one distinct advantage that is only very rarely experienced otherwise — flying it would provide a *direct tail wind over the entire distance of the flight*. But there was also an important disadvantage, for in flying along on only one pressure contour, the plane could be traveling many hundreds of miles north or south of its shortest route as its path curved northward into the pressure ridges and southward toward the pressure troughs. Pilots would sometimes bring up with me the advisability of flying the pressure pattern, but I never knew of any pilot who actually attempted that kind of flight.

As soon as I knew the navigator's selected track and the pilot's estimated departure time (ETD), I would set to work preparing a "prog" (prognostic chart) which was a forecast of the change in pressure-height contours that we previously had analyzed on a constant pressure chart. This of course was before the days of jet travel and the planes being flown from Westover were mainly the four-engined, propeller-driven C-54 Skymaster cargo planes. Built by Douglas and first put into service in 1942, the C-54 was the Army Air Corps' (and later the Air Force's) version of the civilian DC-4. In the Navy, it was called the R5D. The C-54 usually flew at 16,000 or 17,000 feet, depending on whether the craft was eastbound or westbound. In some cases, depending again on eastward or westward flight direction, they would be at 8,000 or 9,000 feet. With a cruising speed of well over 200 miles per hour and a range of almost 4,000 miles, the C-54 could fly non-stop to Europe, but most of the crews would take the trip in hops, usually setting down for refueling at Lagens in the Azores. Nevertheless, most of the flight-level wind forecasts we made for them were for the entire flight to Europe. The charts we used to calculate the wind forecasts for these flights were either the 500-millibar chart, for the 16,000 and 17,000-foot flights, or the 700-millibar chart, for the 8,000 or 9,000-foot flights.

For the non-meteorologists who read this, the millibar is the name given to a unit of pressure that meteorologists have found convenient to use. Air pressure at the surface of the earth is roughly about 1,000 millibars. Thus the 500-millibar level, or 500-millibar chart, is the height where one would find the weight of the atmosphere is only about one-half its weight at sea level. The altitude at which the 500-millibar pressure is found varies over a limited range in the horizontal and likewise varies with time at the same point in space as pressure systems migrate, but is roughly about 18,000 feet above sea level. The 700-millibar pressure, on the other hand, being greater, than the 500-millibar pressure, is found at a lower altitude, at about 10,000 feet. Its height varies similarly as do the heights of the other levels mentioned, the so-called standard levels.

With these constant pressure charts, mere lines drawn on a sheet of paper, we are portraying pictures of invisible skyscapes — undulations of mounds, swales, hills, valleys, lofty mountains (highs), deep canyons (lows) — that go on in all directions beyond the outlines of the map being viewed, unending and captivating to a meteorologist's eye.

The weather features still being shown on the routine television news programs, the highs and lows, are not those found at the 500- or 700-millibar pressure surface or at any other upper air level. Except for a broad brush depiction of the jet stream, they are almost exclusively the features found at the surface of the earth, or on the surface weather chart. Showing the weather systems on the surface chart is a good way of letting the viewers know where precipitation and other weather events are occurring, but to the real meteorologist, that is, the person who actually constructs the forecast from raw data, the surface chart by itself is of little value. The real weather forecaster, the person who formulates the weather prognosis, will want, among other things, to see the 500-millibar chart. To the real weather forecaster, the 500-millibar chart, that picture of the weather systems as they appear at about 18,000 feet, is an indispensable tool. To many commercial TV weathercasters, who may not be meteorologists but simply readers of weather forecasts prepared by others, this could be somewhat irrelevant.

To sort of complete this picture of the various weather charts, there is also a standard chart describing the weather systems at about 5,000 feet, this is the "850-millibar chart." And when you are cruising along in a jet-propelled airliner you are usually at an altitude of about 30,000 to 35,000 feet, or, as the meteorologist would say, you were flying at approximately the 300- to 200-millibar levels, respectively.

From the progs we prepared, which were projections of the future movements of the height contours that characterized a particular pressure-height chart, we literally manufactured the winds that the flight crews should encounter as they proceeded on their many hours of flight over the Atlantic. The first step was to produce the prog chart, usually for a time period of 12 or 24 hours in the future. In those days before computer-produced prog charts, this was a laborious drawing procedure requiring a careful consideration of the propensity for deepening or filling of pressure systems and the movements of ridges and troughs. The forecaster making those charts drew upon all his knowledge of the theory of pressure system dynamics. Just as important as his drawing pencil was his "pink pearl" erasure which was liberally employed as he proceeded to construct contours with spacing between them that he considered to be just right. The spacing and the character of the contours' curvature, whether cyclonic or anticyclonic (respectively, in the Northern Hemisphere, counterclockwise and clockwise) would determine the speed as well as the direction of the wind. Once satisfied with his "chart of the future," the forecaster laid a clear plastic scale (the "geostrophic wind scale") across the contours and read his predicted wind speeds along the projected plane's flight path. The corresponding wind directions were obtained with another scale by simply reading the angular value of the contours.

All this may seem quite taxing, but life at Westover in 1948 wasn't all drudgery. For the first time in my career, we were allowed to wear civilian clothes when off duty. And I was looking forward to a weather familiarization trip. These flights, in which the forecaster was to observe the weather along the route for which he had prepared the forecast, were to be made on the long over-water routes commonly flown out of the base. It seemed to me the idea had great merit, in addition to having immediate confirmation of my flight-weather prediction, at our landing I'd change into civvies and, like an ordinary tourist, enjoy the sights at some place I'd many times heard about but had never seen. The station was abuzz, in fact, with the stories told by a recently returned forecaster who had taken a familiarization flight to Paris. He regaled us with the sights he had seen, including a visit to a Paris nightclub where the waitresses wore bow ties and high-heel shoes — and nothing else. Curiously, the familiarization flights ended at about the time he returned, which was a good thing. It was quite certain this farm boy would have been completely out of his element in Paris.

On one otherwise uneventful day, the first lieutenant who was operations officer popped into the map analysis room and shouted: "**forecaster!**" Except in cases of attempted humor, this was not the normal form of address when a pilot or others requested assistance from the weather forecaster. Looking up from my map, I could see this visit was not to be thought of as an attempt at humor. The lieutenant was extremely agitated. "Tell me what to tell them," he blurted. "The president's plane is running out of fuel!"

As a still essentially inexperienced forecaster, the lieutenant's message was just about the last thing I would have wanted to hear. I was quite aware of the need to maintain the Westover weather station's reputation for being the home of cool, calm and reliable weather forecasters. Composure was the key. "Sir," I asked, "where are they, what's their altitude and how much time do we have?"

"They're at 16,000 feet," he replied. "Incoming over the Atlantic and we don't have *any* time!"

"I'll need just a few minutes to compare their winds with the winds at lower altitudes," I said, as the lieutenant left to monitor further messages from the plane which I assumed to be the Independence, President Truman's C-118, a plane that resembled a C-54 but was sleeker and had better performance. Over his shoulder the lieutenant said he'd be back in five minutes for my response.

Five minutes was not the time interval that I would have liked to assign to myself for the calculation of a wind prediction having life or death implications for the passengers of the plane belonging to the President of the United States. A working condition such as this had not been anticipated in any of the many lectures at Chanute's weather forecasting school.

Although it is certainly not a requirement, there have been any number of good weather forecasts that have benefited from a smattering of good luck. And, luckily, I had just finished drawing my 700- and 500-millibar progs. The 500 prog told the story. The president's plane was bucking a strong head wind. Further, a drop-down to the 700-millibar level, to about 10,000 feet, wouldn't help. Head winds were almost as strong at that level as at the plane's present altitude.

At warp speed I whipped out a section of an 850-millibar prog to check out the wind situation at that level, at about 5,000 feet. The 850-millibar level was not routinely flown at by planes venturing out on long hops over the Atlantic. It was simply too close to the water. And while over the relatively warm Gulf Stream, convective currents could produce an uncomfortable level of turbulence. Also, in the back of our minds was the thought that, in the event of a problem, there'd be little gliding space.

The winds at the 850-millibar level were decidedly favorable. They were lighter and, having a more northerly direction, had the additional advantage of being more of a cross wind and less of a head wind than the winds above. With the operations lieutenant breathing down my neck, I picked off the last zone of wind from my 850 prog while explaining that a drop-down to that level was clearly what the president's pilot must do. The paper with those forecast winds was snatched from my hand and within a minute or two the lieutenant was back in Base Ops radioing the recommended flight level and winds to the plane.

Later in the shift, after things had quieted down in Base Ops, I sauntered over to the lieutenant's counter to see what had happened. If the Independence had gone down in the Atlantic I knew the end of this buck sergeant's budding forecaster's career was close at hand. But everything had gone well. The lieutenant, now a model of composure, told me there was "no sweat." They were not landing at Westover but had contacted us merely because, I supposed, that with all the flights leaving Westover in support of the Berlin Airlift, it was well known that we were the east coast "weather map factory" and *the* place to contact for weather information. And furthermore, he said, the president was not on board, only the vice president (actually the vice president-elect). He said that as if to imply the consequences of the plane going down would not be as great if the person on board was Alben Barkley instead of Harry S. Truman. I didn't want the plane to go down even if the highest rank on board was just a private, and I am sure the lieutenant really shared that view.

And that was the end of it. The Berlin Airlift was in full swing. It was on to other wind forecasts, more flight folders to prepare, a struggle to find suitable alternates on the Continent. For a little while, I wondered if Vice President Alben Barkley ever knew how close he came to trying on a life jacket. At the time, it seemed to me he had not been told of the head wind crisis, for if he had, I thought I might have received a thank-you note on vice presidential stationery, though I was perfectly aware I was only doing my job.

Years later, as a matter of mere curiosity, I tried to reconstruct the date of that flight. I concluded that it was most likely near Christmas of 1948 when the vice president had accompanied Bob Hope on his Christmas show, entertaining the troops in Europe. If that were the case, not only would the Veep have been on board but also Bob Hope and Irving Berlin. And, of course, Bob Hope never went anywhere on these troop-entertainment trips without an honor guard of beauties. That necessary contingent was composed of Jinx Falkenberg, who brought her husband (Bob Hope and Irving Berlin had brought their wives); and the curvaceous Radio City Rockettes. That plane was certainly carrying important cargo.

I had inherited a predilection for the Republican Party, but as time went on I would come to think of Alben Barkley, a Democrat, as one of my favorites as a vice president. His sense of humor and sense of history was something to admire.[2] He could have himself ascended to the presidency in 1948, but would not think of opposing his friend, Harry Truman, who will go down in history as one of the very best of presidents.

---

[2] See Alben W. Barkley, *That Reminds Me,* Doubleday & Company, Garden City, New York, 1954, 269.

This Douglas VC-118A Liftmaster, or DC6A, was the presidential plane used by Presidents Kennedy and Johnson. Built as a cargo plane, it was later converted to be Air Force One, a designation originated by President Eisenhower. Except for a slightly wider body, this plane was identical to President Truman's C-118, the *Independence*. President Truman's plane was artistically finished with markings that made his ship resemble an eagle. This one is on display at the Pima Air and Space Museum in Tucson, Arizona.

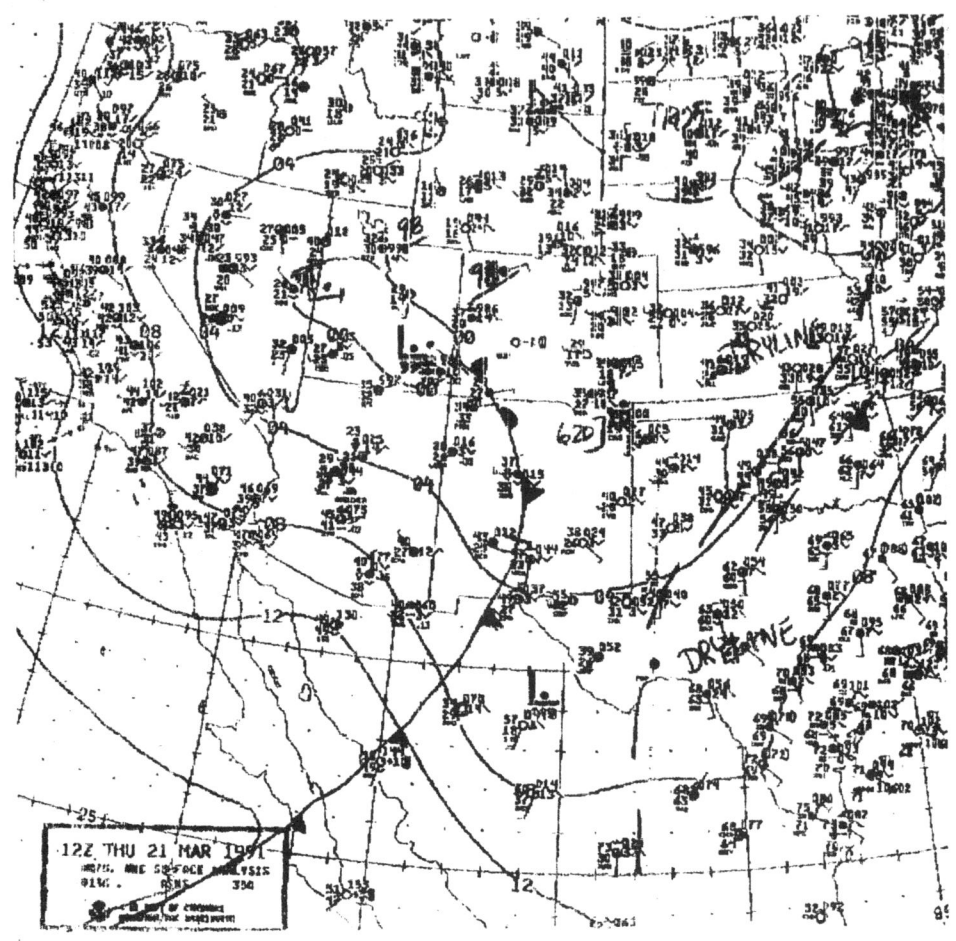

This is the western US section of the earth's-surface weather chart. This chart and the upper air charts of the entire US seen on the following four pages have been produced by computer. They all depict the atmosphere at precisely the same time. It happens to be noon (noted as "12Z"), Greenwich time, or 0800 hours EDST on the first day of spring, 1991. Received by satellite, the data were always rather hard to read. The station plots were only a little larger but very legible on maps plotted by the weather observer at the weather station. His plots, for each station, had to fit under a dime.

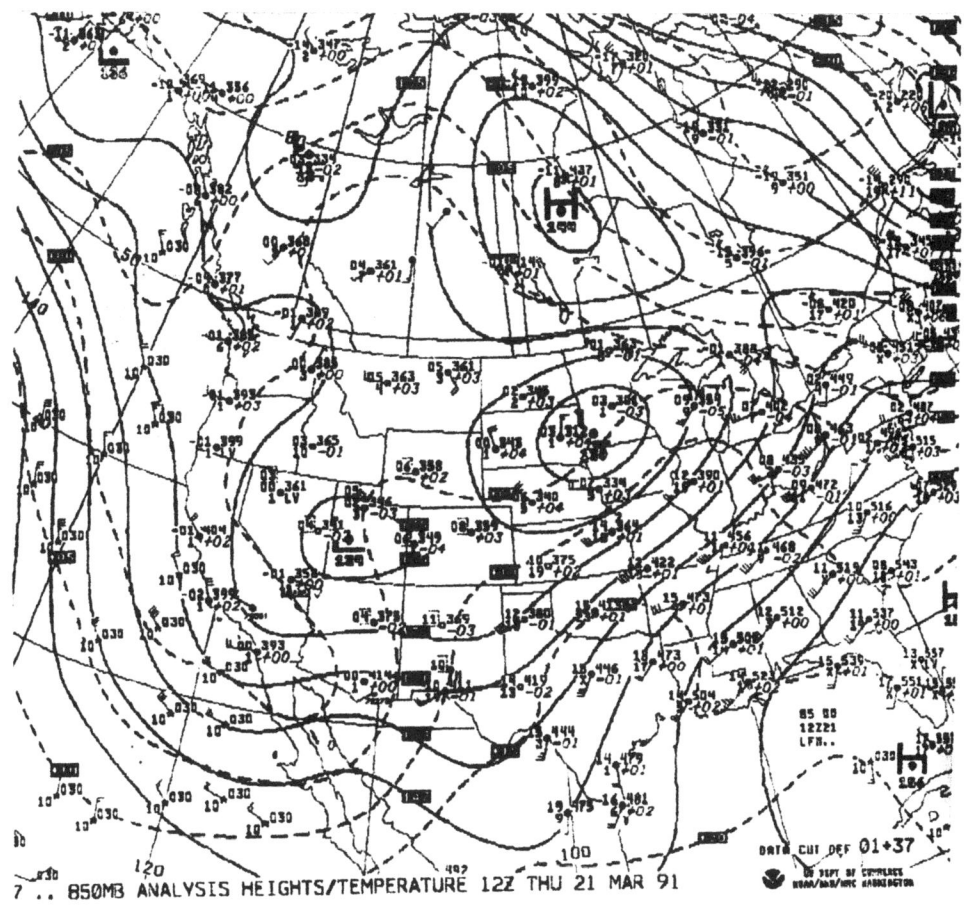

850MB ANALYSIS HEIGHTS/TEMPERATURE 12Z THU 21 MAR 91

Wind directions and wind speed (in knots), temperatures and dew points in the Celsius scale, and heights and height changes of the 850 millibar pressure surface are plotted on this "850 Millibar Chart." The data comes from balloon-borne radiosondes, sent aloft daily at midnight and noon Greenwich time, and a few pilot reports. Lines are contours of height. They expose the locations of high and low pressure cells. The 850 millibar (MB) pressure surface is at a height of about 5,000 feet or about 1,500 meters. Changeover from feet to meters on US upper air charts was made in the 1950s.

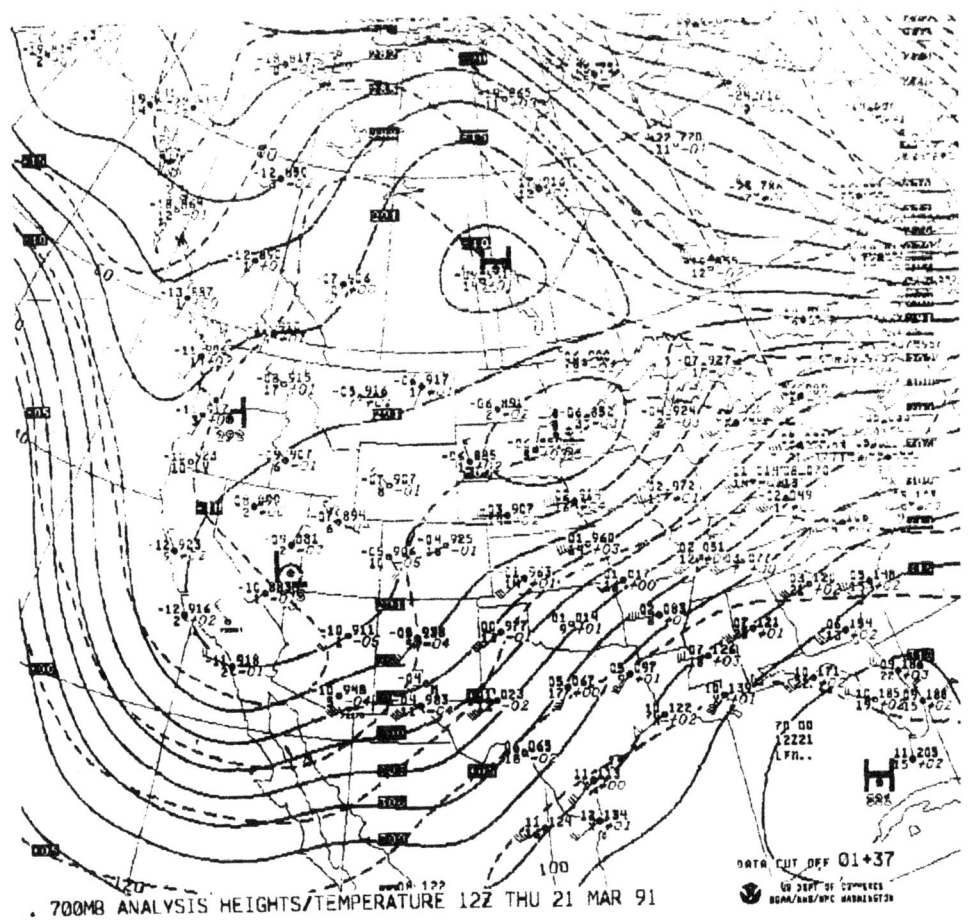

700MB ANALYSIS HEIGHTS/TEMPERATURE 12Z THU 21 MAR 91

The 700 Millibar Chart. The 700 millibar pressure surface is found at a height of about 10,000 feet or about 3,000 meters. It is plotted with the same information that goes on the 850 Millibar Chart and analyzed similarly. Representations like this of upper air charts are not provided to the general public on the weather portions of the TV news programs. As a way of judging the veracity of the prediction, many meteorologists would like to see these charts displayed when the weather forecast is presented.

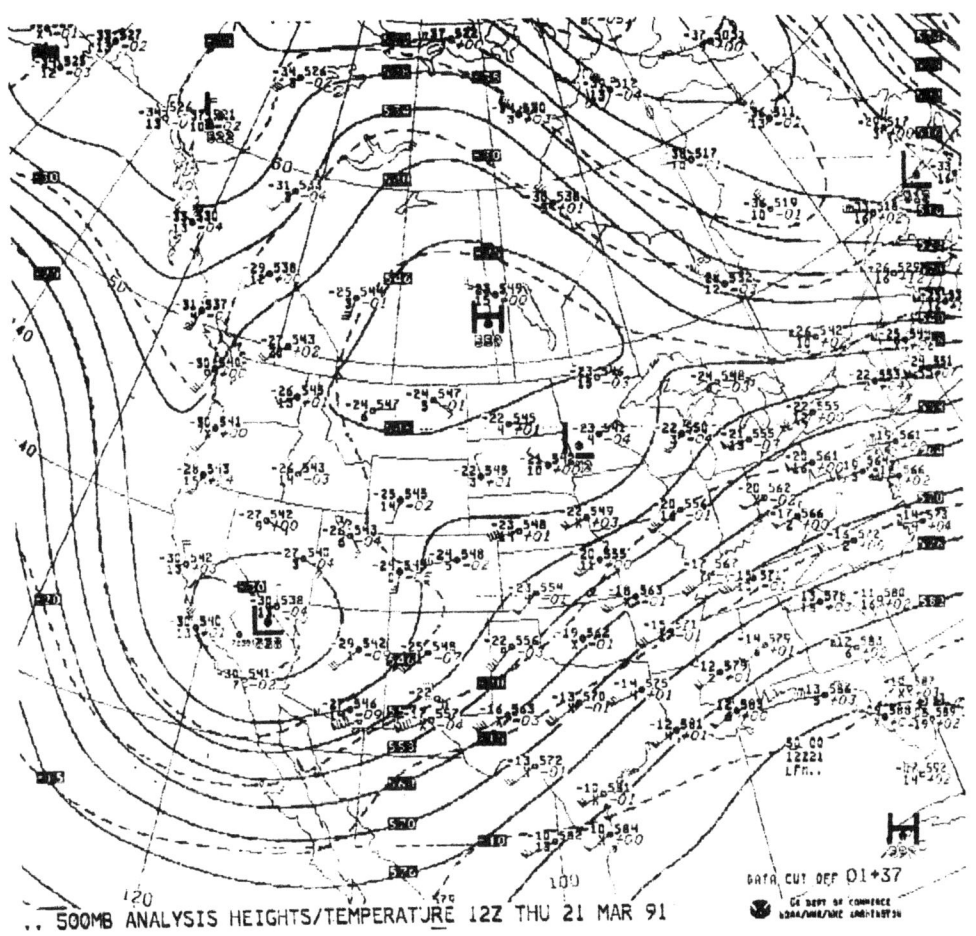

The 500 Millibar Chart, plotted and analyzed like the 850 and 700 MB charts, depicts the atmosphere at about its midpoint of pressure height. (Pressure at the earth's surface is roughly 1,000 millibars.) The 500 millibar height is at about 18,000 feet or about 5,500 meters. This is the basic chart of the weather forecaster. His calculations of future atmospheric (weather) changes flows from his predictions of the changes in the 500 millibar chart.

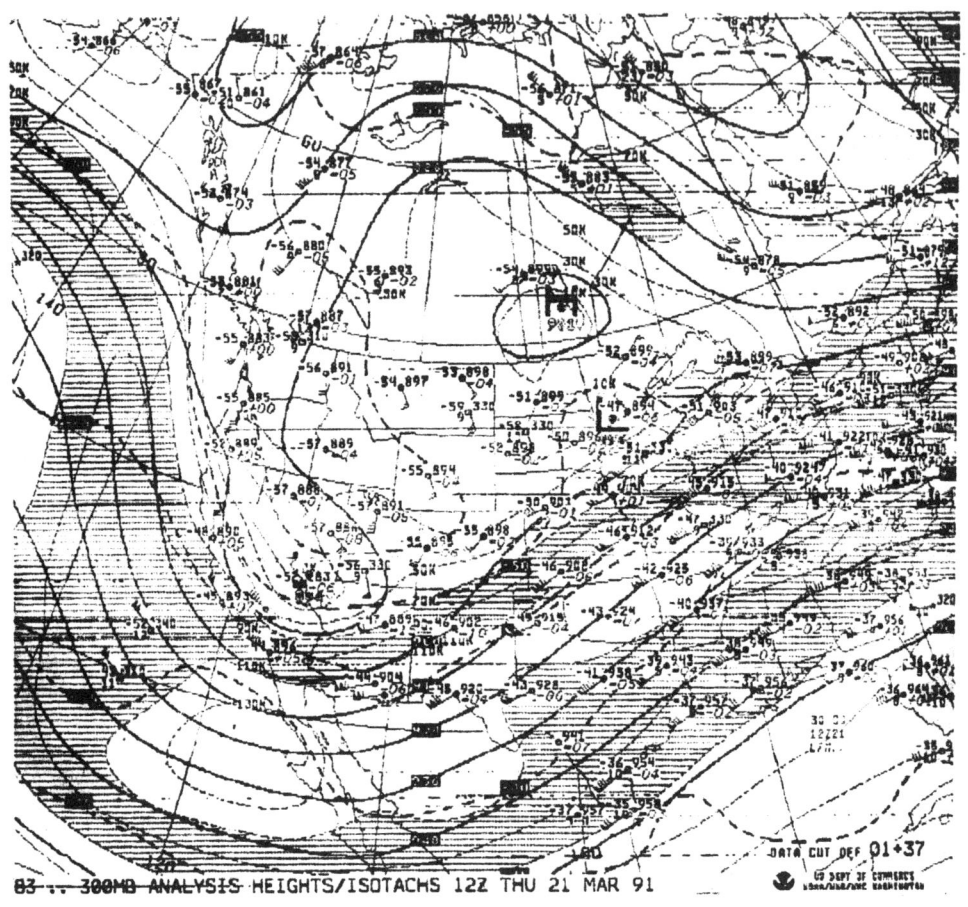

The 300 Millibar Chart. The 300 millibar pressure surface is found at an altitude of about 30,000 feet or about 9,000 meters. Isotachs, lines of equal wind speed, are shown on this chart. Shaded areas delineate the location of the jet stream. Note that the jet is split into two segments in the southern US. The perfectly horizontal streaks are caused by interference from snow that had accumulated on the 12-foot satellite-signal receiving dish.

Chapter 13

A Lonely Job

During those days at Westover's 8th Weather, the weather package provided the pilot not only included forecast winds by longitudinal zones but also vertical cross sections of the clouds and frontal systems to be encountered along his route. Like the weather maps, the professional appearance of the vertical cross sections depended greatly on the artistic ability of the forecaster.

Because the clouds drawn on the cross section were to be the clouds the pilot would find *directly* on the plane's path, an extreme precision in forecasting not only the generation and dissipation of clouds but also their movement was required. For a flight that would extend to about a half a day in duration, a complete picture of *all the clouds* to be encountered was then and always will be a demand beyond the capability of the science. For the clouds that we were not so sure about, we tried to compensate for the unachievable *complete* picture by shading in with hash marks those clouds that we were confident would be near the flight path but quite possibly not directly on it.

Although our main forecasting routes were those over the Atlantic, as indicated in the last chapter we also had occasion to provide forecasts for planes heading for sites in the North. On one December day in 1948, we received word that a B-17 had gone down on the Greenland ice cap, that vast white wasteland that extends up to about 10,000 feet. Rescue attempts were launched from Westover, with weather for the area supplied by Westover's Base Weather. The rescue mission centered immediately on the old reliable C-47 Skytrain. It was believed the C-47 would be light enough to keep its wheels from crunching through the ice, but the question did cause some concern. As luck would have it, the 47 landed and broke through the ice. With two aircrews now on the ice and the method of their rescue still to be found, supplies to include tenting for shelter and radios for communication were airdropped to them. The downed airmen set up camp.

The rescue attempt then centered on two gliders. They were to be towed to the site, loaded with the two crews and then snatched off the ice by a low-flying aircraft. The gliders were landed on the ice but they could not be retrieved. We now had four aircraft on the ice cap.

The third plan involved our tried and true C-54. It was an idea that most of us thought was also destined for failure. But to our surprise, it was a resounding success. A colonel based at Westover obtained a C-54 at Wright-Patterson AFB that had been fitted with skis and a new kind of prop, an electrically-operated reversible one, and JATO (jet assisted takeoff). He asked the downed men to tramp out a stretch of snow upon which to land that would avoid any crevasses that might lurk beneath, landed on that "runway," turned the '54 around, had the men clamber aboard, and lifted that bird off with the help of the JATO units. He was awarded the Distinguished Flying Cross.[1]

Though we had no ice such as the stranded Greenland crews had experienced, we did have our share of snow at Westover, and our share of the challenging tasks of predicting snow as the nor'easters visited us with the white stuff. Predicting snow depths is one of the forecaster's most challenging tasks. At Westover, the forecasters became well known for their skill in that regard. Just so there is no mistaking my motives, please note that their reputation was established before I became involved in forecasting, and I can claim no credit for the success of the forecasts.

Perhaps out of curiosity or, more likely, sensing the makings of a controversial story, a reporter from either the Springfield or Hartford, Connecticut newspaper (the story may have run in both papers) compared the verifications of the snow forecasts made by the US Weather Bureau with those made by the Westover forecasters. These papers would call Westover Base Weather for the weather forecast which they published. Both cities were on the north-to-south-running Connecticut River. Their proximity, Springfield being just three miles away and Hartford only 25, as well as a slight propensity for air channeling along the stream, would result in the occurrence of similar weather or, in the least, similar weather forecasts for each city. That comparison brought on a flurry of high-level interest at the Weather Bureau — and at Air Weather Service headquarters.

Printed in black and white in a newspaper for the public to read was the assertion that the Air Force weather forecasters were more accurate than their Weather Bureau counterparts. That impression, of course, could not be allowed to continue. More precisely, the opportunity for such comparisons could no longer be countenanced.

And so the Westover Station Weather Officer (the officer in charge) received a letter from Brigadier General Donald Yates, the Commanding General of Air Weather Service. I was in the weather station "front room," the place where the forecaster briefed the aircrews, when the SWO read the letter to the forecasters then on duty. In essence, the letter stated that both weather forecasting organizations used the same meteorological information

---

[1] My thanks to former S/Sgt. Tom McClung for his recollections which filled in some gaps of my memory of this event.

and there was, therefore, no reason why one should be more capable than the other. Therefore, the AWS forecasters would no longer provide weather forecasts to civilians. And that, as far as I know, was the beginning of the "no weather forecasts to civilians" policy that the AWS would then follow. We allowed ourselves a certain upwelling of pride with the reading of the letter's closing statement wherein the general extended his compliment for the excellent work being done by the Westover forecasters.

This relieved the Westover forecasters from having to take time out of their critical aircraft support duties to answer calls, many of them generated by mere curiosity about the weather, from civilians. The Air Force weather forecasters were not there to help the housewife decide if she should hang out her washing on Monday or if a local club would have good weather for its picnic. But were they really better than the weather bureau forecasters? We knew the answer to that one!

The National Weather Service forecasters (formerly the Weather Bureau) in fact do a masterful job these days in interpreting the various real-time and prognostic computer-produced charts that have since come to the fore and in creating weather predictions from them. Not many people know that the forecasts they have become accustomed to from their favorite television weather personality have actually been prepared in almost all cases by a National Weather Service forecaster who is receiving considerable guidance from the computer-produced progs and other computer assists.

Back then in 1948, the sophisticated computers that now provide the forecaster with what now is considered indispensable guidance were not yet in existence. The quality of the forecast then depended entirely on the skill of the forecaster who sat on a high stool at a map table in a base weather station. *With no help from anyone,* he would rely on his ability to not only properly analyze the plotted data but also to project forward the changes that he saw would be occurring in those data. And from those changes arrive at weather and wind forecasts for his terminal, for weather and winds to be encountered along any number of flight routes and for the pilots' destination weather. Absent from the state of present-day weather prediction was the time to be saved in using computer products. With phone calls requesting forecasts, pilots and navigators in and out of the station having questions about their weather and winds enroute, the need to spend time on the analysis — where data were sparse on certain sections of the map, could the analysis be done another way? And if so, heaven forbid, on that long half-day flight over the Atlantic, could there be a head wind component where a tail wind was now being predicted? It was an incredibly busy environment. In the end, after sifting through all the available information, it was decision time. In the course of any particular shift, there were many, many decisions to be made. And these were the forecaster's alone to make. There was no questioning the fact that the safety of the flight crews and their passengers depended on the advice he gave. He was not just telling the public that it was going to be a day for bumbershoots. It was a lonely job.

A safety poster that was then displayed in Westover's Base Ops played on the relationship between the flyer and the forecaster. I regret that I do not know the name of the artist. The poster was a cartoon in the style of the cartoons then being drawn by M/Sgt. Jake Schuffert, and this one may well have been his.[2] In this cartoon poster, a pilot and forecaster are pictured standing before the weather map in a typical weather station briefing room. The pilot wears a flight suit. A fifty-mission-crush cap perches precariously on his head, and a cigarette droops from the corner of his mouth. The forecaster, with pencil balanced over an ear, is bare-headed. And this is not referring only to the fact that he wears no cap. There are only two strands of hair on his head, the lone survivors of the previous many that became casualties of the years of briefing pilots such as the one standing before him. He has obviously just tried to provide that pilot with his "weather words." While twirling his circular slide rule, the pilot says: "Skip the lecture, Curly, I've got a date with an angel." Double-entendre doesn't get much better.

The chief warrant officer (CWO) who was my chief mentor held the record for the best actual time of arrival (ATA) versus the estimated time of arrival (ETA) for the C-54 flights from Westover to Lagens, Azores, that over-water flight of 2,200 miles and of 9- to 11-hours duration. On one flight, the difference between ATA and ETA based on his wind forecast was only four minutes. The ETA was calculated by the navigator using the predicted inflight winds. I admired greatly this warrant officer's forecasting skills and was deeply aware of the fact that, with due respect for the great instruction I had received in forecasting at Chanute, all I really knew about practical forecasting I had learned from him. He was understandably proud of his record to the Azores and often checked with Base Ops to see how the other forecasters had fared in comparing ATA's with ETA's. Fearing that one day he'd find out I had badly bungled a wind forecast, I wished he would stop making those checks.

One day, after checking the ATA's versus ETA's at Base Ops on a C-54 that had flown on my wind forecast, my chief mentor walked over to the map table where I was drawing contours on a 500-millibar chart and soberly informed me that I had bettered his record by one minute. It appeared to me he was having a difficult time believing it, and not more so than I. "It was just luck," I said, and continued with my map analysis. I had

---

[2] Jake Schuffert, a hero as well as cartoonist, had flown 50 bombing missions during World War II including the hazardous raids on the Ploesti, Romania oil fields during which he had bailed out of a crippled B-24.

taken great pains in crafting the prog that was used for that record-breaking wind forecast. Under the assumption that concentrated diligence could possibly overcome lack of skill, it was a habit that I tried to follow in making all the progs. But I also believed you had to have luck, and it was beginning to seem that I might have it. This was at a time when I still was not sure I was cut out for a career in meteorology, and it buoyed my spirits. Perhaps Andy Anderson was right, way back there in basic training when the field assignments were being made, only a little more than two years ago but seemingly more like half a lifetime, perhaps he was really on the mark when he insisted I take meteorology. In that short period of two years, I had become impressed with the importance of the aviation forecaster's job but was still not sure it was the career for me.

Arguing for a continuation of the career was the fact that we weathermen were generally regarded as an elite group. Being one of them fostered a certain elan. In the weather stations and in the weather headquarters, *esprit de corps* abounded. I doubt that any of us then would have voluntarily accepted a transfer to another command if it were offered, even if the transfer were to be accompanied by a promotion. We still had that "special outfit" feeling that had been so common in the old AAF Weather Service.

And our optimistic outlook was bolstered by an event that may seem trivial to present-day weathermen. It was this: *we were told that we would be allowed to wear civilian clothes when off-duty.* Way back in early February of 1946, when I arrived at the Ft. Sheridan induction center, I had been required to ship my civilian clothes home and, since that time, which seemed a short eternity, the military uniform had to be worn, both on and off duty, even when on furlough. Being allowed to put on comfortable civilian clothes in which to kick back while off duty was a pleasure that could not be overemphasized.

Other changes followed. It was about this time that the official designation of air bases became known as Air Force Base rather than Field, Base Weather Stations became known as Weather Detachments, Station Weather Officers began to be called Detachment Commanders. It was also the time when Marie, who had been in Michigan to be with her mother during the arrival of our second child, presented me with another lovely daughter, who was named Willa.

And, as if the Air Force was in a hurry to assert its new identity, the changes continued. In that year of 1948, Air Weather Service reorganized itself and the Westover weathermen were placed under another command. They were no longer Westover Base Weather of the fabulous 8th. The station now was Detachment 12/23L of the 12th Weather Squadron with headquarters at Mitchel Field, New York. That reorganization coincidentally brought the weather stations at Westover and Selfridge into the same squadron, another lucky break, for it was then possible to be transferred to Selfridge with comparatively little ado. It had been a dream of mine for a long time to be assigned to Selfridge, and early in the new year of 1949 it happened. I was transferred to Selfridge's Detachment 12-20L with call letters MTC (chosen to identify its location with the nearby town of Mt. Clemens), where I would find the flying customers to be vastly different from those I had so far been associated with.

The year is 1948. With his barracks in the background, the author enjoys having his photo taken while wearing civilian clothes for the first time since he enlisted in the Army Air Force in early 1946. Note the Western style belt with metal end that would come back into vogue in the late 1990s.

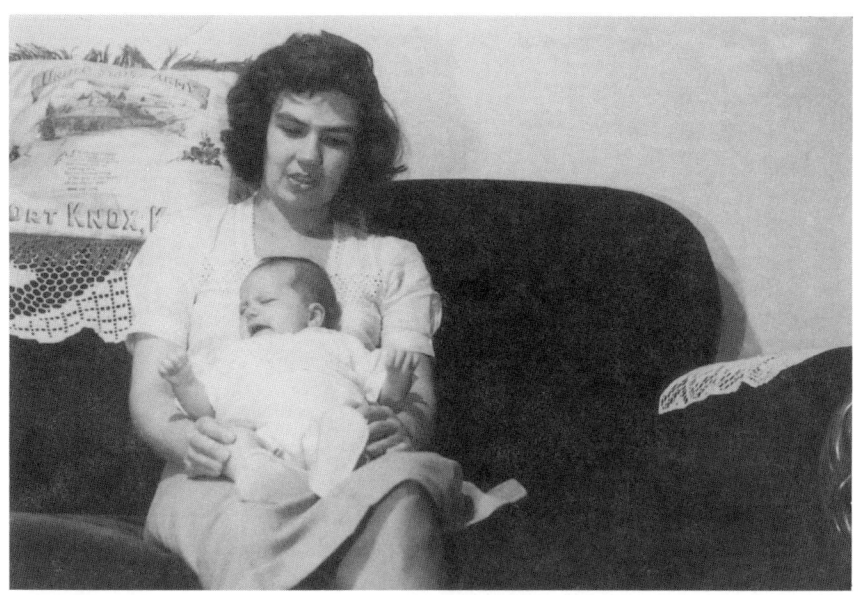

Top: Marie Cogut holds baby Willa on the sofa of her parents' home in Royal Oak Township, Michigan (later City of Madison Heights).
Bottom: As a young lady in the 1980s, the baby in the photo above stands on the wing of a restored T-6 in its hangar at the Detroit City Airport. Its military days long past, the T-6 is privately owned and painted in colors never seen in its active duty days

Top: Marie stands in front of a B-17 restored in World War II battle colors that has landed at the airport in Lansing, Michigan in 1999.
Bottom: A C-54 of the Military Air Transport Service at Wheelus Field, Tripoli, Libya in 1951. Assigned to the 29th Weather Squadron at Wheelus, this aircraft made weekly flights to bring mail to the 29th's weather troops at its Detachment 29-2, located on the sands of Dhahran, Saudi Arabia. The Dhahran denizens referred to it as the "T-Flight" (for Tripoli). A C-54 fitted with skis and JATO units rescued the B-17, C-47 and glider crews that were stranded on the Greenland ice cap in December 1948.

Chapter 14

Brown Shoe Air Force

Selfridge Air Force Base, Michigan was first known as "Joy Aviation Field" after Henry B. Joy, the Michigan Realtor who developed the airfield in 1914. Activated as a military unit in 1917, the airfield was then named after Lt. Thomas Selfridge, the first military pilot and the first military air casualty. Selfridge Field would become known as the "home of the generals." When I arrived there in January of 1949, it was said that of the officers who had once been stationed at Selfridge, 145 ultimately achieved general officer rank. An interesting statistic but not of much real interest to any weatherman.[1] There was, in the Air Weather Service, only one general officer, the commanding general of the organization. No one who chose a career in Air Force meteorology had any real hope of ever being promoted to general. But that, after all, wasn't why one chose that career field. It was, rather, the opportunity to engage oneself in the challenge of the perfect weather observation and the pursuit of the faultless forecast. (Or, perhaps, to hone whatever artistic talents one might imagine he had while drawing weather maps?) Promotion into the more modest levels of the military rank structure would of course have been nice, but with the harsh realities of the big one (WWII) fading somewhat from memory, and military "downsizing" (a term not then used) in vogue, even that was not much to be expected.[2]

Not only because it was situated close to both our parents' homes, but also because its sturdy old brick buildings gave it an aura of permanence suggestive of its historical importance, Selfridge AFB, on the shore of Lake St. Clair, about 10 miles northeast of Detroit, evoked a warm, homey feeling for Marie and I. The weather station was located in the south end of the base's most impressive building, the brick Base Headquarters building which also housed Base Operations where the pilots prepared and filed their flight plans.

---

[1] Among the generals who were once stationed at Selfridge were Carl "Tooey" Spaatz, first chief of staff of the Air Force; Emmett "Rosie" O'Donnell; Curtis LeMay; and Jimmy Doolittle. Charles Lindbergh and Captain Eddie Rickenbacker also were at Selfridge. *Selfridge Air National Guard Base Diamond Jubilee Celebration, June 1992.*

[2] With 4,198 persons assigned to it, the low point in AAF Weather Service manning was reached in June 1946, a great reduction from the peak of 19,000 in 1945. By January 1949, manning had recovered somewhat, reaching 8,300...*Air Weather Service: a Brief History 1937-1991*, William E. Nawyn and Rita M. Markus, May 1991, P. 5.

Just as at Westover, the weathermen at Selfridge were true professionals. Friendly, helpful, when I walked into that station in January of 1949, they did everything they could to have me checked out in as short a time as possible. (In this, it is conceivable, of course, that there might have been a somewhat selfish motive. For if I checked out quickly and was given responsibility for a shift, they might enjoy longer breaks between shifts.) They had the base carpenters make up a "shingle," a green-painted board with hooks attached upon which "Sgt. Cogut" was printed in white letters. I was told to hang the shingle whenever I was on shift, under the lettering "Selfridge AFB Base Weather," that had also been printed in white lettering, on the green-background board permanently attached to one of the walls. I might have flinched at the green coloring of my shingle, taking it to be a reflection on my newness in the field, except that it just happened to be the preferred color scheme — all of the forecasters' names were printed on green shingles. Color schemes at Air Force bases seemed to flow from whatever color paint the supply sergeants then had on hand. If they received a large shipment of gray paint, we would have thought Selfridge was a Navy base.

Selfridge was an Air Defense Command (ADC) base, meaning it was home to a fighter outfit — the 56th Fighter-Interceptor Wing and its 56th Fighter-Interceptor Group. The fighter pilots might be flying the Lockheed F-80 Shooting Star, the Republic F-84 Thunderjet or a single-engine prop-driven star of WWII such as the F-51 Mustang. Back at Westover, which was a Military Air Transport Service (MATS) base, where the primary mission was, of course, air transport, I had become thoroughly familiar with the capabilities of the big four-engined, propeller-driven transports. I had seen jets before, at Chanute and at Scott, but those were just transient occasions in which I was a mere spectator. At Selfridge, I came face to face with the operational requirements of the jet-propelled fighter, whose designation had been changed in June of 1948 from P for pursuit to F for fighter. The nearest thing to jet-propelled flight at Westover had been a JATO demonstration in which jet assisted takeoff units were attached to the underside of a C-54's wings, where they emitted dark smoke clouds as they pushed that fast-becoming-anachronistic craft up off the runway in a steep, fighter-plane takeoff angle, much faster than it had ever risen before, as if it too could become a part of the exciting new age of jet travel.

Of the pilots who were not flying the fighters, many did not have flying as a primary duty. They would keep up their flying skills and qualify for monthly flying pay by putting in four flying hours a month, often flying the twin-engined C-45 or C-47. These two aircraft were also often used by the weekend warriors. Being close to a large city, we had quite a few weekend warriors.

As the jets began entering the Air Force's inventory in significant numbers, to distinguish between the two classes of aircraft we began referring to the propeller-driven aircraft as "conventional." There was a certain comfort in dealing with the propeller-driven aircraft. Should the weather turn sour, we knew the pilots of those planes could tool around the intended landing field until the weather improved or could fly to an alternate base where ceiling and visibility conditions were more suitable.

With the early jets, however, it was imperative that the pilot never descend from his cruising altitude unless he knew that the weather at his destination airfield was to remain good enough for landing. If the cloud ceiling at the jet pilot's destination was too low or the visibility too restrictive, the jet that had descended for landing might use up too much fuel in that low-level dense air, where the jet engine was not very efficient, and the pilot might then be unable to reach an alternate landing field. Especially in the winter months, this was an important concern at Selfridge where the station condition we all wished to see listed in the remarks section of our hourly weather observation as CAVU (ceiling and visibility unlimited) was often absent. To reinforce that view of the jets' limitations, we were told of a flight of jets that had descended for landing at an airfield in Iowa where they were caught in a zero ceiling and visibility situation with a sudden snow shower. The entire flight, we were told, had crashed as fuel ran out.

Snow showers were more prevalent in our Great Lakes area than in the prairie states where that ill-fated flight had run into trouble, and this became a cause for concern. As water vapor from the comparatively warm Great Lakes was injected into the cool air, the almost ever-present and rather shallow stratocumulus clouds formed. During exceptionally cold outbreaks, this mechanism could also produce cumulus and even cumulonimbus (the thunderstorm cloud) with towering turrets. Unlike the warm-season cumulonimbus whose tops may exceed 60,000 feet, these cold-season Great Lakes thunderstorms often were only about 10,000 feet high. Nevertheless, they presented a clear danger to aircraft operations, particularly when the vertical temperature structure would indicate that the thunderstorm would produce snow rather than rain. Then the ceiling and visibility at a CAVU field could deteriorate in minutes to below landing (or takeoff) minimums.

While sudden weather deterioration was a nightmare, a prolonged period of good weather could also raise concerns, for that was going to change. The question was — when would it happen? When the weather has been good for many days and you predict it to deteriorate, and you "hit" that forecast, going against the persistence of the weather adds to the satisfaction of the verification. One of my most pleasurable forecasting moments occurred with such a change during what had been an unusually long good-weather pattern.

It was an early evening of a September day. All day long the weather observer who used the tall Macomb County government building in nearby Mt. Clemens as a convenient visibility checkpoint, because it was exactly three miles away and the cutoff between VFR (visual flight rules) and IFR (instrument flight rules) was also three miles, had been reporting the CAVU condition as he recorded his observations on the WBAN (Weather Bureau, Army and Navy weather observation form) and then transmitted it to the world (or at least to all stations set up to receive that weather circuit) over one of our teletype machines. In fact, Selfridge had been VFR for almost two whole weeks. The weather had never been better.

Just after sunset a reserve lieutenant colonel, a weekend warrior, walked into the station with his pilot clearance form in hand. He needed four hours of flying time and planned to get the four hours by flying around Michigan in a C-45 that night. He pointed out that he had to fly VFR and would not fly if field conditions were IFR. Whereas I usually took great pains to touch every obviously pertinent meteorological base when briefing any pilot, I would go beyond that when briefing weekend warriors. Their inability to fly on a daily basis, as opposed to the pilots on active duty, made them more vulnerable when deteriorating weather was encountered. Keeping that thought in mind, I briefed that LC very slowly, repeating points up to and beyond the point where he could perhaps think I was insulting his intelligence. As compared to Selfridge's fighter pilots, he was an unusually patient man. He stayed with me, giving me the impression, unlikely as it might have seemed, that he was absorbing every word.

The gist of the forecast was that the sky would be clear and the visibility completely unrestricted until 2100 hours (9 p.m.), at which time, as I carefully explained, small thin patches of ground fog would form around the field. I cautioned that he should then be especially watchful and remain close enough to the field to quickly land if that should be necessary. The fog patches would be thickening and, by 2200 hours, fog would restrict visibility to below three miles, the VFR minimums. It was unfortunate, I said, that he would not be getting in his four-hour flying quota that night. With a bemused look, but still ever so polite, he thanked me and left for his takeoff. I knew he was just trying to be nice to this sergeant. He really didn't believe much of what I said, and that was understandable, after all, we'd been CAVU for about half an entire month.

That being a weekend, weather station traffic was somewhat lighter than usual. Compared to the rather hectic traffic of the normal weekday, there were few interruptions. I had plenty of time to re-examine the forecast, to analyze the moisture and wind conditions, two critical elements in the formation of ground fog. Also called radiation fog, ground fog would depend on the extent of radiative cooling of the ground surface and on a flow off Lake St. Clair that would be light enough to bring in moisture at ground level while the air aloft remained dry enough to enhance cooling to the dew point or to the point of saturation. And that ground-level wind also had to be low enough in speed to eliminate fog dissipation due to mixing with the upper drier layers. As I checked these elements at about 2030 hours, I could see they were all on track for the predicted fog formation and then went on with my map analyzation duties, with answering the phone, etc.

In a little while, busy with map analysis and the preliminary forecast of Monday morning's flying weather, I had forgotten the briefing. Then, at about 2210, the colonel walked into the weather briefing room. He was carrying two cokes obtained from the coin-operated soft drink machine in Base Ops.

"Sarge," he said, "I bought you a coke so you and I can celebrate a toast."

It had been a warm evening. The frosty cold, hour-glass-shaped bottle would feel good to the touch. Handing me the coke, he continued: "Here's to the very best weather forecast I've ever had. Thin fog patches formed, as far as I could tell, and I was watching it closely from above, at exactly 2100 hours, and the field went IFR at exactly 2200 hours by which time I had landed. And that was exactly what you said would happen."

I had flinched a little at his use of the word "sarge," it was not the form of address that was considered appropriate by the military, unless of course, the person being referred to happened to be Beetle Bailey's boss. We were to be called "sergeant," not "sarge," a nicety that had always been a matter of complete indifference to me. In any event, anything that LC said was going to be completely OK.

We then stood there, leaning on the forecaster's briefing counter upon which the weather observer placed new yellow sheets of hourly teletype reports from stations that were routinely reporting their weather from locations as far east as the Atlantic, as far west as the Pacific, and from Canada to the Gulf of Mexico. We sipped the cokes as the LC eagerly absorbed the analytical details of the evening's meteorology. With such a captive listener it seemed entirely appropriate to delve into things you wouldn't bring up with, for example, a fighter pilot. Such things as the saturation vapor pressure, the surface inversion, the distinction between radiation fog, advection fog and steam fog — all those things you had felt the need to explain to the pilots from time to time, pilots who never seemed to have the time for it or the inclination to listen to it.

I wanted to reciprocate and buy him a coke in return, but that was against regulations. Presumably, it could have been taken as an attempt to curry favor with one of higher rank — a ridiculous thought, perhaps, in this day and age but not so in 1950.

That would be only one forecast of many thousands. The 10-cent coke could be thought of as a mere courtesy easily dismissed, but the colonel's thoughtful appreciation, coming at a time when I was still in the beginning stages of refining my forecast procedures, would make the presentation of one cold, refreshing drink rank in later memory with the Legion of Merit that would be awarded with much more pomp decades later in a place called Vietnam. And I thought of my former associates at Westover, none of whom would have experienced the luxury of this relatively quiet Sunday night on a fighter base where the traffic had been light enough to provide the time necessary for an unhurried analysis and formulation of a forecast; and yet, those Westover forecasters had produced the immensely difficult and critical forecasts in support of the Berlin Airlift, one of this nation's most noble undertakings, and had done so under great pressure and with great reliability. To my knowledge, they had not received so much as a certificate of appreciation, or even a cold drink to commemorate the effort.

We had a warrant officer at Selfridge, a weather forecaster who, like me, must have been raised on a farm. In any event, he had a penchant for gardening which he pursued with diligence on his days off. Matt Mleziva and his wife, Dorothy, purchased a home in nearby New Baltimore with a plot of land for gardening. One day when Marie and I and our three little girls went to visit them — our third and last daughter, the lovely Pamela had been born while I was at Selfridge — I found Matt working in his garden. Marie and the girls went directly into the house to visit with Dorothy, and I walked around the side of the house where I found Matt, busy with his planting. He apologized for wanting to complete the planting before stopping for our visit and I urged him to do so, he was nearly done and we could visit when he finished. As I stood on the edge of his plot, I could see that Matt took his gardening work quite seriously. Before beginning the planting he had obviously spent a lot of time working up the ground with a hoe and rake. I picked up a clump of his soil and felt it crumble softly in my hand — the texture of Angel food cake — just the kind of ground that all serious gardeners strove to acquire. As I walked around to the side of the sown rectangle, I could see that his planted rows were as straight as tightly stretched strings.

Matt was just finishing with the planting of the last row when two boys appeared. They began running through his garden, chasing each other, kicking that tenderly worked earth in all directions. Matt glanced at me and I at him. I knew that as a forecaster, he had to have developed a lot of patience, but would he put up with this mayhem? He did, continuing with his planting. I started to tell the boys to leave the garden but then thought better of it. Matt and I had never seen each other's children. These were obviously his boys and he would not take kindly to my attempt at discipline.

Keeping up a conversation about the weather, our normal topic of conversation, began to become difficult. Matt was now focusing with visible irritation at the increasing violence being inflicted upon his plot of land. In between staring at the carnage, he began casting quick sidelong glances in my direction. Soon, we weren't talking at all, only gazing at the young barbarians. Finally, the time had come. Laying down his hoe, Matt looked at me with studied directness, and this friend, with anger now no longer being masked, blurted out the long-delayed question — "Are those *your* kids?"

Misinterpretation of the apparent "facts" was something that could also happen in the forecasting arena, and it was important to guard against it. There was a class of pilots who would read the latest reports posted on the clipboards and assume that all the weather that was occurring along their proposed route would be reported by those stations. Telling these "weather sequence readers" that there was weather along their proposed route that was not being reported by those weather stations was usually a waste of time. The reports they were reading came from weather stations that just happened to be located at the Air Force or Navy bases or, more often, at towns selected by the Weather Bureau for weather reporting. There was a lot of unaccounted weather in the voids between stations, but to those pilots, the weather they saw being reported was the only weather there was, and that weather was never going to change. Frequently, these same pilots, because they were unqualified for IFR flight or simply because they chose not to fly IFR, would insist on having VFR weather. They were often eager to begin boring holes in the sky and, I suspected, might be inclined to take unnecessary risks. Like the pilot in the Westover safety poster, they probably had a date with an angel.

The degree of caution that I extended to the weekend warriors I also extended to the other pilots whom I judged to either be generally inexperienced in flying or unfamiliar with the area they planned to fly through. One day a pilot who seemed to fit that description walked up to the briefing counter and the experience became unforgettable. He would have been one of the afore-mentioned sequence readers except for the fact that he understood very little of the code that, to save transmittal time and space on the teletype paper, all weather obs were transmitted in.

The fact that this pilot wore the gold bars of a second lieutenant put me immediately on guard. Saying that he had to fly VFR, he requested to see the latest hourly weather sequences for the stations along the route from Selfridge to Traverse City, Michigan, which is 200 miles northwest of Selfridge, on the opposite Michigan shoreline. He would be flying the T-6, the single-engine, prop-driven WWII trainer. At his request, I read to him the latest reports from Flint, Lansing and Traverse City itself. These stations and all others in the Lower Michigan peninsula were reporting ceilings and visibilities that were well above VFR limits. "It's a no sweat flight," he said as he handed me his clearance form for signing.

I was tempted to say: "Not so fast!" but we weathermen had been reminded on any number of occasions that we were a service organization, a fact that we were never to forget. So I pointed out the cold front that was then, as accurately as I could place it on the surface chart I had just analyzed, somewhere over Lake Michigan, and northwest of Traverse City. Ceilings and visibilities, I said, could easily drop to below VFR minimums as the front completed its crossing of the lake and made its landfall on the Michigan shore. The front appeared to be slowing in its eastward movement so its ETA at Traverse City could not be very precisely calculated. Given the T-6's cruising speed of 145 mph and west-southwest winds at his proposed flight level with a head wind component of about 15 mph, his time enroute would be about one and one-half hours. With that information, the lieutenant, somewhat reluctantly, walked across the street to the cafeteria/PX complex. He would shop in the PX, drink coffee in the cafeteria and be back when the next hourly reports were in.

He was back at the forecaster's counter in exactly one hour and listened as I read to him the new station reports. The front had not yet made landfall. I reiterated the expected downturn in the weather with the front's arrival. He returned to the cafeteria/PX complex to while away more time and was back again as the next hourly reports came banging in on the teletypes.

Conditions had not changed. The front was still over Lake Michigan. I began to wonder if I had misanalyzed that last surface chart. Perhaps, in the worst case, there was no front at all. Now, *that* would have been the embarrassment to end all embarrassments! But that was not likely, I was sure there was a front out there over the lake. But, perhaps, it was a front that carried no weather. There *are* dry fronts. Because there were, at that time, no earth satellites and very few weather radars, it was not possible to positively assess the strength of the front. The only weather radar in the entire area was the one we had at our Selfridge weather station, the AN/APQ-13, an item that had been salvaged from a B-29 bomber. This radar had an effective range of only 50 miles, much less than the 150-mile range of the CPS-9, the radar that would be installed at Selfridge Base Weather a few years later, in the early to mid-1950s. Even that radar would not have helped in this case, for the front was more than 200 miles away.

I made another comparison of the discontinuities in wind direction, temperature and humidity between the Wisconsin and Michigan sides of the lake. The front was there all right and the density difference of the two air masses it separated convinced me it could be producing weather.

Finally, after his third return from the cafeteria/PX, we saw the clear demarcation of the front in the data being reported along the Michigan stations. It had passed Traverse City and the observer there had not taken a special observation which meant the station had stayed VFR with its passage. Traverse City could have been in one of those "sucker holes" dreaded by the weather forecaster — it could have been in a spot that had better weather than almost anywhere else along the front. But the other stations it was passing had also stayed VFR. With a final note of caution, to the effect that with cold air pouring down from the northwest *behind* the front, convective cells could erupt over the lake and drift over land, I signed the clearance. As he began to walk away, I advised that he could circumnavigate any weather that might exist along the front, but if he saw cumulonimbus being generated in the cold air over the lake, he should approach Traverse City with caution and, if any of those cells should happen to have reached the airport, make a 180 to land somewhere else. Then, from the weather station window, I watched the good lieutenant climb up into a beautiful VFR sky.

At about the time he should have landed, Traverse City transmitted a special observation — a thunderstorm was in progress. Ceiling and visibility were below VFR minimums. I scrambled over to Base Ops. "Get in contact with the pilot of that T-6 and tell him to land somewhere else," I said. But the operations sergeant couldn't raise him on the radio. I went back to Base Ops several times before the end of my shift. There was no contact with the T-6. Finally, we knew he'd be out of fuel. "He's down somewhere," the sergeant said. Down? But where? Safely landed? Or crashed?

That evening, at home, I waited for the message that the plane was missing. We had no phone, so it would be a message hand-delivered by someone from Base Headquarters. Or, I'd hear it over a commercial radio station: "We interrupt this program to advise that an Air Force training plane with one person aboard..." Or, I'd read in the *Detroit News* that a certain Air Force pilot who had been on a routine training mission had crashed in remote northern Michigan. But none of that happened. In fact, there was no word at all about the plane.

Days went by. Then, about two weeks later, like a ghost out of the past, the lieutenant bounced into the station with a clearance form. It was at shift change and though there normally would be only one forecaster present, at that changing of the shifts there were two of us. The other forecaster offered to provide the lieutenant his weather briefing. "No thanks," he said, "I want to talk to the sergeant over there." He was pointing at me. He was going somewhere where the weather was absolutely CAVU all the way — truly a no-sweat flight. I feigned a matter-of-fact attitude. It took only a couple of minutes to inform that pilot that his weather was VFR all along his route and would stay that way for the duration of his flight. I signed the clearance. He started to leave, then turned back.

"By the way," he said, "the next time you tell me you believe the weather can deteriorate, I'm really going to listen." He then proceeded to describe, as he phrased it, the hairiest thunderstorm he ever saw. It had rolled into Traverse City, he said, and upon seeing it, he turned back and landed, safely, at the closest airfield.

I counted myself lucky with that one. The episode illustrated the quandary the forecaster often found himself in. It was important to evaluate the meteorological situation on its own merits, to remove from your thinking the desires of the pilot who could be overly anxious to be somewhere at a specific time. You had to be totally objective or you ran the risk of leading the pilot into thinking it was a waste of his time to absorb the weather picture you painted. It was especially bad to be pessimistic. Once the pilot had sensed he was dealing with a habitual pessimist, he'd dismiss all of that forecaster's advice, even when the apparent pessimism was based on good evidence.

Unwarranted pessimism could easily devolve on the forecaster through the loss of a plane to weather causes even though he had handled the situation in the most professional manner. I had a friend at Selfridge to whom this began to happen. For various reasons beyond his control, pilots who had received weather information from him had crashed on takeoff. And one day when the weather was otherwise perfect, he had been on duty when three planes were damaged on the ground due to what could be called freak high wind gusts that erupted suddenly and then died out within just a few minutes. And a plane to which he had provided wind forecasts over the pilot-to-forecaster radio, disappeared over Lake Michigan. We began to think that if the accumulation of these unfortunate events were not giving our colleague a pessimistic outlook he would have been much more resilient than we were. Not long after, he was transferred to a remote Pacific island where the plane traffic would presumably be light. I hoped in those new surroundings his luck would change.

Knowing the exact condition of the ceiling and visibility at their base was something that constantly occupied the attention of the observer and the forecaster. This was a critical requirement. One afternoon, the duty observer, Sergeant George Popadines, stood outside the weather station looking south toward the Mt. Clemens county building, our prominent three-mile-visibility, or VFR visibility, marker. To his left, the flight line was clear and dry. On his right, just 50 feet away, rain poured down on the drive leading to the weather station. Back in the station, Sergeant Popadines asked the station weather officer how to record that on the official weather recording form, the WBAN. The SWO thought about that for a minute and said, "I *don't know*."[3]

On another afternoon the observer and I were standing on the south steps of the weather station, peering at the fuzzy sky that enveloped Mt. Clemens. The county building was fading from view in haze and smoke drifting in from Detroit, and above us, the ragged bottom of a low stratus ceiling crawled at snail speed over the field. Just when we were remarking that we were going to have an exceptionally easy shift because the weather was simply too bad for flying, we heard a prop-driven fighter coming in low. In a matter of minutes the plane (memory suggests it was a Thunderbolt, but it could have been another make) zoomed in below the clouds and landed. Being surprised that any plane would come in with the ceiling and visibility as low as it was and wanting, as we always did, to debrief the pilot to see if he concurred with our observation of the ceiling and visibility, we could hardly wait to talk to the pilot. In those days before weather satellites and reliable weather radar, the pilot report was an important tool in the weatherman's attempt to construct the total meteorological picture in that part of the sky for which he was responsible. The thought that I might also bring up with the pilot the hazards of flying on a day like this briefly ran through my mind. As that pilot walked into the weather briefing room, however, I quickly scrubbed the idea of a weather lecture. For he was none other than Colonel Francis Gabreski, one of America's famous fighter aces. It was well known that during World War II he had shot down 28 German planes, and later, after the Korean War broke out, he would achieve an additional 6.5 victories as a jet pilot in that conflict, making him the nation's leading living fighter ace. Only two other Americans, both killed in plane crashes, Major Richard Bong with 40 kills and Major Thomas McGuire with 38, would have more victories. For one of the first times, I hardly knew how to begin the weather conversation. The great difference in our rank, he a colonel and I a buck sergeant, was in itself enough to give me pause, but in addition this was a man I had read about and admired since the time I was in high school. I first asked if he had an important reason for flying into our base, but immediately realized that was an inappropriate question and quickly segued to: "Did the ceiling and visibility look low when you came in, sir?" I didn't elaborate with our assessment — that the ceiling and visibility looked almost alarmingly low

---

[3] Ted Cogut conversation with George Popadines, 1 September 2001.

to us. Gabreski broke into his famous smile and said it looked pretty low. Years later I would meet him again when I was stationed on Bermuda, briefing a century-series fighter mission that he led. His jet fighters were to air-refuel between Bermuda and the Azores; and Gabreski was just as cool contemplating that event which also could have had its hazards.

In May and June at Selfridge, it was not so much the clouds as the May flies that made life interesting for the weatherman. Looking something like a small dragon fly, these gossamer-winged nighttime fliers flew together in large numbers and were sometimes thick enough to actually reduce visibility. On the midnight shifts they congregated beneath the light above the weather station door and accumulated to a depth of a half-foot or more on the concrete stoop. When the observer opened the door to make his weather observation, he would have to clear a path with a broom and often a shovel. In the process, a few would fly into the station. This would make the forecaster go into attack mode with the station's official fly swatter, putting in jeopardy his reputation for professional appearance. Here, I am not referring to his personal dress, which at such times admittedly might have seemed a bit disheveled, but to the unscientific wiggles that appeared in his isobars, as he swatted with one hand while drawing pressure lines with the other. If he happened to make a lucky hit, the May fly would decorate the map with a brown splotch which might have seemed puzzling to the pilots who knew that brown shading on a weather map was the indicator for a dust storm, just as green shading was the indicator for rain or snow and yellow shading for fog. (When thunderstorms or lightning were reported, their symbols were drawn in red.) While, for operational considerations, we usually preferred not to see a cold front passing the station, in the May fly season we welcomed it, for the front, at least for a little while, would sweep the May flies away, pushing them out over Lake St. Clair.

Not long after the Traverse City thunderstorm episode, I received an unbelievable promotion to staff sergeant; and not long after that the Air Force presented us all with our blue uniforms which were adopted in 1949 but not issued till late 1950. The promotion was unbelievable because advancement to the first three grades had been frozen throughout that immediate post-war period. I had fallen into a bit of luck again. The Air Force-type rocker chevrons with the centered star, which came to us before we received the blues, having been sewn onto the sleeves of our old Army olive drab and khaki uniforms in 1949, were transferred to the blues. The new uniform was more smartly tailored but became the object of jokes. Our comedians who could not pass up the chance said we'd enlisted in the Greyhound bus service. Along with the clothing change, our shoes were changed from brown to black. One aspect of the uniform change that many of us regretted was the discontinuance of the AAF shoulder patches, the orange and blue weather vane sleeve patch and the circular weather pin that enlisted men had worn on epaulets and overseas caps. Nevertheless, I liked the new uniforms. For one thing, the AAF brass buckles and lapel insignia (the US on the wearer's right and propeller with wings on the left), which had required a considerable amount of polishing, were replaced with an aluminum-colored buckle and US lapel pins of the same material. These were not meant to be polished. And so we passed through that period in American military history when Air Force personnel wore Army uniforms decorated with Air Force insignia. To this day one can still find veterans who say, with pride: "I served in the brown shoe Air Force."

The Graveyard Shift at the Weather Station

"Hurry with that cold front! They're planning another attack!"

Still thinking he might be cut out for a career in the art world, the author drew this cartoon about the May flies that plagued Selfridge in May and June.

Briefing an over-flying pilot over pilot-to-forecaster radio, channel charlie, 1950. The sign board reads:

DETACHMENT 16-8L
16TH WEATHER SQDN
2103rd WEATHER GROUP
STN WEATHER OFFICER
CPT W.E. SMITH
STATION CHIEF
TSGT P.E. DOWNEY

The shingle with the duty forecaster's name hangs below the sign board, obscured in this shot by the slanting display upon which the yellow teletype sheets of hourly weather reports were posted. The reports are all but invisible in this shot because of the glare from the overhead fluorescent lights.

Chapter 15

Jet Streams and Other Things — We Call Off the Bet

Through its rawinsonde section, the Selfridge weather detachment provided precious upper air information which was transmitted via teletype to other weather stations and was plotted as one station of many on upper air charts used by forecasters around the world. With the rawinsonde information as his basis, the forecaster would predict flight-level winds, icing and turbulence zones. Through examination of the thermodynamic diagram plotted from the same data, he would assess atmospheric stability and the opportunity for cloud formation including the cumulonimbus (thunderstorm) cloud.

To receive the radio transmission of upper air data — pressure, temperature and humidity — from the heart of that rawinsonde system, the radiosonde, which was ascending into the atmosphere tied by waxed string to a balloon the size of a one-car garage, rawinsonde sections used the pre-World War II SCR-658. Located outdoors beside the upper air shack, the SCR-658 was a manually operated radio direction finder. Like the postman, the SCR-658 operator did his duty out in the elements — in the rain, sleet, hail, snow.

As described in Chapter 10, with two cranks the operator kept the antennas aligned on the radiosonde signal. One crank changed the vertical aspect of the antenna, providing the elevation angle of the radiosonde, the other controlled the horizontal, providing the azimuth angle. To be sure that he was perfectly aligned on the moving signal, the operator viewed two sets of green blips in a control panel window similar to an oscilloscope. When the antennae were perfectly on target, the high points of the blips would coincide. Inside the upper air shack (rawin shack), the elevation and azimuth angles were plotted to obtain upper air wind speeds and directions which were assigned to specified heights that the plotter had calculated from the radiosonde data.

As the balloon ascended into the high atmosphere, it began to expand in the increasingly less-dense air. Eventually, stretched to its limit, it would burst and float down to earth on its paper parachute. The radiosonde

137

package would occasionally be found, and still is, usually by hikers who travel the back country.

The rawinsonde system was capable of providing data to very high altitudes, even higher than the normal operating range of the new-to-the-scene jet aircraft. This information was provided twice a day on that world-wide noon and midnight Greenwich time schedule. At Selfridge, the rawinsonde team conducted their radiosonde flights from their operating location, the rawin shack, which was about a half mile from the weather station. When finished with their run, they would call their data to our weather observers who then plotted it on the Skew T - Log P diagram and transmitted it via teletype to merge into a collective with similar data from other rawin stations across the nation.

Prevailing winds frequently carried the balloon-borne radiosonde across Lake St. Clair and over Canada. One day, I received a phone call from a farmer in Ontario who found one of our radiosondes.

"It fell right in front of my horses as I was plowing," he said. He described what had fallen: a red paper parachute, a deflated beige-colored neoprene balloon, a white box-like affair which was the radiosonde itself inside of which and accessed by a hinged door was a water-activated battery, an aneroid cell for measuring pressure, a printed circuit board (commutator bar) with a moving arm actuated by the aneroid cell that would switch in temperature, humidity and pressure circuits, a humidity sensor located near the top of the box, and a ceramic temperature sensor that protruded out from the box. Attached to the bottom of the box was a white two-inch by nine-inch cylinder that housed the radio transmitter. "It spooked my horses when it fell," the farmer added.

Thinking of how it was becoming the "in-thing" in this country to sue for whatever cause and perhaps that was also happening in Canada, I asked the farmer if he or his horses or plow had suffered any injury. "I did not and they did not," he said, and then, reversing the thrust of the question: "I think the horses may have stepped on the instrument, but it's not hurt bad." He then wanted to know if he was required to send the instrument back to us, and I received his heartfelt thanks when I offered that he should take it to his house and place it on the mantel as a souvenir of one exceptional day of plowing. As for the need to return the radiosonde, that policy would change from time to time. The problem was that, once it was returned to us, it had to be sent back to the manufacturer for recalibration, a cost that usually made its retrieval unattractive. I don't remember if the policy at the time specified the return of that radiosonde, but given the circumstances of its descent and the farmer's extremely cooperative attitude, in any event I wanted him to have it.

With the increasing number of debriefings of jet pilots, on a daily basis we were learning more and more about the flying characteristics of jet aircraft and also of the winds and weather those aircraft were encountering. One day, an F-80 pilot placed his Form-175 clearance on the briefing counter for my signature, saying: "Tell me where the icing is." And I did so, while saying that he only had to do so-and-so to avoid the icing zone. "You don't understand," he said, "I *want* to fly into the icing." That kind of remark made to any aviation forecaster who had spent his career trying to steer pilots *away* from any possibility of icing, was sure to command instant attention.

My response: "You're kidding, of course." But, no, he was serious as could be. He *meant* to fly into the icing.

"I have a theory," he said, "that the jet won't ice up." Now *that* was an interesting theory.

It clearly was his assumption that the vastly different flying characteristics of the jets protected them from icing problems, even while flying in clouds with great icing potential. I sent him into the only area where there was a possibility of icing but advised that this would not be a good test, as the icing, if there was any, would be very slight and perhaps not even noticeable by him, which was the way it turned out to be.

In a few days, he returned for another briefing. He wanted, again, to fly into an icing zone. I then did something that I had never done before or since. I sort of lied about the weather, telling him where icing could occur but not where the worst icing would be. Due to a great streak of luck that would stay with me throughout my aviation forecasting career, I had not lost a plane due to weather. I didn't want his plane to be the first. This pilot was persistent in his efforts to defy the icing gods. On succeeding shifts he would return, always without a flight plan for that would be made after I responded to: "Tell me where to go to find icing." This went on for several weeks. He was making the same request, I surmised, of the other forecasters. For my part, I continued to send him off into the particular areas of the sky where icing would be unlikely or, at the least, in the form of what we called "light rime" which I believed would not be a significant hazard. Eventually, he stopped coming into the station. I hoped it was because he had been transferred to another base. He had survived all the flights he made on my briefings, and I hoped he had the same luck with the other forecasters he approached. I have since concluded that pilot was a bona fide test pilot, operating out of our base to be where icing could easily be encountered, ferreting out the capabilities of the jet plane for the benefit of pilots who would follow him. But the wonderful ability of the jet to fly at altitudes that would put it above most of the weather tended to lull pilots into thinking jets were immune to the effects of bad weather. Incidents in the years to come would quash that kind of thinking.

Being the first truly operational jet in the United States, the F-80 did have to undergo its days of testing. One day in 1949, a colonel and the other pilots of his flight of F-80s walked into Selfridge's Base Weather and requested a flight briefing — *to Europe.* I could visualize a four-engined transport taking off for Europe easily enough — *but an F-80?* It had only been three years since that day at Chanute where I had seen my first F-80. It was a sleek little plane, but it never seemed to me to be an ocean-crosser. I briefed the stateside leg and was shortly joined by Captain William Smith, the station weather officer (weather detachment commander), who had been forewarned about the mission and had prepared for the over-ocean portion. Captain Smith signed their clearances and the F-80s took off for Goose Bay, Labrador, the first hop of their flight. We then believed that this flight, led by Colonel David Schilling, a World War II ace of the ETO who had 22.5 victories, was the first west-to-east jet crossing of the Atlantic. I reflected that it was being done 22 years after Charles Lindbergh had made the very first west-to-east crossing by air in his single-engine, prop-driven Ryan monoplane, *the Spirit of St. Louis.* Recently, however, I read that the flight on that day might have only been the *second* west-to-east jet crossing with the first being made a year earlier, near the beginning of the Berlin Airlift.

The arrival of the jet age raised the question of whether the Air Force weather forecasters had kept pace with that significant advance in aircraft capabilities. Routine analysis of upper-air pressure systems, winds, temperature and moisture, had, up until that time, been conducted only as far up as the 500-millibar level, or to a height of about 18,000 feet. But the jets were flying at the 300-millibar level or about 30,000 feet and even up to the 200-millibar level or about 40,000 feet. For the forecaster of that time, these were essentially uncharted skies. Even though it had been known that B-29 bombers had flown into head winds that were much stronger than expected while on those bombing runs over Japan that would eventually end World War II, it was not something that we thought much about. The B-29s had pressurized compartments for the crew. For propeller-driven planes, they flew at rather high altitudes, often at about 30,000 feet, and that was where they encountered rivers of air of impressively high speeds. From time to time our rawin section also would report exceptionally strong winds, some as high as 200 miles per hour, at the 300- to 200-millibar pressure surfaces. And our first reaction was to suspect that the lads in the rawin shack had made an error. At those times we would ask them to recheck their calculations. The answers always came out the same.

To assess forecasting capabilities at the jets' cruising altitudes, planners at weather headquarters set up a test. At Selfridge, and perhaps at other Air Force bases as well, for several weeks forecasters were required to predict 24-hour wind speeds and directions for selected stations at the 300-millibar level. These forecasts, done anonymously, were transmitted over the teletype by the weather observer. The transmission time usually sneaked up on the forecaster while engaged in a briefing, in answering phone calls or in other on-the-spot forecasting duties. It was a rare occasion, in fact, probably never, that he had an opportunity to carefully construct a 300-millibar prog from which to pick off those 24-hour-forecast winds. Unless the predictions were done that way, the way they had been done at Westover for the lower altitudes, there'd be little chance they'd leave a lasting impression for accuracy. I recall that WOJG Matt Mleziva,[1] he of the destroyed garden, was very interested in the verification of the various forecasters' predictions, and I assume he was probably verifying better than the rest of us though he had said that I was that culprit. The forecasts were officially verified at one of the weather headquarters, probably at Air Weather Service itself. In a twist of that famous Grantland Rice quote: *At a receiving teletype somewhere an unknown scorer indicated whether we had won or lost and wondered how we played the game.* Not long afterward, we would all be rotated through a new weather forecasting updating school which was probably precipitated as a result of that high-level winds forecasting test. We would call the course "SHAFT" for Specialized High Altitude Forecasting Training, though I suspected "training" had unofficially been appended to the title by some wag in order to fit the acronym he wanted it to have. In orders assigning us to the school, it was also referred to as "High Altitude Forecasting Course."

The year 1950 was a year of important happenings. Among them were these: Marie and I embarked on a comprehensive house-rebuilding project; the North Koreans invaded South Korea; I would be sent TDY (temporary duty) to the High Altitude Forecasting Course and; as mentioned in the last chapter, Pamela, an Air Force squadron commander, was born. I should probably state that she did not immediately take on that responsibility.

Having been raised on a farm, it was inevitable that I should, sooner or later, use gardening as my tension-relaxer when off shift. Through an ad in the *Detroit News,* my father-in-law, Walter, found a house for us on an acre of land, which we promptly bought on the land contract plan. This was quite a find as it was only about a mile from Selfridge's back gate, and of utmost importance to a farm boy, the land was rich. Tomatoes, sweet corn and beans could be grown with little more effort than to throw the seeds out and watch the plants pop up out of the ground. To solve the problem of having too little land for growing all we wanted (and needed, for even a staff

---

[1] WOJG, Warrant Officer Junior Grade, was one of two warrant officer ranks, the other being CWO, Chief Warrant Officer. WOJG was later changed to WO, Warrant Officer, and the CWO rank was expanded to CWO-2, CWO-3 and CWO-4.

sergeant had barely enough, some would say not enough, money to raise a family on), we planted pole beans at the corn hills and watched as the beans climbed the corn stalks, thereby saving precious planting space and making it easier to pick the beans.

But the house was too small. Still, it was all we could afford. Walter had a solution. In his Sunday morning routine, he read every ad published in the *Detroit News*, and one Sunday he found a house for sale that was to be torn down to make way for Detroit's first freeway, the John Lodge Expressway. The house, one of several dozen that were also to be torn down, could be purchased for an unbelievable $200! Being a neophyte when it came to evaluating houses to buy, I took Dad with me to offer advice and also to let him share in the experience of seeing someone buy a house for a mere $200. In 1928, he had had a two story brick house built in Hazel Park, a suburb of Detroit, and it had cost him $10,000. Like Walter, Dad was of that generation where to be a machinist, a farmer and also a carpenter was not exceptionally unusual. Dad was impressed with this bargain. He picked out a house that had been built entirely of white pine. The trim boards, the sheeting, the full-dimensional 2X4s, even the full-dimensional 2X8 floor joists, were all made of white pine, that choicest of builder's woods. In 1950s house construction, white pine, because of its scarcity and cost, was reserved for trim work and cabinetry.

The contractor from whom I bought the house had it dismantled and trucked to our yard. Suddenly that yard was filled with old but excellent white pine lumber. It was one thing to have the lumber, but quite another to get started on a house-enlargement project. Where was I to begin? Dad solved that one. One day while I was on forecasting duty at the base, he came out with shovel in hand and began digging the foundation that would extend the house. With the digging completed, he brought in a cement mixer, bags of cement and a load of gravel and began mixing up batches of concrete. In a few days, we had a foundation upon which to build the house extension. Just before I left for Chanute's High Altitude Course, with the help of brother-in-law Robert, the floor and walls of the addition were up and Walter began building the roof that would enclose both the new and the old structure with height enough to provide an upstairs bedroom for one of the girls.

As I left for Chanute, I felt quite pleased with myself. I had had the good fortune, of course, in being the son of a man who knew construction work, that I could claim no credit for. But I had obviously employed great skill in the selection of in-laws(!).

Every weather detachment that I ever served in had at least one forecaster who was also a pilot. Every one of them were down-to-earth, regular guys. The fact that they flew routinely made them more acutely aware of the weather effects on aircraft, an asset that I thought quite valuable. But in addition, they could be counted on if you needed a hop. The pilot-forecaster at Selfridge, a first lieutenant, was sent along with myself to the High Altitude Course. He elected to take his wife with him, but Marie and I decided that with decisions to be made in the house-construction project, she and the three girls would stay home during this, my fourth technical school. This would also save a little money that was needed for building materials.

In its introduction to a lesson plan under the subject heading *Constant Pressure Analysis and Prognosis,* Chanute's US Air Force Technical School, Department of Weather, provided a rationale for the High Altitude Course which implied that the forecasts for those high altitude "uncharted skies" may not have been as accurate as desired:

> The material in this paper is by no means new. It is simply a collection of old concepts and ideas gathered together for the specific purpose of increasing the accuracy of upper air forecasts in so far as wind direction and speed are concerned. Basically, the principles used in making a forecast at any level are identical! Hence, if in the past, the forecasts issued for the 10,000 ft. level have been acceptable there is no reason to expect that the forecasts for 50,000 ft., or or any other level, are to be inaccurate, or, in other words unsatisfactory.

The High Altitude Course would present us with the anatomy of the jet stream, a topic for which there seemed to be almost no information as recently as when I attended the Weather Forecaster Course in 1948. The High Altitude Course would also focus on improving upper air analysis and prognostic chart construction skills

Routine class procedure involved drawing standard-level pressure charts for the 300- and 500-millibar charts. These were placed on a light table, where the current analysis was superimposed on the 24-hour-old charts. By drawing lines through the intersections of the current and 24-hour-old height lines (lines that delineated the high and low pressure systems), one could see the changes that had occurred over the past 24 hours. These charts were a complex myriad of lines. We called them "spaghetti charts." As I analyzed them, several lines of the weatherman's song often ran through my mind:

> Our pressure lines are intertwined
> Our fronts are underground,
> The winds that blow
> From high to low
> Have blown me off the track.

In the analysis and construction of progs, for those of us who had served in the "map factories" such as Westover, attending map analysis and map prog classes was a case of reinventing the wheel. In that pre-computer age, we had already mastered the construction of prog charts to the extent that was then possible. About a decade later, when the sophisticated computers came about, progs that made use of a number of models would be possible as would the handling of vorticity, the tendency for any point in the atmosphere to undergo spinning motion. We did pick up important information about the jet stream, primarily the polar jet which would be found usually in the height range of 25,000 to 40,000 feet with the lowest heights in winter, and in a latitude range in this hemisphere that would be farther north in summer than in winter. There was also that largely new-found phenomenon called "high altitude clear air turbulence" that could be set off with the wind shear occurring near the jet stream. As the class progressed, it became apparent that the class designers also had a sort of mentoring concept in mind. With the students coming from all sorts of weather stations, from those where the traffic was heavy and the station was in operation 24 hours a day to the light-duty stations with daytime-only flying where only a small number of progs had been constructed. The students from the light-duty stations would presumably pick up information from those who had been in the map factories.

When we were about halfway through that 90-day course, my classmate, the lieutenant from Selfridge, said he needed to get some flying time and invited me to fly along with him back to Selfridge. Being away from Marie and the girls for about a month, I didn't have to be asked twice. He never told me so, but I suspected his selection of Selfridge for the destination of his flight was made to make that visit with my family possible. At Selfridge and at every other weather station that I would be assigned to, in an atmosphere of warm camaraderie, you could depend on the consideration of your fellow weathermen. The lieutenant certainly exemplified that fine tradition.

We took off from Chanute in a T-6, the single-engine, two-seater trainer of the World War II era. With the seats arranged one in front of the other, I sat in the back where I had a good view of the ground and sky but no communication with the pilot. As far as I could tell, I didn't need communication — I was enjoying a picture-perfect flight. When we descended to land at Selfridge, our flight path took us directly over the house, giving me a thrilling bird's-eye view of the big new roof Walter had completed in my absence.

The weekend flew by all too quickly. Soon, it was Sunday evening — time to head back for Chanute. While getting our weather clearance we kept the duty forecaster occupied with stories about the High Altitude Course which he knew he'd also be sent to attend. Fully aware that he was presuming to give a weather briefing to two forecaster-fliers who were then enrolled in a course which was providing them with the very latest in the science of flight-level wind forecasting, and perhaps thinking those two forecasters would certainly be checking the winds themselves and that he'd be insulting their intelligence by drawing their attention to it, as I recall, he did not dwell on the description of the wind field, and more importantly, on any perception he might have as to the way that field would be changing.

We looked at the winds aloft chart, which hung, along with many other charts, from spring clips attached to one of the station's walls. Except for the surface chart and an upper air chart which we produced ourselves, these came in over the FAX. The chart depicted a slightly quartering head wind on the track to Chanute. That chart was plotted with winds only. Had there been temperature and pressure-height values on it, we might have seen clues as to the tendency for deepening or filling of the rather weak trough of lower pressure. The trough, apparent in a wind-shift line, lay slightly west of our route. The 850-millibar chart, which would have provided temperatures and pressure-height change for an altitude of 5,000 feet, which was close to our flight altitude, had not yet been received over the FAX. Constant pressure charts such as that one (and the 700- 500- 300- 200- millibar charts), were produced from radiosonde data that were transmitted only twice a day, at midnight and noon Greenwich time. The most recent 850 chart, therefore, was constructed from data that were then almost 12 hours old — a lot of changes can occur in weather charts over a 12-hour period. In any event, the trough would not be important wind-wise unless it should deepen and tighten the pressure gradient ahead of it, but it would provide impetus to the scattered evening thunderstorms that lay along the route.

We took off into a typical Midwestern summer's evening. Warm, but it cooled off nicely when we reached cruising altitude. The T-6 could not climb over the tops of the cumulonimbus, but we would circumnavigate them. I settled into my seat, checked my parachute straps and got comfortable.

When you are flying through an area with scattered thunderstorms without the benefit of radar, as we were, your ability to circumnavigate them becomes more difficult after dark, and it had turned dark shortly after takeoff. Probably helped by the occasional lightning flashes which illuminated those cumulonimbus cells, the lieutenant was doing a great job of thunderstorm-avoidance. In evading the cells, he obviously had to change course several times. At about the time when it seemed we should have been near Fort Wayne, Indiana, we dropped down toward a cluster of lights and found we were over Battle Creek, Michigan. We were north of course with a stronger and more direct head wind than the winds aloft chart had shown — the trough had deepened. (That wind phenomenon would later be dubbed the low-level jet, after its big brother, the high level jet.) With the wind situation now quite obvious, we headed directly for Chanute, on an uneventful course which was then in the lee of the thunderstorms and, slowed by the winds, in time landed.

As I climbed out of the cockpit, the lieutenant said there was only 15 minutes of fuel left and that for the past half hour he had been worrying about how he would get me out of the T-6 when, as he feared, we would run out of fuel. As much as that could be said in the rank-separated military environment of that time, I thought of the lieutenant as a good friend and knew he was genuinely concerned with my well being. We were standing on the apron in almost complete darkness — there either was no moon or a cloud was obscuring it. This saved him from seeing the ghastly expression on my face as he matter-of-factly explained, without so much as a hint of a joke in his voice, that he had decided he would roll the aircraft over and have me fall out. I was over-joyed that that had not been necessary and grateful for his great job of dealing with the thunderstorms and the unexpectedly strong head winds.

In time, I think we all come to know certain of our characteristics that separate us from others. Clearly, I was quite different from those heroes who could accept the thought of dropping out of airplanes when they were high enough to sort of estimate if their bodies would accelerate through 78% nitrogen, 21% oxygen and a few other trace gases at a rate approaching 32 feet per second per second. And also, if they were lucky, they might be able to demonstrate if their bodies could be supported adequately by strings tied to a piece of cloth.

Speaking of heroes and parachutes, my generation had been raised with Lindbergh's cross-Atlantic flight held up as a feat worthy of emulation. I too had been captivated by his spectacular achievement, but it had always been far more interesting to me that Lindbergh had been forced to *parachute out of airplanes four times* before he ever attempted his historic flight. No one could question that he was a genuine hero.

In a couple of weeks, we all stopped drawing the spaghetti charts, returned to our home bases, and got back into our forecasting routines. One day while off duty, I stopped in at the weather station to check my mail. I had my mail delivered there to have a good reason for chatting with the duty forecaster — about the weather, of course. While talking, we watched the wind recorder's pen arm jump from about 5 or 10 miles per hour to 50 miles per hour. Staying at 50 mph for less than a minute, just as suddenly as it had risen the wind dropped back to its previous 5 or 10 miles per hour. With no weather system to increase the general pressure gradient, and consequently the wind, the forecaster had no wind warning in effect. Had there been such a warning, precautions would have been taken. In any event, the gust picked up an F-80 and slammed it along the apron where it came to rest against two other parked aircraft. The result: three damaged aircraft from a wind gust of essentially instantaneous duration. It had been produced in a downrush from a single cloud that was just starting to build into the towering cumulus stage and would dissipate before reaching the cumulonimbus (thunderstorm) stage. Some years later this phenomenon would be given a fancy name — the microburst. Back then, we hadn't a name for it but knew what it was and knew just as much about how to predict it in the long term as is known now.

In recent years, because of aircraft problems associated with it, weather information broadcasters on the news programs have made the public aware of the wind shear phenomenon, the sort of mechanism that had damaged the three planes at Selfridge. Planes can get into trouble when pilots attempt to land or takeoff with wind shear close by. Most of the time the possibility of wind shear is "telegraphed" by the presence of towering cumulus or cumulonimbus. And the phenomenon can actually be diagnosed with a fairly new kind of radar that recently has been installed at many airports, the Doppler radar, making it possible to warn pilots when wind shear is actually occurring. Such technological advances are making air travel much safer than it used to be.

One day, in January of 1949, an old friend walked into Selfridge Base Weather while I was on forecasting duty. It was Andy Anderson, my buddy from basic training and the Weather Observer Course, from way back in 1946. Andy had remained a weather observer, had been sent overseas, and was now on his way home to Durand, Michigan and out of the service. He was about to become, as it was then popular to say, taking the image from a novel about a businessman in the new post-war world, the man in the gray flannel suit. The fact that the end of that three-year enlistment was just around the corner was not up-front in my consciousness. The truth of the matter was that I had been captured by the challenge of forecasting, and I seemed to have a hidden desire to see where it could lead me.

Andy said he was stopping by to call off our bet — the bet we had solemnly made in those first few months after our enlistment in the Army Air Force. It was the bet flowing from his assertion that I would not reenlist and my response that I just might.

"I don't want you to reenlist," Andy said, "just to win that twenty dollar bet." A reader today will immediately assume he was just being flippant. Who would reenlist for a paltry twenty dollars? But he was deadly serious. It was true that we had been granted a few promotions since our basic training days, but twenty dollars was still a lot of money to a GI. We were paid once a month, on the first of the month, and it was a common practice in our household to search for stray coins beneath the sofa cushions during the third and fourth weeks of every month. Of course, I wouldn't have reenlisted for twenty dollars, but Andy was not so sure of it. We agreed to call off the bet. Andy still was not satisfied. Before leaving he asked, "Now you're *really* not going to reenlist?"

Actually, I just couldn't make up my mind. And to avoid coming to terms with it, I had decided to take advantage of the provision that allowed one to extend his enlistment for one year. I would put myself in a holding pattern, seeing if I was cut out for that forecasting business, delaying the career decision until the next year. "No," I said, "I'm not going to reenlist — at least not at this time. I extended my enlistment one year."

Andy sort of flinched. The thought of extending one's time in the service, even if for as little as only one year, was anathema to him. "So," I said, "I have another year to go. And at the end of next year — who knows what I'll do?"

On that note, with an expression of extreme pity telegraphing that he was sure I would reenlist, Andy left. I never saw him again. It was a somewhat sad and typical aspect of service life that you would make good friends, serve with them in unforgettable situations, perhaps even in combat situations where the bonds formed would be akin to family ties; and one day, with their barracks bags slung over their shoulders, they'd walk away and you'd never see them again. Andy was one of those soldiers who served with dedication and, seeing no future in the service, couldn't wait to return to civilian life. I could not forget that he was the one who had talked me into entering the meteorological field. And now he was leaving it. Unlike others whom I would meet years after they had made similar decisions, I believed Andy would not be looking back.

Top: Weather observer Sergeant George Popadines services wind equipment atop the weather station building. Standing at a typewriter mounted on the typical tall weather station typewriter stand, the author types the daily forecast, a routine requirement at the end of the midnight shift.
Bottom: Sergeant Popadines checks temperatures and humidity at the instrument shelter outside the Selfridge weather station.

Top: The small object barely visible between the balloon and the tallest tree is a radiosonde. Tied to the balloon by waxed string, the radiosonde transmits temperature, humidity and pressure to rawin section personnel who track the radiosonde with a radio direction finder to obtain elevation and azimuth angles from which to calculate winds aloft.
Bottom: The setting sun paints a cumulonimbus, or thunderstorm, cloud. Cumulonimbus was abbreviated "cb" on the WBAN (Weather Bureau, Army, Navy weather recording form) and on the teletype weather reports.

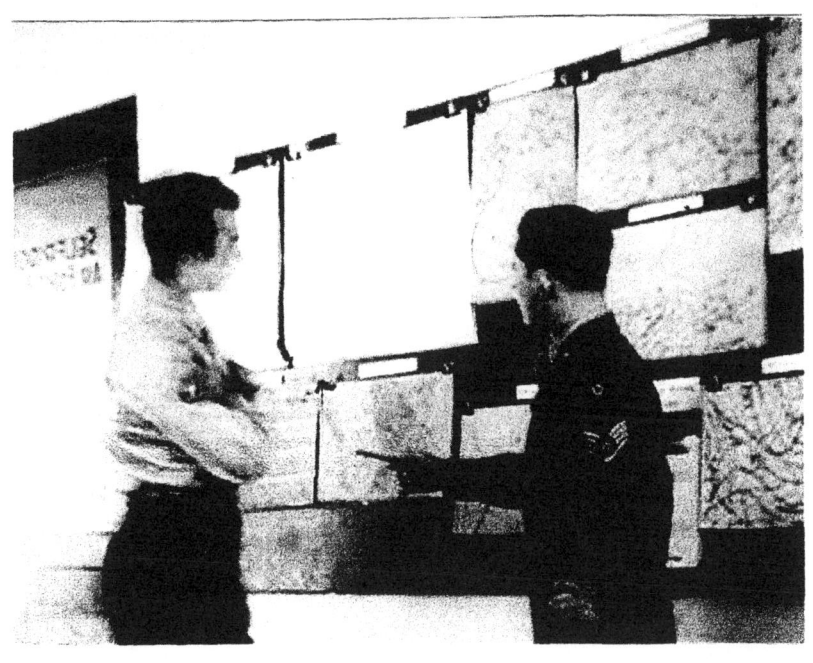

On a day in 1950, weather observer Airman Ray Doty takes note of the FAX chart pointed to by the author. FAX charts (gray) were transmitted from Suitland, Maryland. The white charts are the local area surface map and its forecast map, prepared by Selfridge observers and forecasters. The door opens to Base Operations. Its sign reads:

SELFRIDGE
AIR FORCE BASE

BASE WEATHER STATION
2103rd WEATHER GROUP
2059th WEATHER WING
AIR WEATHER SERVICE
RESTRICTED AREA

Top: A weather observer tears off sections of teletype paper which he will post in the adjacent weather briefing room. Three teletypes are to his left.
Bottom: This FAX machine, demonstrated for us in weather observer school in 1946, was used well into the 50s. Photos taken in 1950.

Chapter 16

Saudi Arabia — No Tree,
No Grass on Ground to Spy

In January 1951, after a train trip from Detroit back to my old stomping grounds of Westover, I took off from that base, a passenger in a Douglas C-74 Globemaster. My destination: a place named Dhahran in the kingdom of Saudi Arabia. I had asked for the assignment, but it still came as a surprise. I had actually volunteered for it, but that was more than a year earlier, near the end of the one-year extension of my three-year enlistment. Shortly after volunteering I was alerted for Saudi Arabia as expected, but that trip was canceled when I was told I would have to extend my enlistment to have enough time to complete the tour. This I refused to do as it would obviate the $300 reenlistment bonus that we were counting on to pay for more building materials. About a month later, I finally committed myself to another three-year reenlistment, and near the end of the first year of that hitch, someone in personnel assignments at Air Weather Service headquarters, then at Andrews Air Force Base, Maryland, remembered there was a foolish forecaster at Selfridge who actually had once volunteered for Saudi Arabia. That personnel person knew the volunteering had happened the year before and technically was no longer in effect, but one had to look hard to find anyone who'd ever volunteer for Saudi Arabia. So this was a bird in the hand that had to be taken advantage of. And so I was sent to Dhahran to be the replacement of the forecaster who had gone there the year before. At Air Weather Service headquarters, there probably seemed to be poetic justice in the fact that I would be replacing the forecaster who had gone there in my place because I would not reenlist when it would have been more convenient, from the assignments perspective, for me to have done so.

That somewhat rash decision, to volunteer for Saudi Arabia, was taken because I knew I was overdue for an overseas tour and I remembered the stories of the troops coming back from the Far North and of how cold it was and how they had to shovel tunnels to get to the instrument shelter from the weather station, and also how devastatingly cold it was on my first nights at Ft. Sheridan and Westover Field. In short, I wanted to be sent somewhere where it was not very cold — from my study of climatology I knew Saudi Arabia would not be cold.

I had fully been expecting my first overseas assignment to be to Korea where that savage war, a war then being called "a police action," was entering its seventh month. Of course, I could have volunteered for Korea after that war started, but aside from the big negative that it could be bitterly cold there, I had always adhered to the notion that it was bad to tempt fate. If it was meant that you were to go to a war, so be it. Go, and do your duty to the best of your ability. But let the war gods decide that you should go.

Probably the worst part of the Saudi Arabian assignment was the fact that it was an isolated tour. No family members, excepting that of the brigadier general commanding, were allowed. That may have been a purely practical decision for there were no quarters for other families, and, for that matter, nothing but a few Quonset huts and tents for approximately 100 US Air Force personnel, the entire complement then at Dhahran. The good part was that the tour was only 12 months long.

The piston-driven C-74, the Air Force's largest transport with a loading capacity of 25 tons or 125 troops, had four engines, would cruise at over 250 mph, and had an interior that seemed spacious enough for football practice. As I settled onto the canvas side-bench seat, I felt a certain comfort, not unlike the hop I had taken that time in the B-17, a comfort in the steady sound of the C-74's mighty Pratt & Whitney engines, each having a 3,000 horsepower capability. I wondered if this one could have been used in the Berlin Airlift where the C-74 had flown large tonnages to the beleaguered city. Two bubble canopies at the pilot's compartment gave the big ship the look of a monstrous bug.

We landed on the Azores, at Lajes, formerly called "Lagens," which seemed a little unnecessary. With its range of over 7,000 miles, the big bird could have taken us all the way to Africa, but there were passengers to deplane at Lajes and passengers to pick up. Among the gaining passengers was a British sailor, who was going, as I remember it, to meet his ship at a North African port. At our stop in Port Lyautey, French Morocco, the US Navy prepared a meal for us at its mess hall. It was not sumptuous, but nevertheless an astronomical improvement over the box lunches we had had on the plane. Turning to me as he stood ahead in the mess line where the Navy cooks plopped large spoons of vittles on his loaded mess tray, the English lad remarked: "You Yanks eat like lords!" He was just being a normal, polite Englishman, the chow really was no better than that ladled out in the Army or Air Force mess halls and I had never heard anyone praise the food in those halls, even when it might deserve to be praised. It was a cultural thing. The GI's I traveled with, probably as a result of having had the de-elevating experience of KP, would have found it hard to praise anything at all about a mess hall.

From Port Lyautey we flew on to Tripoli, Libya, landing at Wheelus Field. The 29th Weather Squadron headquarters was located at Wheelus. It was the parent unit of Detachment 29-2 with station call letters HZDR at faraway Dhahran, Saudi Arabia, the place I was heading for.

At Tripoli, we transferred from the C-74 to the old tried and true, and smaller, C-54. After a number of hours over the Mediterranean, we landed for refueling at Nicosia, Cyprus where the customs inspector sprayed the entire interior of the plane, including its not-so-happy passengers, with DDT. Then, airborne again, we made the short crossing of the eastern Mediterranean and over the desert made known during the first World War by Lawrence of Arabia. We penetrated deep into the airspace of that Arabian desert, farther than Lawrence had gone, out over vast seas of sands called the "Rub al Khali," a place also aptly described as "the empty quarter."

My first impression of Saudi Arabia came to me through the windows of that C-54, flying at about 9,000 feet. We would drone on over that desert for a thousand miles. I looked out from time to time. It was nighttime. There was nothing to see, not the warming cluster of light of the occasional small town that you saw as you flew over the prairies of the American Mid-west or West, places also sometimes called "deserts," not even those solitary lights of a farmer's or rancher's house to break the visual void. No headlights of a lone car or truck. No indication at all that any humans might be looking up at our navigation lights as they plodded through the sands, following the stars.

Then, as we were approaching the shore of the Persian Gulf and the airfield at Dhahran, the orange glows of natural gas flares, being burned off in the oil fields, lit the desert for miles around. Until we landed and I was bounced by jeep to that "house" with the half-round integrated wall and top that looked like the half slice of a huge corrugated drainage pipe, Quonset hut 113-A, my home for the coming year, the oil field flares and the unending sands they illuminated were the only noticeable features of the landscape.

I thanked the driver of the jeep, pushed open the screened door of the Quonset, threw my bags on the nearest unclaimed bunk and learned very quickly that, if you didn't enjoy raising a knot on your head, you did not walk directly to the side wall (really the roof) of a house with a half-round top.

If I had expected a welcoming committee, I would have been disappointed. The Quonset, home to the enlisted weathermen, was empty. After I got to know the habits of the weathermen who lived in that unpretentious dwelling, I would know better than to be surprised by their absence — they were all either on duty in the weather station or they were hanging out there. The weather station was the only building on the airfield with air conditioning, and even if they were not scheduled for an observing or forecasting shift, the weathermen could usually be found there — just keeping cool.

But they might also be at the NCO Club. At the club, they'd be drinking beer, the only place it was allowed under the strictures of the Saudi Arabian culture, or they might be sipping "baksheesh cokes," an imitation Coca Cola made from a syrupy extract. These resembled the true cola in color only. Sweet, and with no fizz at all, their price was perhaps their only attraction. Baksheesh was the Arabic word that we used for "free."

Although we had flown for nearly half the night, and it was still nighttime, keyed up by the strange surroundings, I was too alert to immediately hit the sack. In my imagination, this was the part of the world that was the setting for Scheherazade's stories — of the *Thousand and One Nights* — of Sinbad the sailor, of Aladdin and his magic lamp. Simply going to bed would not be the proper introduction to these storied sands.

With such thoughts, I stepped out of the Quonset under an umbrella of stars. There was not a cloud in the sky, a condition I would find almost routine during the coming tour. With that nighttime relative humidity down to about 15% (it would be about 5% and even less in the middle of the hot afternoons), the stars seemed to leap out at you. Having spent all my life in America's Mid-west or East, where the higher humidity masked their brilliance, I had never seen stars so bright. Shuffling over small sand dunes, I made my way toward a little cluster of lights that would be coming from the 24-hour-operating weather station.

Next to the weather station stood the spartanly furnished civilian air lines terminal. Like the Quonset huts, it was made of corrugated steel, but bigger. Looking into the terminal as I trudged through the soft sand, I saw a snack bar with red plastic-and-chrome chair and table sets such as would be seen in cheap diners back home. I would have been surprised to have had someone tell me that I would be frequenting that establishment — it was not a particularly attractive place — but, as I would learn in the coming months, at Dhahran airfield, that was the *only* place to buy so much as a sandwich. During that tour, I would spend some leisure moments there, trying, with no prospect at all of success, to recapture a moment from the past through the mere act of ordering a hamburger and a coke. You could tell that the art of making hamburgers was still in the formative stage at Dhahran. The hamburger would be almost as thin as a piece of corrugated box paper (and, if I had ever tasted cardboard, I think it would have been difficult to sense the difference), the coke, like those at the NCO club, would be served in a paper container and would be sweet and flat, made of the same syrupy extract. Too sweet, it was usually not possible to down the entire make-believe coke. Nevertheless, during that tour most of us would return again and again for another dose, always hoping that the next one might be the real thing.

The weather station was made of stone, an attractive piece of workmanship. I stepped onto the stoop, knocked the sand off my shoes and opened the door to hear the old familiar sound of a tapping radio key —a sound from the distant past, a sound I had not heard since my days as a weather editor at Westover. A man in civilian clothes, khaki-colored shirt and shorts, had been tapping out the station's last weather observation, sending to whoever wanted to receive it at some far off place. Seated next to him was another civilian, similarly dressed, who, with headphones on, listened for incoming weather messages which he typed out on a manual typewriter.

A captain was the forecaster on duty. In a sort of self-conscious way, telegraphing the fact that no one could make reporting for duty at Dhahran for a 12-month tour a pleasurable event, the captain made a strong effort to welcome me. "You'll like forecasting here," he said. "There are no thunderstorms."

Now that indeed would be good news. I remembered well the flight in the T-6 from Selfridge to Chanute, the circumnavigation of cumulonimbus cells, and the many forecasts I had made where C-45s or C-47s were on their way from Selfridge to Bolling Air Force Base at Washington, DC and how it was a problem, often, to find a day where those relatively low-flying craft could top out over the cumuliform clouds that built in the orographic lifting over the Alleghenies. What would be particularly nice about not having thunderstorms was that *there'd be no need to predict them* and to issue the wind warning when they were predicted so that parked aircraft could be tied down. I was skeptical, but did not want to spoil the warmth of the welcome by indicating doubt. He could have said, "There'll be no snow," which I would have accepted more quickly. Or, "There is no fog." Because of the dry air that, too, would have been easy to believe.

I wanted to believe him, but I had been struggling with forecasting decisions for two and a half years, which some would say was not very long, but it certainly was long enough for any forecaster to develop a habitual attitude of healthy skepticism. Continuing with his obvious attempt to put me at ease in regard to facing the prospect of 12 months at Dhahran, the captain said he himself only had a few months left to sweat out before rotating back to the states, implying that he was proof one could survive the tour.

The captain's remarks to me were refreshingly different from what most newcomers to weather stations heard. At most stations, in utter seriousness, the veteran forecasters would tell the new forecasters that they had just reported for duty at a place *that was the most difficult area for forecasting in the entire world.* I learned to take those remarks in stride. It was true that terrain differences gave each station its own unique weather pattern and that the effects of terrain on the migratory weather systems had to be analyzed and mastered; but unless you were the character in Al Capp's Li'l Abner comic strip who went around everywhere with a dark cloud over his head, *every* station you reported to couldn't be the most difficult in the world. I appreciated the captain's efforts to put me at ease with respect to this forecasting assignment, and I hoped his thunderstorm non-occurrence prediction would prove to be right — at least until after he had left Dhahran. It would save us both an embarrassment. With that thought, I trudged back over the sands to my Quonset home.

Next morning, while it was still somewhat dark, I walked across more sand to the corrugated steel building that served as the shower room. There was no hot water system, but I was able to shave with warm water, nonetheless — the water pipes had been installed only slightly under the ground surface, within the layer that would be heated by the sun, and they still retained some of the heat that had been acquired from the previous day's residence in the hot sands. As a finishing touch, I patted on a dab of after-shave lotion.

With the sun now showing over the dunes, I stepped back onto the sand and saw there was not much more that could be seen by daylight than had been seen by starlight. Our little settlement was surrounded by a desert similar to the Mojave in southeastern California. In every direction the sand marched off to infinity. Here and there, isolated small yucca-like plants barely managed an existence. Not a tree was to be seen anywhere. Nor was there even a single blade of grass. Above, the bright night stars had been replaced by a silvery haze, the drifting telltale residue of a desert dust storm that had erupted days ago somewhere over that 300,000-square-mile expanse of sand. There was no wind at ground level but the storm's remains, high aloft, were being carried to us on the upper winds.

In the few minutes that I had been standing there, flies began gathering and soon transformed my person into what had to appear to others as an animated cloud, a black amorphous creature with indistinct and wildly waving appendages. I was the Al Capp character, but instead of a dark cloud hovering above, I was entirely within a black cloud. I was swatting at the flies, aiming at first with measured deliberation, then wildly waving as if they were the May flies at Selfridge. They looked exactly like the common stateside fly, but in behavior they were vastly different — showing not the slightest fear of humans, they completely ignored all attempts to wave them away. I don't know if this was true, but I was later told that their non-fear of humans arose because they were protected by His Majesty, King Ibn Saud, and no one in Arabia dared bother them (Ibn Saud, 1888-1953, the founder of Saudi Arabia). This could well have been one of those stories that was saved up at most stations and then trotted out one day to be used on the new guy. It doesn't sound true, but the reason for the protection, also told to me, was somewhat plausible, it was that the flies assisted in the pollination of the date palms that grew in isolated oases. Not being able to scare them off, I ran to the inside of the Quonset, outdistancing the black cloud by only a foot or two, and slammed the screen door tightly shut. There, I was told the Arabian flies particularly liked a certain then popular brand of shaving lotion, which I had used that morning, and for that reason no one there used it. For the rest of the tour, I did the same.

But now to the weather. During that previous night, old Thor must have had as difficult a time going to sleep as I had had. He must have been watching as I trudged over the sand to the weather station and heard the good captain speak of the happy absence of thunderstorms. Thor could have been expected to take that remark as a challenge. For it was on that very first afternoon of my long Saudi Arabian tour that Thor unleashed a parade of thunderbolts that rolled, and rolled and rolled across the Dhahran sky until there could have been no one at all, no one anywhere on the airfield, no matter how preoccupied, who would have failed to understand what it was. Even the troops who were off duty taking in a movie in our small rock-walled theater building came to eventually understand what it was. Though, at the first few rumblings, they had scrambled out of the building, quite sure we were being bombed by Joseph Stalin in retaliation for the US involvement in the Korean War. (We were only a little over 900 miles from the Soviet Union.)

When the storm hit, I was reconnoitering the airfield, trying to get the lay of the land and suddenly found myself groping through a dust and sand storm that was being fiercely driven by the down rush from the cumulonimbus. There was not much water vapor in the air layers closest to the ground, and the rain that fell into those layers cooled them, making them more dense. Now being heavier than its surroundings, the cooled air fell to earth in a free-fall frenzy, deflecting into furious walls of rushing dust and sand. With both hands I tried to shield my face from the stinging, sharp-edged bits of flying quartz.

When I saw him the next day, the captain, who had given me the happy news of a no-thunderstorm tour, said the storm was so unbelievable that for a few minutes he had been certain we were being attacked by German 88-millimeter guns, guns that he obviously had once seen too much of. He could have been forgiven for his statement that thunderstorms did not occur at Dhahran. As the months rolled by, I was to learn they indeed were rare.

We provided forecasts for the various airliners that landed at Dhahran — TWA, KLM, Air France, etc. Since we were "the only game in town" (or anywhere on that desert), the prohibition against giving weather forecasts to civilians was waived. On April 11, 1951, while forecasting on the day shift, I waited with some anticipation for a TWA captain who always gave me exceptional post-flight reports. Most pilot reports were verbal, the pilot's best recollections of the flight. But his were drawn on a map, and in great detail. By longitudinal zones, he recorded winds and drew sketches of the clouds he had encountered. To a forecaster at Dhahran, a pilot report was essential, and a pilot report from a pilot who took pains with his presentation, as this one did, was priceless. We labored under the handicap of an extreme paucity of data. In the direction from which the weather systems usually came, out over the vast Arabian desert, there were no weather reports for a thousand miles. And when radio reception was bad, even those beyond a thousand miles simply could not be received. I was slightly taken aback therefore, when, this time, instead of immediately laying out his weather folder for my inspection, the TWA captain burst into the weather briefing room with the question: "Did you hear the news?"

"Radio reception has been bad," I said, "And our newspapers and letters are always at least a week late. What's up?"

"President Truman just fired MacArthur! Can you believe it?"

My TWA friend heard this over his radio while in-flight from Europe in his Lockheed L-1049, the "Connie." Five-star General of the Army Douglas MacArthur was the military brain of World War II who captured the admiration of the American public by island-hopping our forces in the Pacific, avoiding the casualties of taking the islands one-by-one. And in Korea, he had employed a similarly successful tactic when he took an end run, landing the 1st Marine Division at Inchon, a considerable distance north of our Army that had been pushed back into what would have been a no-win defensive position at the Pusan perimeter. The tactic turned the war around. But MacArthur's independent style would bring about fears of an expansion of the war and ultimately caused the President to take his drastic action. MacArthur's dismissal *was* hard to believe and it reminded me that we were having a very safe tour while our comrades were engaged in desperate fighting in Korea. (Some time later I would learn that one of my classmates of the weather forecasting class and his men had been overrun by the Chinese communists who would enter the war on the side of the North Koreans.)

Most of our "news" came over short-wave from Radio Moscow, at night when radio reception was good. This was obvious propaganda information, not exactly unbiased. Broadcast by people with no accent at all, who sounded as if they could have just walked off the streets of Detroit, their often-repeated line was: "The heroic Chinese people's army had pushed back the imperialist American aggressors." They said that even when it was very obvious to us that the battles then being fought were being won by our forces.

Radio Moscow did serve a purpose, however, the broadcasters strained attempts to seem serious when providing ridiculous information made for some laughs during the long, dead night shifts.

Dhahran was only about four hours flying time from the Soviet Union and Stalin's intentions in regard to the US participation in the Korean War were unknown. On that bare desert landscape, it was clear that concealment in the event of an attack would be impossible. As preparation for the possibility of attack, out on the dunes we practiced defensive fires with a tripod-mounted 30-caliber machine gun and were ordered by the base's first sergeant to dig an aircraft-attack defense trench beside our Quonset hut home.

When you dig a trench that is to be used for protection against strafing aircraft, it is best to make it a zig-zag affair such that, even if the enemy pilot is lined up perfectly with the trench's longitudinal axis, there is always a protective wall of earth you can crouch behind. Planning the layout of such a trench was a snap for we weathermen. We laid it out to match the design of the zig-zag blue line that identified a ridge of high pressure on our weather maps. *Digging* the trench, however, was an entirely different matter. The soil at Dhahran, though seemingly only a soft sand on the surface, which was abundantly evidenced by the sand that seeped into your shoes every time you walked between the Quonset and the weather station or anywhere else on the airfield, became a hard, cemented, desert-soil caliche before the depth of one shovel blade was reached. We had picks as well as shovels, but these were of little use. Many a gallon of sweat had to be removed from our bodies before a noticeable dent could be made in the caliche. Though we worked hard and were determined to finish the job as soon as possible (for no one knew just when the Soviet fighters would come in on a strafing mission), at the rate of our progress we'd be rotated out of Dhahran, and our successors would be rotated out of Dhahran and their successors would be rotated out of Dhahran, before the job could be finished. We could see that project was going nowhere.

Then, an interesting coincidence occurred. On a day when the motor pool sergeant decided to go to the NCO club for a baksheesh coke, several of our weather troops happened also to be there. And although he had known that the weathermen had a reputation for friendliness, he was still surprised to see how many baksheesh cokes they brought to his table, and he was equally surprised at how interested they were in his motor pool operation. They listened in rapt attention as he described such things as the distributor points gap setting on a jeep or how often the trucks had to be serviced in that dusty climate. He never had a better audience. Then they invited him to come up to the PIBAL launching site for an opportunity to let go of a fully inflated balloon and watch it soar up into the desert sky. The motor pool sergeant wondered why he had never known balloon blowers to be so accommodating; and as a return on that favor, he offered to demonstrate how one of his trucks, the one equipped with air compressor and jack hammer, could be used to dig an aircraft-attack defense trench. With his truck, it would be a mere piece of cake and, incidentally, would confirm, again, the resourcefulness of the weathermen.

That truck was soon parked in the slight shade cast by Quonset hut 113A and when not on shift at the station, we were all soon initiated into the teeth-jarring rhythms of the jack hammer. Curious about the beehive of activity at 113A, other denizens of the Dhahran Quonset complex began strolling by to see it. They took due note of our five-foot-deep community-improvement project that had been laid out in that attractive zig-zag pattern, mimicking a ridge line on a weather map. At first, we thought our visitors could be looking for story material for a *Better Homes and Garden* article, but it soon became clear that they had a more pressing motive. Their homes-away-from-home had also been targeted by the first sergeant for a landscaping effort, and our project was the example to be followed.

Although we were only a few miles from the Persian Gulf, most of us, I believe, were surprised to learn that our weather regime was not of the maritime type. It was predominantly continental over much of the year, dominated by the winds that blew from the Rub al Khali. In the spring and summer, temperatures rose impressively. On June 6, my daughter Leta's birthday, the air temperature rose to 119 degrees Fahrenheit, which I was told was a record high. Of more importance was the persistence of the heat. For the rest of that month, there was only one day in which the maximum did not equal or exceed 100 degrees (on the 14th, the mercury stopped rising at a mere 99 degrees). Out on the apron, mechanics told us they fried eggs on the solar-heated aircraft wings. There were long periods during which absolutely no clouds could be seen, not even the high, thin cirrus clouds that, in most other places in the world, usually manage an appearance even when the middle and low clouds, the rain producers, cannot. During the month of May, there were 10 days with not a wisp of cloud; in June, there were 15. As you stepped outdoors on those perfectly clear days, the high albedo (reflectivity) of the light-colored sand bounced the sun back at you, making sun glasses a virtual necessity. And the heat radiating from the sand and the sides of the buildings made you think of the hot blast from a suddenly opened oven door. To this Mid-westerner, that was a new experience.

Speaking of ovens, the bakery at Dhahran was operated by Italian nationals. They had complete charge of the bakery and mess hall. For the most part, they were men who had been stranded after World War II in Mussolini's African-Italian colonies. They did a first-class job of operating the bakery and mess hall. Instead of having food items tossed at your tin tray while you moved along the slow mess line, the procedure in most mess halls, you walked into a dining room, sat down at a white-table-clothed table, and munched on warm, fresh-baked Italian bread while a waiter brought your order. The meal usually began with minestrone soup and ended with dessert that the Italian waiters called "sweets." Now that was a mess hall! That kind of treatment made you forget, for the moment, the drab landscape and sheer boredom that existed everywhere on that airfield except, perhaps, at the weather station, where the forecasting challenge always seemed to be more than a match for the pervading ennui.

Italians were also employed in receiving and sending our meteorological messages. Wearing earphones, they sat before the radio consoles, listened for the radio signals and simultaneously pounded them out as meteorological messages on their manual typewriters. They hadn't the foggiest idea of the meaning of what they were typing, they knew little English and the content of the strings of numbers in the meteorological codes would have been a mystery to them, but one could be assured that whatever came out as an intelligent signal would be faithfully typed for us to use. The only messages they would fail to provide us were the ones that were missed because of radio blackouts, probably caused by sandstorms over the desert.

The Italians had their own club. Beside it, a bocci ball court had been laid out on the sand. It was a rare day that, in passing by it, one wouldn't see a half dozen or so Italians playing bocci ball. They seemed happy and it seemed to me the war had not ended too badly for those Italians who had survived to secure employment at Dhahran. They also seemed better adapted to the climate than we were.

But the champions in acclimatizing to the hot and dusty weather were of course the natives. They knew how to cope with the heat. Even if it was only no more than a foot wide, our Arabian friends always walked in the shadow cast by a building where, incidentally, they would also be somewhat sheltered from the blowing sand. A number of them wore the old, US Army olive-drab overcoat that I had always hated to wear in my early Army days. These probably were gifts from GIs who had no doubt discarded them with pleasure when the lighter-weight blue coats became a part of the official Air Force uniform. Those heavy old GI overcoats were not only effective barriers to the airborne sand, but also induced perspiration which promoted evaporational cooling. This personal air conditioning trick probably worked well for I would see the Arabs wearing those long, woolen coats even during the hottest part of the afternoons. Just thinking about carrying that horse blanket for any length of time, not to mention wearing it in the searing Arabian sun, was enough to make anyone sweat.

Worse, perhaps, than the hot afternoons were the nighttimes when temperatures were too high to accommodate restful sleep in our non-air conditioned Quonset. As an example, in the early morning of May 12, the minimum temperature was 73. In the succeeding days, the minimums rose and sleep came only in short, fitful periods. We waited for months, it seemed, for the minimum to drop below 75 again. On the last day of June, a day when a dust storm reduced afternoon visibility to one-half mile, the lowest nighttime temperature was 90.

In that year of 1951, Dhahran was quite a different place, and a much safer place, than the way it was when our Gulf War airmen and soldiers arrived there. At the Dhahran airfield, the half dozen or so Quonset huts, a few small stone buildings (the most prominent being the weather station, small theater and smaller library), a corrugated steel terminal building and hangars made up the enclave.

Except for the American oil workers' ranch-style homes at the nearby ARAMCO (Arabian American Oil Company) facility, there was nothing outside the base other than some mud-walled, adobe huts that probably were on the sands even before Ibn Saud took control of the area.

One day, in an attempt to alleviate the monotony, we drove a jeep out beyond the airfield. We were "going to town," to the mud-walled village of El Khobar, to observe the local commerce. (Since the name is the same, I'm assuming this miniscule protrusion on the desert sands later became the location of the Khobar Towers that was bombed with the loss of 19 American servicemen in June of 1996.) At an open-air (as they all were) stall, a Bedouin shopper was haggling over the price of a piece of goat meat that hung on a hook attached to a log supporting the makeshift roof. When the number of rials had been agreed to, the seller brushed off the flies that covered the carcass, took out a large knife, and hacked off a chunk which was handed to the buyer. This was merchandising in an uncomplicated way. There were no scales and no fancy wrappings to deny the (protected?) flies a taste of the meat. Having observed all there was to see of that "shopping center" in about 15 minutes, we drove the jeep as far out into the desert as we dared, thinking it would be fun to be able to say we had gone where no American had gone before. When we reached that imagined spot, we stopped, walked up and down the dunes and came upon a sheet of paper just barely sticking out of the sand. I bent over to look at it. The paper was a Sears and Roebuck order blank.

On another trip outside the airfield in yet another attempt to alleviate boredom, we flew in a C-47 to Bahrain Island which at that time contained nothing much more than some tented street vendor stalls where Arabs hawked rugs, jewelry, articles made of brass and other items. Here I took a crash course in bargaining but without much success. Before I learned how to terminate an unwanted bargaining discussion I was harangued into purchasing several trinkets that I did not really want. There are now, I am told, high-rise buildings on Bahrain. That would have been impossible to even imagine in 1951.

One of our missions at Dhahran was that of providing our Saudi hosts with training in airfield operations. The Saudis had been introduced to aircraft operations much earlier, we apparently were to bring them up to speed as they sought to create an air force. To accomplish that end most effectively, individual Saudis were evidently assigned to become a shadow of that American whose skill it was his duty to acquire. Two Saudis who had received their college educations at Cairo University were assigned to me. I thought that one would have been enough, but assumed that two were assigned because it may have been concluded that I was more elusive than the other forecasters — and would require more watching than one person could effectively accomplish. Like with all the uneducated Saudis we met, we would soon develop friendships. While I was on shift, these two stayed within two or three feet of me, scrutinizing my every move. If I picked up my trusty pink pearl eraser to erase part of a line I had drawn on a weather chart, they would ask why I had concluded the first attempt was in error. I would respond by saying that a well-worn eraser was the sign of a good forecaster, but any attempts at telling jokes would escape them — they were absolutely absorbed in the seriousness of their assignment.

One day I decided that the training was too much of a one way street, and I struck a bargain. I would go beyond the mere surface transmission of meteorological knowledge and teach them everything I knew about weather forecasting if, in return, they would teach me the Arabian language. It was agreed, and after a few weeks the weather troops would come to me when they sought light entertainment, saying: "Say something in Arabic."

Though it is nearly impossible for an American to reproduce the throaty sounds of the Arab language, some of the basic sounds of their words seemed to me to be particularly suited to convey the meaning intended. In these examples, my "best shots" at what I knew of the phonetics and meanings, you may see what I mean.

The Arabic word (in phonetics)	The translation (my interpretation)
flooss	*money*
wahjid flooss	*lots of money*
kelossed	*all worn out*
sidiki	*friend*
shoof	*to look*

165

It soon became apparent that my accomplices in learning had acquired a lot more weather forecasting procedures than I had learned about their language. I think they may have gone around me and spied on some of the other forecasters while I was off duty.

If that low-latitude sun and the dust and sand storms were a bit much to bear, the desert night with its stars made brilliant by the dry atmosphere was an inspiring sight. So when I learned that a shipment of binoculars had come in to our little Quonset hut PX, I hurried over and bought one. When not working the swing shifts, a number of us would see what we could see through the 12X50 binoculars. We sat in front of our Quonset home in deck chairs someone had commandeered from the NCO club, listened to Patti Page and Connie Francis on 45 rpm records, marveled at those stars and, in that day before Neil Armstrong, guessed at the makeup of the moon.

Some of my friends said I planned to rent the binoculars to the troops who wanted to use them to find their rotation dates (their dates of return stateside). Master Sergeant Robert Lafferty, one of the forecasters, went so far as to write the following poem that was published in the *Dhahran Duster,* our mimeographed airfield newsletter (at Dhahran, my nickname was "X"):

X did buy a spyglass
To see for miles around,
He wanted to see a blade of grass
Agrowing on the ground.

He wanted to see a cloud on high,
And hoped to see some rain.
He spied and looked and shoofed the sky,
But he did look in vain.

He focused the glass and wiped the lens,
And tried to find a tree.

He ignored the laughs and knowing grins,
But ne'er a tree did see.

No tree, no grass on the ground to spy,
Nor any cloud above it,
What he saw with naked eye,
He only saw more of it.

Top: The C-74 had a large cargo space, the largest cargo transport plane of its time, and had a range of 7,000 miles. *Photographed at Wheelus Field, Tripoli, Libya, 27 October 1951.*
Bottom: The Lockheed Constellation, the most beautiful airliner ever built. This TWA version of the plane is on display at the Pima Air and Space Museum in Tucson, Arizona.

Top: Quonset hut 113A was home to Dhahran's weather troops. Not air-conditioned, they were usually there only when it was necessary to sleep.
Bottom: At Dhahran's Weather Detachment 29-2, forecasters Captain (later Major) S.P. Eckrem (left) and the author discuss the latest surface (ground level) weather map. The heavy line barely visible on the six-hour-old map hanging above is the ITCZ, the intertropical convergence zone, along which turbulent weather including thunderstorms would be expected to occur. *Official US Air Force photo, 5 September 1951.*

Chapter 17

Single-station Analysis and Other Trials

Dhahran was a place that reminded you why, in the Weather Forecaster Course, they taught a class called "Single-station Analysis." Actually, while we were in that class, we thought the effort expended in teaching it was quite a waste of time. After all, where could anyone be sent where data were so scarce that those procedures had to used? To concoct forecasts for a station where you only had the weather observations you produced yourself, to be isolated from weather data taken at surrounding stations, might have been the case during World War II, but the war was then more than two years past and, in the post-war, weather data seemed abundantly available. But it was not so at Dhahran.

As you scanned the weather maps being plotted by the weather observers at Dhahran, as noted in the previous chapter you would see no data at all for a thousand miles in the direction from which the weather usually came, from across the vast Rub al Khali. To today's stateside forecaster, that is a situation that can scarcely be imagined. It would be the equivalent of a forecaster at Washington, DC attempting to formulate a forecast for that city when the nearest upstream weather information available to him was a weather observation from Omaha, Nebraska! As a consequence, we were forced to place unusual dependence on the weather observations we ourselves produced, on single station analysis, and always hoped that the next pilot who landed had been conscientious, had paid close attention to the cloud systems and weather he flew over, under, or in, and that he or his navigator would take the time to thoroughly brief us in regard to it.

Local hourly weather observations, supplemented by special observations when the weather changed significantly, were taken by our observers just as they were everywhere else at Air Force bases. This was indispensable, giving the forecaster his "starting point." But the most important local weather information came from the PIBAL, pilot balloon runs, also taken by the weather observer.

A PIBAL run was taken every six hours, timed to coincide with the noon and midnight, Greenwich time,[1] radiosonde schedule and supplemented

---

[1] Greenwich, a borough of London, England, lies at 0 degrees longitude, the prime meridian. In the weather services, Greenwich time is used almost exclusively and the letter "Z" is its indicator. Thus, 00Z and 12Z on a weather map is midnight and noon, respectively, Greenwich or London time.

with an additional run taken midway between those times. It was a team effort requiring the cooperation of two to three weather observers. Preparations began with the observer who made the hydrogen in a cylinder about six feet long to which chemicals had been added. Once emptied of the gas, the solid residuals, clinkers, that formed in the cylinder had to be pounded out with a steel spud. In the next step, the observer inflated a balloon with just enough hydrogen to balance a 30- or 100-gram weight. Thus the balloons used for the PIBAL were of two sizes and were called either 30 gram or 100 gram. (A 10-gram balloon was also in the weather station's inventory and was used by the weather observer to measure ceiling heights.) As soon as the balloon was released, an observer tracked it visually with an optical instrument called a "theodolite" that was similar to a surveyor's transit.

For night runs, the amount of hydrogen injected into the balloon was increased just enough to also suspend, in a balanced state, a paper Japanese lantern that held a candle. Immediately prior to launching, the candle was lit and the observer then tracked the candlelight as the balloon ascended. Hydrogen being readily ignited, the candlelight launch was handled with extreme care. (The flammability problem was later solved to an extent by using a flashlight bulb energized with a water-activated battery. And at most stateside stations, helium, an inert gas, later replaced hydrogen.)

As one observer tracked the balloon or the candlelight, at one-minute intervals another observer recorded its elevation and azimuth angles. Then, inside the station, from a table based on balloon ascension rates, elevation angles were converted to horizontal-distant-out and were plotted against the azimuth angles to ultimately produce wind speeds and directions at thousand-foot height intervals.

We forecasters studied those PIBAL winds intensely, milking all the information we could out of them with the hope of finding new techniques that might help us formulate the forecasts. And, of course, we employed those single-station analysis techniques that had been taught in the Weather Forecaster Course, the techniques that some of us then thought were a waste of time.

To the pilots as well as the forecasters, the feared weather event at Dhahran was that fierce dust and sand storm, the *shamaal*. The worst of these erupted as the tail ends of strong cold fronts swept through the desert. During the most severe shamaals, it was not unusual to have pilots report that visibility was restricted up to heights of about 10,000 feet. On such days, the suspended dust would change the sun into a white disc, almost perfectly like the full moon, and shadows all but disappeared with daylight dimmed similar to that when the sun is in total eclipse.

On June 7, the day after our record high temperature of 119 degrees, we experienced a particularly fierce shamaal. At the peak of its fury, strong, persistent north and northwest winds were averaging 37 miles per hour, and blowing dust reduced visibility to three-fourths of a mile. If you were bold enough to walk outdoors, and few were, you had your hands in front of your face to shield it from the sharp-edged sand fragments that would make your skin feel as if a million red-hot needles were being flung at it.

During that month of June, blowing sand or blowing dust, or a simple case of suspended dust left behind by a previous storm, reduced visibility on 15 days. The shamaal that began on June 7 reduced visibility for seven consecutive days. On the 8th, visibility was down to one-fourth mile. No pilot would plan a flight into a field with those conditions. In addition to the hazard of trying to land (or take off) with virtually no visibility, the wear on engine parts would be significant and the sand storms adversely affected radio transmissions.

Analysis of our PIBAL data showed us that the probability of blowing dust or sand was enhanced when the winds increased in speed upward through about 6,000 to 8,000 feet. Statistical data have a place in day-to-day weather forecasting, but without a tie-in to the current meteorological dynamics, without corroborating evidence, without any reliable indication of the character of the pressure systems that were approaching from the desert interior, that aptly named empty quarter, it was sheer agony to forecast restricted field conditions due to blowing dust or sand and, also, in the alternative, to forecast the absence of it. We placed heavy emphasis, therefore, on corroborating reports of dust storms that pilots might have spotted in their flights over the desert.

The commercial airlines — Air France, Royal Dutch (KLM), TWA, Middle East and others — constituted the bulk of our traffic and provided us with most of our post-flight weather reports. The airliners were usually going to or were from Karachi, Bombay, Cairo, Athens. They were primarily the sleek Lockheed Constellation, the L-1049. With its streamlined fuselage and triple tail fins, the Connie remains the most beautiful airliner ever built. Powered by four Wright engines, it could cruise at nearly 300 miles per hour with a range of over 4,000 miles.

From his flight scheduled one day a week, we'd also get weather reports from the ARAMCO (oil company) pilot who flew a Navion along the oil pipeline (called "The Tapline") that stretched out about 1,000 miles over the northwest desert. He was looking for oil stains on the sand which indicated that bolts had been removed from the pipe connections or holes had been shot into the pipe. Admirably suited for this job, the Navion was a single engine plane with a cruising speed of about 150 miles per hour, slow enough to inspect the pipeline for leaks. Designated by the Army as L-17, it was used by the Army in Korea as a FAC (Forward Air Control) and in other assignments. A variant had seen service in World War II, and it would see service in Vietnam.

We also received valuable weather reports from our Air Force's B-29 weather recon plane and from other Air Force flights on their way to or from Asmara, Eritrea; Beirut, Lebanon; Cairo, Egypt; Teheran, Iran; Tripoli, Libya. The last mentioned was the most frequent and the most eagerly awaited, as it was the 29th Weather Squadron's C-54 that brought our mail every Tuesday. We called it the "T Flight," for Tripoli and Tuesday. (As an indication of how small our world really is, in 1998, on a golf course in Arizona, I would meet retired Senior Master Sergeant Carl Erwin, originally from Albuquerque, who had been the flight engineer on that plane and, in yet another coincidence, he had also been stationed at Selfridge.)

There were no jets at Dhahran, which made it sort of a step down from my days of forecasting at Selfridge. But this was also an advantage because the piston-engine planes, which we still called "conventional," were flying at those lower altitudes where most of the weather was and therefore could provide us with valuable insight as to what weather lay out there beyond our view.

One of the main flying routes out of Dhahran was the one to Bombay, India, about 1600 miles to the southeast. In June, with the onset of the summer monsoon, the forecast of landing weather at Bombay became quite difficult. Thunderstorms that were brought to the sub-continent from the Indian Ocean on the winds of the summer monsoon persisted, it seemed, for weeks, lowering ceilings and visibilities to near zero. Finding alternate landing fields with dependable weather was a tough job, reminiscent of the struggles to find suitable alternates on the European continent during the Berlin Airlift.

As a break from both the Dhahran weather and the utter monotony of the airfield, five-day rest and relaxation (R and R) tours could be taken to places that were garden spots by comparison. These R and R sites were Beirut, then a peaceful city and about 1,000 miles to the northwest, and Asmara, about 900 miles to the southwest. I had been thinking it would be nice to go to Asmara, the capital city of the then UN trust territory of Eritrea in northeast Africa. Asmara was at an altitude of 7,300 feet. If all relevant meteorological elements were equal, on an afternoon when the sky was clear and solar heating at a maximum, the dry adiabatic lapse rate would prevail and would restrict the temperature at Asmara relative to Dhahran, lowering

Asmara's temperature by 39 degrees.[2] Thus on the day that we had our maximum of 119 degrees, the equivalent Asmara temperature was 80.

When my time for R and R arrived, I turned it down. By that time, I had been on the Arabian sands for about half a year, and I rationalized that the gods who were in charge of such things knew that, throughout that tour, I had been clever in giving the impression that I strove to achieve a 100 percent dedication to duty. Their opportunity to get back at me for those occasions when I would have failed in that pursuit, a failure such as that imagined interlude of luxurious relaxation in cool Asmara, could easily unfold as a little-noticed plane crash in the highlands of Eritrea. My friends, however, had a more simple answer: "Better keep an eye on X," they said, "he's been standing too long in the Arabian sun."

One day, an Air Force pilot who had just returned from Teheran in his C-47 told me that, on his flight up there from Dhahran, he had experienced easterly flight level winds at his altitude but had been given a forecast of westerly winds by one of our forecasters. Teheran was about 600 miles north-northeast. He had flown at about 10,000 feet, so I dug out the 700-millibar chart that would have been current when he left. When I showed him that analysis, which clearly indicated he should have had westerly winds, as he had been given, he left and returned in a few minutes with his navigator who carefully explained his several wind calculations. They convinced me that our 700-millibar map analysis had been wrong.

The problem was that we had been assuming there was lower pressure over the Soviet Union whose border was some 150 miles north of Teheran. Since the Soviets did not provide us with their radiosonde reports and there were no other such reports on the Teheran run, we had a wide gap of data in the north and no way to know that, in reality, a flight at 10,000 feet toward Teheran was really a flight toward higher pressure, in other words, a flight into easterly winds. The weather systems during that period had become fairly stagnant with little day-to-day change in the upper winds. So when I drew the next 700-millibar chart, my analysis showed easterly winds.

The forecasters ranged in rank all the way from staff sergeant (my rank) to major. There were several captains, and the schedule happened to be such that I was being relieved at the end of shift by a captain who believed lower pressure over the Soviet Union looked more plausible and that a flight to Teheran at that time would be a flight into westerly winds. But the navigator's serious demeanor had convinced me the winds should be easterly.

---

[2] The dry adiabatic lapse rate, often referred to by meteorologists, is the rate of cooling that a parcel of non-saturated air will undergo as it is lifted — 5.4 degrees F per thousand feet of altitude.

I think the captain believed the navigator had miscalculated, which of course was possible. So, when he drew the next 700-millibar chart under those fairly stagnant conditions with the pressure systems changing but little, his analysis showed a low to the north giving westerly winds on the Teheran route.

The Arabian heat was probably getting to me for I continued in my bull-headed approach to the synoptic situation. On the next day, back I came with a high in the north and easterly winds on the Teheran route. The other forecasters were enjoying this diversion, fully aware that we were both flaunting an important weather station procedure — the obligation to follow continuity from one map to the next. Our battle of the highs and lows went on until the weather systems got moving again, which then provided an entirely different upper air configuration upon which we agreed. If that hadn't happened, the detachment commander probably would have confiscated our upper air analysis pencils. (Upper air charts are now drawn by computer and computer glitches, or faulty data that the computer would not recognize as faulty, now can result in charts that on occasion can also be erroneous in certain areas even though based on much more data than we had at Dhahran.)

It pays, sometimes, to be standing in the way when progress is being made. During a visit of the 29th Weather Squadron's inspection team, which was headquartered at Tripoli, one of our sergeant forecasters pointed out that no one from this isolated duty station had been promoted, a situation that he, and we, believed to be quite improper. There was some implication in the complaint that the troops at Tripoli *were* being promoted; and while duty at Tripoli was not exactly what most troops would volunteer for, when we thought about comparing it to our station, in our moments of slight exaggeration we saw Tripoli as a place not far removed from the French Riviera. In a few weeks, that sergeant was promoted and shortly afterward I came out on promotion orders. I was made a technical sergeant and that was a surprise, coming as it did several years before I thought it could happen. Several other enlisted men in the station also were promoted.

Our total complement at Dhahran airfield consisted of only about 100 American servicemen of all ranks and military occupations. The dull sameness of the days and the expectation of seeing only the same faces, day after day, seemed to wear away on all of us. Amid the boredom there grew a need to take advantage of every change in the day-to-day routine, no matter how slight, as grist for discussions aside from the usual weather talk. When one of our forecasters was dinged by our little library for having lost a book he had checked out, his name appeared on the mimeographed daily bulletin, and that event filled the discussion void for at least a week. The lost book, a novel by Philip Wylie, was titled: "The Disappearance."

One of the most popular changes in the daily routine was the arrival of the commercial airliners that parked just outside the terminal and weather station. All airline passengers in those days seemed to be more appropriately dressed than they are now. Probably the best dressed passengers were those who had boarded the Air France Connie in Paris. The women looked as if they had stepped out of some glossy Parisian fashion magazine. These passengers would step down from the aircraft onto portable stairs and walk to the terminal with the false prospect of being refreshed by one of those syrupy cokes. They would of necessity pass by the mini-reviewing stand, the terminal rail, where the weather troops who were off duty and the duty weather observer, whose job it was, of course, to be outside in any case because that's where the weather was, gathered as a sort of welcoming committee. Sometimes, the observer would be accompanied by the forecaster who, quite naturally, had a duty to respond to the observer's request that he confirm the weather readings. This was a perfectly proper welcoming committee that made comments in a manner that would accurately be said to be dignified. Just as if they were only comparing weather data, they would note, for example, that their observed statistic was once again proven — that the female passengers with open-toed shoes could not walk the 75 yards to the terminal without having to take off their shoes to pour out the sand.

Those brief glimpses of passengers unloading from and then boarding the commercial airliners were all there was to be seen of the female sex. There were no women on the airfield. Outside the airfield, if a camel caravan were passing by, it was possible to note the probable presence of a woman. These Arabian women would be dressed in opaque, black, ground-length, sack-like robes which covered their entire bodies, even their heads. Two eye slits were cut in the head-covering portion to enable them to see their world. None of them wore filmy veils or engaged in the sinuous dances that we had come to expect from the scenes in Hollywood movies.

One day, one of our observers, who was fond of photography, met one of the caravans and decided that was a picture-taking opportunity that should not be missed. It happened that he snapped a picture of one or more of the black-robed, enshrouded women. This was taken to be a grave offense, and the local chieftain had the observer escorted to the Dammam jail. There he languished, as I recall, for a week or two before our US military could secure his release. The women he photographed, it should be noted, were all perfectly dressed in their non-revealing Arabian garments.

The extreme challenge of preparing weather forecasts based to a significant extent on single-station analysis was probably a blessing in that it kept our minds riveted on the forecasting problem and prevented us from dwelling as much as we might have on the pervading boredom. Outside of our little group, however, boredom began to take its toll. A mechanic on a

C-47 climbed aboard his aircraft one day with the presumed purpose of checking out the running of the engines. While in the cockpit, he decided, apparently, that it would be a fine day to fly back home. Even if he knew how to fly, which seemed doubtful, home was about 8,000 miles away and the 47 had a range of about 1,200 miles. He had little opportunity to prove his flying skills for the aircraft was barely off the runway when it nose-dived into the ground. We were told our mechanic was killed in the crash.

The dull day-to-day sameness, the heat, the dust, the sand — all were taking their toll. A young airman from one of the administrative units was luckier than the mechanic. He attempted to hang himself in the corrugated steel shower building, but was discovered in time to save his life.

Isolation such as that at Dhahran would affect you in subtle ways. I recall that, after being on those sands for about a half year, I began to dread meeting my associates. As I started out from our Quonset home or from the weather station, I would look out and try to time my walk on the sand such that I'd meet no one I knew. I simply did not want to have to engage in a conversation with someone who had nothing whatsoever to say that was new. If such a conversation were to occur, I felt I could predict almost exactly what that other person might say. We'd been there long enough to have heard everything anyone had to say. There were no conversational surprises left.

One fine morning, from out of the settling dust that was trapped in the morning temperature inversion, an acquaintance from another Quonset, who was about to complete his 12-month tour, walked up to me and declared: "I can guess your serial number."

I, too, had at that time been on that desert for most of my tour, long enough to have reached the stage where answers become more abrupt and, when discussions are initiated, the diplomatic edge all but disappears from your side of the conversation. "No," I said. "You cannot."

But of course there was some logic to his approach. The old Regular Army serial numbers such as I had were composed of a two-letter prefix followed by eight digits. When I enlisted, my serial number was RA16165522. The "RA" for Regular Army, was later changed to "AF" for Air Force but the numerical portion was kept as before. The first two numbers designated the part of the country from which you had enlisted. (Incidentally, I always thought the four pairs of numbers in my serial number, a fairly good poker hand, augured for a good future in the military.) However, it was beyond the realm of statistical expectation that my acquaintance or anyone else could make a successful outright guess of an eight-digit number. Of course, he didn't guess my serial number. But he was persistent. Undaunted, he set about interrupting others, telling them, also, that he could guess their serial numbers. He was able to guess the first two digits in a case or two because he knew what part of the country the man came from, but that was the extent of his success.

The serial-number guesser extended his operation to the weather station where he was seen hanging around at all times of the day and night. The weathermen on the different shifts began comparing notes and concluded that, in his search for the perfect serial-number guess, our acquaintance was forgetting an important daily requirement — sleep. Soon, his actions were being watched by nearly everyone in the weather detachment. In trying to keep track of how long it had been since he retired to his Quonset, they, too, began to lose sleep. Finally, on one of the midnight shifts, the observers found him sprawled out on a map table, in deep sleep. They calculated he had gone eight days without sleeping, and it took them the better part of the shift to wake him.

For those who had the stomach for it, one could leave the airfield and attend the public punishments meted out by the official executioner. One of our camera fans brought back photos of severed human hands and a severed human head that had been hung on a fence near the execution site. Hands were chopped off with a sword as punishment for stealing; heads for a more severe infraction. Upon viewing the photos, I reflected that even though we left valuables lying about in the Quonset unsecured and often with no one there, nothing had ever been stolen. And mentioning the executioner's name in the presence of our Arab friends would bring on an expression of terror that you could recognize even though it seemed that a strong effort was being made to suppress it. No one could argue that that kind of justice wasn't effective, but I would rather risk losing items to theft, if need be, than to know hands would be severed because of it. Knowing my feelings in that regard, I won't have to disclose how I feel about the practice of severing human heads.

While I was at Selfridge, I had been asked to write a column for the Selfridge newspaper which I called *Cloud Chasing with Weather*. It occurred to me that a similar column might help to elevate morale at Dhahran. By sheer coincidence, while I was mulling over that possibility, Captain Joe Hopkins, our detachment commander and a veteran of both the Marine Corps and the RCAF, made me an offer I couldn't refuse. With his captain's bars gleaming, he told me I was to write a column for the detachment that would be published in the airfield's newsletter, the *Dhahran Duster*. (If memory serves, Captain Hopkins was later transferred to Tornado Alley, a bomber base in Texas.) With the invaluable assistance of the detachment clerks, a particularly zany column began to appear in the *Duster*. Its title, taken from the dreaded sand storm, was *Shamaal Talk*. In it we told of the exploits of imaginary weather characters who were all possessed of superhuman abilities. As a takeoff of the radio soap opera *John's Other Love,* we wrote of "John's Other Forecast." the protagonist, of course, was completely unflappable (and unfathomable) under even the most adverse weather conditions. His stellar achievement was the rescue, through an astute handling of an immensely complicated meteorological situation, of an aviatrix, daringly (for that time) called "Sexy Suzy — the sky girl of the Middle east," an idea that was also a takeoff of a radio program of that era.

One afternoon in early summer, as I stood out on the sand watching the observers track a PIBAL, a handful of letters was given me. These had just arrived on the T-flight, the weekly C-54 mail run from Tripoli. All of the letters but one were from Marie. Continuing the tradition that began in my basic training days, even on the Saudi desert Marie and I corresponded daily, our letters making up a significant part of the T-flight's mail load. I smiled in amusement at the letter she had inserted into one of the envelopes. It was from my draft board. They were officially informing me that if I didn't report for induction into the military there would be serious consequences. I made a mental note to tell Marie in my next letter that, contrary to what they might think, being in the Air Weather Service was the same as being in the military. To my astonishment, the last letter of the handful was from Dad.

As I looked at the angular, European-style handwriting written in the blue ink of a fountain pen, I became vividly aware that this was the first letter my father had ever written to me. Fluent in five or six languages, English, which he would have learned as a 20-year-old in Henry Ford's night school for immigrant employees, was the last he had mastered. I opened the envelope slowly, not wanting to know what I had suspected his letter contained.

There were missing letters in some of the words. The missing letters were the letters that were silent when the words were spoken. I never thought much about how he pictured the English language. But now I recalled that he always pronounced the "w" in "sword" and in "answer." The letter was almost entirely a hum-drum account of what was happening. Among other things he said that he had baby-sat my three girls while Marie and my sister Delores went shopping. This greatly pleased me for I wanted my daughters to know something of their grandfather, a modest man that I had always looked on as a truly heroic figure. In the last line of the letter, almost casually he said he was entering the hospital for a pain that just wouldn't go away.

At about the same time that I received that letter, my mother-in-law passed away after an operation from which she did not heal. Prior to her illness and even up to the time that I left for Saudi Arabia, she was one of those vibrant persons whom nobody would believe could ever be overcome by a health problem. For weeks, while I had trod the shifting sands, Marie had been at the bedside of her mother while also contending with the childhood diseases which all three of the girls then contracted.

About a month later, with only three months left on my tour, I was told to report to the major, our new detachment commander. I was going home. Dad was seriously ill. I thought of how, when we worked together at the war plant, I practically had to run to keep up with him as he hurried to enter the plant gate. It was as if he thought the Great Depression was at his heels, and he had to get to his diamond-bore machine before it caught him again. Now, if I were to see him before the unthinkable would happen, I'd have to respond to the Red Cross's summons to leave Dhahran. It was a strange feeling. To go before my 12 months were up was tantamount to leaving a job undone. Sensing my thoughts, the major said that, of course, I'd have to go. And knowing that I'd appreciate a memento of my service there, he presented me a Maria Theresa silver dollar, that beautifully crafted coin that was, in addition to rials, used extensively in Saudi land.

I took a C-54 to Tripoli, then a C-74 to Lajes and the states. Back at my old stomping grounds of Westover I visited old friends at the weather detachment. Staff Sergeant Henry Griffith was on duty. He had been a great help to me when I broke into forecasting and was the only graduate I knew of the AAF's weather school at Grand Rapids, Michigan, the weather school that had a short life in 1943. Sergeant Griffith, who would retire as a chief warrant officer, told me they were still using the basic vertical cross section strip charts upon which I had drawn terrain features in '49 and had corrected the error in longitude on the rhumb-line Atlantic route. A few pleasantries. Then onto a C-47 going to Wright-Patterson AFB and a bus to Michigan.

As I opened the door of the house we rebuilt, the one that was so convenient to Selfridge Air Force Base, only a mile from the base's back gate, I observed that the window I had cut in the front door before leaving was still secure and noted that I had not gotten around to installing the picture window in the living room's bay window enclosure. The window would be one of my first projects. An odd size, I'd have to make it myself. As I looked toward that bay window, Marie said: "Now we can take down the tree."

The beautiful fir Christmas tree that I had placed in the bay window eleven months earlier still stood in its stand. Not at all its old self, the last of its needles must have dropped off at least six months earlier. Marie, an exceptionally tidy housekeeper except for this one lapse, had apparently given no thought to disturbing the tree. I gently removed that dead reminder of a happy time. We would not be placing another tree there during the coming Christmas season, for within a week we would move. I learned I was going back to the fighters, but it would not be to Selfridge.

When I had filled out my reassignment wish-list at Dhahran, the document you were handed some six months before your tour was up, I had indicated Selfridge was my first choice. For second choice I asked for Niagara Falls Air Force Station. My third choice was Oscoda Air Force Base, in the northern part of the lower Michigan peninsular near the shore of Lake Huron. I got Oscoda.

Before leaving for that "up north" assignment, we visited Dad. Though appearing gaunt from the ravages of Hodgkin's disease which he had been battling, I then learned, for the past three years, Dad greeted us with great spirit. He beamed as he tried on my new blue uniform jacket upon which were sewn the even newer five technical sergeant's stripes, and he asked if he'd be seeing us at Christmas time. I promised that we would, but it was not to be. The doctors who had advised that the Red Cross should effect my return to stateside in order that I would have a chance to visit him for the last time, had diagnosed the seriousness of the illness quite correctly. We were to visit him three or four times on trips back from Oscoda. Then, one day, 1st Lt. Clayton Wright, a particularly sensitive weather officer, walked up and stood by while I briefed a pilot. After the pilot left and I had turned back to the map I had been analyzing, he slowly approached, telegram in hand, and said, "I'm afraid it's bad news." We would not see Dad at Christmas time.

Four reasons why it was tough to leave on a 12-month isolated tour: Marie and (l to r) Pamela, Willa and Leta. Pamela became a Squadron Commander at Edwards Air Force Base; Willa became a Policy Writer for Michigan's Family Independence Agency; Leta was named Greenbelt, Maryland's Outstanding Citizen of the Year 2000; Marie made the girls' dresses, the suit she wears, and became an expert golfer.

On his days off, as in this case, the author liked to practice the old skills he had learned as a weather observer. Here he follows a PIBAL with the theodolite. Through the head phone, at one minute intervals the duty observer in the adjacent weather station alerts him to read the elevation and azimuth angles to the balloon. As he reads them, another observer, just out of the picture, stands beside him with clip board and records the angles.

Dhahran, Saudi Arabia in 1951. Street signs are in three languages: Arabic, English and Italian. Lower photo: at this place with no trees and no grass, a brave little shrub tries to gain a foothold below a taxi sign. With a round roof like a Quonset hut but somewhat larger, the building to the left is the civilian airlines terminal. The fenders of the facing auto, a 1947 or '48 Ford, have been extended to fit the wider wheels needed to keep from sinking in sand.

Top: Gag ID cards were presented to Dhahran newcomers. This ragged version has obviously suffered from its exposure to the Arabian climate. A picture of a camel is on the reverse as well as a box for the "Chaplain's Punch."
Bottom: Dhahran's principal street intersection in 1951.

Chapter 18

Fighter Pilots, Flying Saucers and One Lucky Forecast

"My plane's shot up!" This outburst, from a lieutenant I knew only as the recipient of a few of my weather briefings, came as quite a surprise. On this day, he had been the pilot of a North American F-51 Mustang, towing targets for the jets to shoot at. He recounted, with a little excitement rising in his voice, that a jet fighter had decorated his F-51 with bullets as he towed a target over Lake Huron. To talk to a weather forecaster at a time like that made good sense. The alternative could have been a visit to the chaplain. But then, of course, the place that he parked his plane was adjacent to the weather forecaster's shop and blocks away from the base chapel. Also, he must have sensed, as did many others, that the very nature of the job of weather forecasting, that occupational specialty often referred to as "an inexact science," caused the forecaster to be acutely aware of the limits that lie in the paths of those who seek absolute perfection. The pilot, therefore, who sought a sympathetic ear might well find it in the weather briefing room.

I didn't know how to respond, and the F-51 pilot's intense gaze telegraphed that he thought he was due some response.

Finally, I fell back on the well-used and usually effective stall: "Say again?"

Actually, as he more completely described the incident, his plane was not badly shot up. As I recall, there were only two holes in it. But if you were the F-51 pilot, of course, you'd be thinking that that was two too many. It began to dawn on me that I had a duty to cheer up this pilot.

Reaching back for one of the maxims of my earliest days in the military, the advice the wise-acres, the "guardhouse lawyers," gave to each recruit: "Next time, when they ask for a target-tow pilot, don't volunteer." But that was not the best response I might have made.

"That's not how it's done," he said, his face contorted in the pain of the required explanation. "They don't ask for volunteers. They give the job to the pilot with the worst gunnery score!"

I tried to think of some nonsensical remark, even if only about the weather, that might calm him, but nothing at all came to mind. He left the station, looking back at me in a way that made clear he regretted having brought the subject up. This was the only incident of its kind that I ever heard of. After he left, I thought I might have turned the discussion toward the humorous. Remembering the look of pure joy on the face of Captain Don Wolfe, our detachment commander, just after he had the opportunity to take the F-51 up on an around-the-field hop, I thought I could have said: "Look at the bright side, there are a lot of pilots who would jump at the chance to fly what many have called the premier fighter of World War II."[1] But it probably was better left unsaid.

When I reported in to that weather station, Oscoda's Detachment 12-8 of the 12th Weather Squadron with headquarters at Stewart AFB, New York, I was not arriving at an unknown base. Located 125 miles north of Selfridge, for a while Oscoda was a sort of satellite of Selfridge. In 1949 and 1950, while I was at Selfridge, an interesting feature of the forecasting shift had been the requirement to provide weather clearances over a pilot-to-forecaster radio for the Selfridge pilots who were undergoing gunnery training at Oscoda.

Nestled in the jack pine woods where Marie's Swedish grandfather, Alexander Nordstrom, had worked as a lumberjack near the turn of the 20th century when the forests there were composed of the prized five-needled white pine, most of the time the base seemed to be enjoying a state of repose on its carpet of evergreen needles. Oscoda had no forecasters in 1949 and 1950, and no weather station. To provide on-site weather observations, an observer from Selfridge would be sent to Oscoda for a week or so of temporary duty where he worked out of a small Base Ops facility, taking weather observations. The pilot who was in his plane on the ground at Oscoda, ready for takeoff on another gunnery mission or for his flight back home to Selfridge, would call me or the other forecasters at Selfridge for his weather clearance over the pilot-to-forecaster radio. This was a unique weather briefing. Instead of being conducted man-to-man at the forecaster's counter, the briefing was conducted with the principals 125 miles apart.

While I had been in Saudi Arabia, Oscoda had come of age. It had been organized as Detachment 1-18L of the 1st Weather Squadron with

---

[1] Captain (later Colonel) Donald J. Wolfe was one of three forecasters at Oscoda who were also pilots. The others were Captain (later Lt. Colonel) L. Dayton Blanchard and Captain (later Lt. Colonel) James Burgess. This was, I believe, Don Wolfe's first flight in an F-51. In a recent communication with Dayton Blanchard, I learned that, while at Oscoda, Jim Burgess checked out in the F-86 and after leaving the service joined Boeing where he was an instructor pilot and, later, a check pilot. In a letter he sent to me in 1966 while I was in the Artillery in Vietnam, Don Wolfe indicated he had flown over Vietnam in a B-52, the plane that we ground pounders looked on as one of our "big brothers."

headquarters at Wright-Patterson AFB, Ohio, and was later brought into 12th Weather Squadron. No longer a satellite of Selfridge, it was a principal Air Defense Command Base — home of the 63rd Fighter-Interceptor Squadron's F-86 Sabrejets. Those guardians of our skies were there to block any attempt of the Soviets to send bombers toward us by way of the over-the-top polar route. Constantly on alert, they had only minutes to be airborne when the alert klaxon sounded. When they scrambled, the base could be a beehive of activity, otherwise there was little air traffic. Pilots flying routine flights at night got their weather information from the flight service center at Wright-Patterson.

By sheer coincidence, on that first day that I walked into Oscoda's Base Weather, Captain Don Wolfe, who would take over as the detachment's new commander, walked in at precisely the same time and also for his first time there. For both of us, it was our first real acquaintance with the station. After the introductions to the observing and forecasting staffs, Captain Wolfe took me aside and asked. "Did you know they are operating this weather station in the daytime only?"

I had to admit that I did. Moreover, it was the very reason Oscoda had been one of the three bases on that wish list I had filled out at Dhahran. After about five years of working round-the-clock, I thought I deserved a bit of a rest from the graveyard shifts. So I asked around and discovered that Niagara Falls and Oscoda were daytime-only weather forecasting stations. These, then, became my second and third choices, respectively. Of course, I asked for Selfridge as my first choice. Selfridge was a fairly busy 24-hour-operating station, but that choice was made for economic reasons, we had a home only a mile from the base. My rational: If I can't get the base of my first choice, it would be nice to sweeten the transfer with duty at a more leisurely operating location.

"From now on," Captain Wolfe responded, "we're going to operate this weather station 24-hours a day." And so my dream of working like a normal human being, for a change, ended before it even got started. And perhaps with the thought of making sure I'd have no time to dwell on this rather drastic modification of my plans, he gave me the two additional jobs of NCOIC (non-commissioned officer in charge) of the Observing Section and Station Chief. These duties were to be accomplished during the time when I was not pulling the forecasting shifts for which I was fully scheduled. In the weather stations I had previously been assigned to, the station chief was not required to routinely pull forecasting shifts, but I would not have had it that way. I enjoyed the forecasting shifts, and even with me on the forecasting schedule we had too few forecasters to provide round-the-clock forecasting. Don Wolfe's solution, which, he explained, would provide the forecasters with more usable time off, as indeed it did, was the 12-hour shift. One shift would begin at 0700 and the other at 1900. Not only had I not succeeded in avoiding night shifts, I had managed to put myself on a schedule with longer night shifts than any I had ever worked, and with it, had picked up significant additional duties. My best laid plans had certainly gone agley, and I would see that Don Wolfe was a man who knew how to get the most out of you and make you like it as he did it. I would come to remember him as one of my favorite detachment commanders

In our line of work, taking the teletype metaphor, we were always looking for the ungarbled word, but occasionally, as in the following, the garbled word also proved useful. About that time in the history of the Air Weather Service, we were told more precise wording should be used in the phrasing of the forecasts. This lost something in the several translations that it went through. I'm sure what we heard was not exactly what was meant, for by the time we received it, it was said forecasts should be worded like this: "A rain shower will begin at 1445 and end at 1500." To have a forecast verify in which the beginning and ending of each of the short solitary showers that fell from Oscoda's stratocumulus, cumulus and the impressive cumulonimbus were predicted to the minute, was a task beyond any forecaster's capability. With the much improved radars that would be available a few years hence, that kind of precision would not be nearly so impossible but would only be attempted when the forecast was for only an hour or two out. Even then, attenuation of the radar signal and changing atmospheric dynamics made the ending of the event more problematic than its beginning. Nevertheless, I'm sure that example of the garbled word had value, for it caused one to focus more intently on the forecast's timing, and it removed the word "probable" from the forecaster's vocabulary. If you were thinking of phrasing the timing of the weather event "on the head," you'd probably not miss it by as much as when you were not trying for such precision. If line-oriented systems were coming through, fronts, instability lines or troughs, we'd time precipitation's beginning with their projected arrivals. This we calculated by measuring rate of movement on our maps, and modified that speed by the accelerations, negative or positive, that we expected to occur. We fixed precipitation's ending with our estimate of line width. The coded forecasts of our base's weather, ceiling and visibility, that we appended to weather observations, entailed a similar approach to precision. When we had requisite moisture and expected a long period of instability, we'd predict "showers" or perhaps "scattered showers" or even "isolated showers" with that family of showers given a time of beginning and ending.

And now to the flying saucers. Knowing that I spent a large part of a lifetime observing the sky in many parts of the world, I am often asked if I ever saw flying saucers or unidentified flying objects (UFOs) as the Air Force came to call them. I suppose that, in what I'm about to relate, I'll become a prime candidate for abduction by little green men with antennae protruding from their foreheads, but I have never seen a flying saucer, not anywhere. On three occasions, however, I had such sightings reported to me. Two of these occurred at Oscoda. Sometimes I think Oscoda's physical setting — nestled under the lonesome jack pines — and the general quietude of the base had something to do with the Oscoda "UFOs." Otherwise, it seems logical that UFOs would have been reported at all the other bases at which I was stationed. In the middle of the night, the time that the Oscoda sightings were made, there was extremely little flying activity, a time when imaginations, for example, might be tempted to wander a bit.

Just as he would on the day shifts, the weather observer's routine on the night shifts was to go outdoors at least once every hour to note the visibility and the condition of the sky. When he first stepped outdoors from the station well-lighted with large fluorescent bulbs, the observer was unable to see. After the pupils of his eyes had sufficiently dilated to provide night vision, he would look for visibility checkpoints and for stars. If, in a particular sky sector, there weren't many stars or perhaps none at all, he would then try to satisfy himself as to the nature of the obscuring sky cover. At Oscoda, as the observer looked upward, he had to walk around a bit to take the pine branches, the aircraft on the flight line and the hangars out of his field of vision. Out there in the dark, separated from the forecaster, the only other life form with which he might be near enough to communicate with, the observer's imagination could easily shift into high gear.

There was good moon glow on the night the observer came running into the station shouting: "Come outside, quick! You need to see this!" It was about an hour before midnight. The observer caught me in the middle of drawing one of those smoothly curved isobars that I had begun to pride myself in accomplishing. I could see he was excited and knew, therefore, that it was urgent, but in order to avoid putting an inartistic wiggle in that line, I finished it before laying down the pencil. This only took, at most, two or three minutes of time. I then followed the observer to his observation point out under the pines.

"Up there, look up there!" He was pointing as he yelled. "It's just below the altocumulus! Can't you see it?"

Altocumulus is a cloud found in the middle ranges of the troposphere. This particular altocumulus deck, at an altitude of about 12,000 feet, was nicely painted by the moon which glowed through it, producing on its pillowed formation a soft blue and red corona.

"What do *you* see?" I asked. "I can see the altocumulus plain enough and the corona. But what do you see?"

What the observer saw, he could no longer see. Something had been there, just below the altocumulus, but now it was gone. It had, he said, colored lights. "Could it have been a plane?" I asked the obvious. No, he was certain it wasn't a plane.

We stayed outdoors long enough to assure ourselves that whatever he saw was no longer there and then checked with Base Ops to see if any planes were flying in our area. There were no planes aloft.

As part of its Operation Bluebook, a project instituted to handle the unexplained sightings that seemed to be receiving a lot of public attention in the half dozen years or so following World War II, the Air Force had instituted a policy of reporting UFO sightings to the flight service center at Wright-Patterson AFB. The observer, therefore, was required to call in the sighting to Wright-Patterson, which he did. The next morning, an officer in charge of intelligence matters went to the dormitory on base where the observer lived (we were beginning to change the character and the name of the enlisted men's lodgings — they were better than before and were losing the image that the word "barracks" connoted), got him out of bed, and had him answer a list of questions about the sighting that went into a written UFO report. The fuss taken at a time when he would have preferred to be sleeping made him feel a little sorry, I think, that he had been so observant the night before. I doubted that he had seen anything more than the moon shining through the altocumulus, made more spectacular by the coloring of the corona, but he maintained it was a UFO.

The second "sighting" occurred at about 0200 hours. A different observer was on duty that night. As I sat at my map table, analyzing a surface chart, he quietly approached, acting as if he wasn't sure he wanted to bring up what he wanted to bring up. "What's up?" I asked.

"I want you to see something," he said, and he started for the door as I got up to see what was the matter. Remembering that the last UFO had disappeared before I could see it, and sensing that this might be another, I had dropped my isobar pencil almost immediately and followed him outside.

After we had walked out some 15 or 20 feet beyond the porch light that shone above the weather station door, we stopped. That was where he had seen it. And, out there in that dark, moon-less night, I saw it too.

It was a small, white spot of light. Mostly stationary, it would shoot off into infinity at an incalculable speed, and then reappear, magically, back where it had been to start with, relatively close by. The most extraordinary aspect of its movements, aside from its incredible swiftness, was its direction of flight. It traveled only on perfectly horizontal or perfectly vertical lines. It shifted from horizontal to vertical flight frequently, and when it did so, it made precise 90-degree turns. At first I thought it might have been caused by one of the mechanics on the flight line who would be carrying a flashlight, waving it as a member of an airlines ground crew would guide in a taxiing airliner. But no human could carry a flashlight as far and as fast as that spot of light moved. Nor was it possible that anyone could make those perfect 90-degree turns.

The flight paths of that spot of light, which could have been laid out by a draftsman with T-square and triangle, were eerie and amusing. Like a clown jumping for the delight of children, it darted one way, stopped, darted another way, stopped — and always with that same astronomical speed. These motions were completely soundless, and the carpet of pine needles underfoot smothered all other sounds save those of our hushed comments. As if it did not approve of our comments, when we talked it moved about wildly; and when we stood in silence, unmoving, it too became motionless. Like a pet dog, it played its game for us and then paused to see if we liked it.

We finally concluded the UFO was not inclined to leave, so we returned inside the station to discuss the need to report the sighting to the flight service center at Wright-Patterson. Just as a reminder, I said, "If we report it, there'll be forms to fill out and interrogations at first light."

"But we have to report it, if it's really there."

"Yes, " I agreed, "but before we do, let's go out for another look." I greatly disliked calling the flight service center for what I believed would end up being a trivial reason. Back outside, the light appeared, as it had before, in the general direction of the flight line and with the same strange movements.

"Think hard," I said, "did you see anything on the flight line before it got dark."

"Only a C-47," was the reply. And that was it. Our UFO was the weather station's porch light reflecting off the shiny surfaces of the C-47. When we moved our heads, we made the UFO move. We returned to the bright lights within the station and, to my knowledge, never told anyone about our close encounter with a flying saucer — until now.

Those two sightings weren't nearly as spectacular as the one reported to us by the public in 1947 while I was on duty at Westover's weather station in Massachusetts. This was near sunset. Just as the lowering sun had begun to cast its orange glow, a number of excited callers reported flying saucers in the west. I stepped outside the station to see if I could spot them.

It was a beautiful sight. The "flying saucers," about a half dozen in number, were at an altitude of about 8,000 feet. Painted a brilliant orange by the setting sun, they presented a most pleasing contrast with the purple-hued peaks of the Berkshires. No artist could have planned his palette with greater skill. It looked as if some flight leader had commanded the saucers to arrange themselves in what seemed to be near-perfect formation; but unlike any USAF formation, they were not moving. It was easy to see why the public, viewing those orange-colored objects that had smooth edges and shapes similar to footballs, could have taken them for space ships operating under a form of propulsion unknown to earthlings.

But they were only clouds, special lens-shaped clouds called "lenticulars" that acquired a virtual stationary appearance from the fact that they were formed by localized atmospheric waves which are set up in the lee of mountains when winds aloft are strong and atmospheric moisture is neither too little to prevent their formation nor too great to cause the lenticular shapes to be obscured by a general cloud mass.

A lot of controversy surrounds the question of whether UFOs have landed on earth. We are all familiar with the news reports in which people are quoted as having seen UFOs. Some of those interviewed have even claimed to have been aboard a saucer. Nevertheless, to my knowledge, there has never been a scientifically proven report of a UFO landing. I think I know why. I got my clue back there at Westover in 1947 when, on one of the night shifts, we received a bogus teletype message from an unidentified source. It was a NOTAM (Notice to Airmen). This was the report:

NOTAM
FLIGHT OF FLYING SAUCERS INBOUND
REQUESTING 1,000 OCTANE FUEL
ONLY 100 OCTANE AVAILABLE HERE
UNABLE TO LAND WITHOUT IT

But now back to Oscoda. By this time in my career, I had been forecasting for close to five years and had acquired considerable confidence as compared to my first year or two as a forecaster. I had developed the philosophy that, after the forecasting shift, you should go home to your family without worrying whether your decisions would all pan out. You should go to sleep knowing you had done your best. But all that was not as easy to do as it might have seemed. When such thoughts arose I'd sometimes be reminded of corn planting time during my boyhood on the farm. I'd drop the kernels into the planting hills and silently chant: "One for the blackbird, one for the crow, one for the cutworm and one to grow." On the farm you did all you could to change the things that could be changed and learned to ignore the things that couldn't be changed. In weather forecasting, however, it would take a great deal of concentration to forget some of the predictions.

Each morning at Oscoda's Base Weather, near the end of the 12-hour shift, the forecaster prepared briefing charts which he took to the alert hangar where he provided the assembled fighter pilots, the F-86 pilots, with a formal briefing. This included a forecast of the ceilings, visibilities, precipitation, turbulence, icing and winds that were expected over the next 24 hours. These forecasts were for Oscoda and for the other Air Defense Command bases that might be used as alternate airfields. One of those briefings became indelibly etched in my memory.

It all began in the evening of a chilly night in the winter of 1952. During the 12 hours of that shift, I was following an instability line that had slipped out of Montana, headed our way. I asked the observer to plot me a sectional surface chart as soon as the hourly observation sequences came in on the teletype, at the beginning of every hour, rather than on the customary three-hour interval. I spotted the line early in the shift as a zone of falling, then rising, pressures. There was no weather associated with it. On one of the succeeding charts, at the location that the line should have moved to, the observer plotted a rain shower. An hour or two later the line was all but impossible to find, there was no weather at its expected location. (The reader will recall this was before satellite imagery and adequate weather radar.) Later in the shift it became visible again and a station not far from it reported distant lightning. But again, there was no weather along that instability line. It seemed that what little weather there was with that line was at best sporadic, occurring in isolation here and there.

At about midnight I calculated the line's ETA at Oscoda to be more than 24 hours out — by that calculation, it would arrive at Oscoda at 0400 hours of the next day. This was not a good arrival time. If thunderstorms are possible, and I thought they were with this line, forecasters prefer that such lines (also, fronts) arrive in the afternoon when daytime heating adds to the instability and the possibility of having thunderstorms is increased. During the cooler times of the day, and especially in the early morning, near the time of the minimum temperature, statistics tell you thunderstorms are less likely.

I mulled around the pros and cons — would there be a thunderstorm or not? There were two additional pieces of the puzzle to contemplate. In northwest Canada, upstream of the instability line, colder air was being advected. That should intensify the line. And it would be passing over Lake Michigan. Any cells that developed would have moisture to feed on.

So, at that time, the decision seemed somewhat clear. Statistics notwithstanding, I would forecast thunderstorms at Oscoda, even if they were scheduled to occur near the time of the minimum temperature. I would forecast the boomers to roll into Oscoda the next morning — at 0400 hours.

Anyone who has worked the midnight shifts, the graveyard shifts, knows how resistant the body is to adjusting to those hours. Sleeping during the daytime when everyone else in your household has a plan or plans for the day, becomes a sort of happenstance. And then there is the body's desire to reconnect with the old ways, to eat lunch at noon, for example, rather than to sleep through that time of day. I'd always awaken at precisely the beginning of the noon hour, ravenously hungry. After lunch, I'd return to bed, trying to get the remainder of the required allotment of slumber that would properly prepare me for the coming night's work. Another peculiarity of the graveyard shifts and especially of those super-graveyards, the 12 hour night shifts, was the impressive and sudden energy drain that swept over you about an hour or two before sunrise, often at about 0400 hours. This would happen no matter how much rest you had during the previous day or how interesting the work at the moment happened to be. And then, in my case, this always happened: After a half hour or so of having to force myself into action, the biological clock would institute a miraculous recovery, timed perfectly to coincide with the need to prepare charts and forecasts for the 0700 briefing at the alert hangar. Then, during the quarter-mile drive to the alert hangar, I reviewed my preparations in a now fully alert mind, searching for any flaws in the forecast I had decided upon. Like electrons in orbits, thoughts rose to higher and higher energy levels until I braked to a halt and entered that briefing room furnished with a lectern and about 50 folding metal chairs.

During the drive to the alert hangar on that particular morning, this forecaster's brain cells debated pros and cons of the thunderstorm forecast, a habitual sort of review ending in a somewhat unsettling conclusion.

...In this briefing, you need to be careful with that thunderstorm forecast. This is winter time in northern Michigan. Any fool knows a thunderstorm wouldn't have much more than a snowball's chance in Hades up here at this time of year...Remember what Mr. Nobel so strongly advised in weather forecasting school — **never forecast the extreme event**...*He knew that would end up a failed effort almost all of the time... A thunderstorm in northern Michigan in the cold season and at 0400 hours is the extreme event if there ever was one... Why even mention the thunderstorm? If you ignore it and it really happens, they'll think it was a strange, unpredictable event.*

You'll not be blamed. On the other hand — if you've worked it out with the data staring you in the face...Unless — you've forgotten something... Well here's the best plan: mention the thunderstorm only in the worst case. Mention it only if you happen to be pinned down in the briefing...

The metal man-door of the alert hangar clanged shut behind me, and I strode into the briefing room, my rolled maps tucked under my arms — the surface chart, 300-millibar chart, 700-millibar chart, 500-millibar chart. I weaved through the rows of folding chairs that were identical to the ones we often took on picnics and, up on the platform, taped the charts to the easel. These were in order, the surface chart first, exposed for all to see as they looked from their seats or mulled around near the easel, and immediately behind it, to be exposed during the flight level winds forecast was the 300-millibar chart. The 700 and 500 would only be shown if there happened to be a question about the winds at 10,000 feet and 18,000 feet, respectively. You didn't want to lose their attention with details that might not be pertinent to them. These were F-86 fighter pilots of the 63rd Fighter Interceptor Squadron, one of whom, Captain Ralph "Hoot" Gibson, was the 3rd jet ace of the Korean War and a future commander of the Air Force's Thunderbirds aerobatic team.[2] You didn't want to put yourself in the position of the harried forecaster in that captivating Westover safety poster — you didn't want these pilots to be tempted to repeat to you the poster caption. You didn't want then to say: "Skip the lecture, Curly. I've got a date with an angel!"

I made some small talk with the pilots seated in the first few rows, and waited for the CO, a colonel, to arrive. The pilots snapped to attention as he strode into the room. Taking his position in the chair left vacant in the first row, the colonel announced "at ease," nodded at me, indicating that the briefing would begin.

I began by talking the pilots through the features on the surface chart, flipped to the 300, discussing the flight level winds, and followed with a description of the current weather in the Great Lakes area. I concluded by reading from my laundry list of terminal forecasts, for Oscoda and alternate landing fields: Selfridge AFB; Willow Run Airport, Michigan; Niagara Falls Airport, New York; Wright-Patterson AFB; Youngstown Airport, Ohio;

---

[2] I would meet jet ace Gibson again about 47 years later, in May 1998, when he and I participated in an Air Force Association golf tournament at Davis-Monthan AFB in Tucson. His skill as a fighter pilot and my presumed skill as a forecaster did not translate into great golf scores. Our handicaps were a little high. I asked him if he remembered the day I had briefed him on a flight to Washington, DC when he said he wasn't looking forward to the trip because he was going commercial air and they wouldn't let you wear a parachute. He said he thought that might have been when they had him attend the Peach Festival. Being only the third American jet ace, Hoot Gibson was a celebrity. His hometown of Mt. Carmel, Illinois bought the Cadillac he had purchased just before going to Korea and paid him quite a bit more for it than he had paid for it so they could put it in a museum.

Milwaukee's Billy Mitchell Field; O'Hare International Airport, Illinois. These forecasts were my expected cloud and visibility conditions through the daytime and into the evening, with the forecast period ending at midnight. That was the routine. Although we would at times go beyond that period, the forecasts were to focus on the daytime and early nighttime when almost all flying would occur. There was no need to mention the thunderstorm, as that would be beyond the end of the forecast period. Things were going good.

As I gathered my charts to leave the lectern, a blond major, the officer who coordinated the briefings, asked: "What's the weather for tomorrow morning?"

The jig was up. I brought out that nonchalant manner that a lot of forecasters had found to be useful on similar occasions, saying: "Actually, I expect a thunderstorm at that time."

"At *what* time?" the major asked.

"At 0400." I couldn't even keep from mentioning the exact time! If I had, there'd be some cover to be found should the thunderstorm forecast be a hit but slightly off in its timing. In the event that it should occur at, say, 0500, 0300, or even 0700. I could even have said it would happen tomorrow. Then a humorous thought surfaced. I might be wrong, but someone at Air Weather Service headquarters would be proud of me. Here was a forecast stated with uncompromising precision. No hedging in this one. No flim-flamming with words designed to obscure true meaning.

Thinking that I had said more than I wanted to, I turned to leave. The major stopped me again. "Excuse me," he said, "I *did* hear you forecast thunderstorms for four in morning, here in the north and in winter, didn't I?"

The major was polite and was no man's fool. Statistically it was a bad call. Like the words of a popular song — it was for the wrong time, the wrong place. I did believe in statistics as a general guide, but one had to find the determinant path to the forecast and then follow it, regardless of whether that led to the extreme event. There were certain physical processes. You identified the pertinent ones, tracked them, and you projected them to an unyielding conclusion.

Still, doubts flowing from those so obvious statistics could not be completely squelched, and I left the alert hangar thinking how bad it could be to face the fighter pilots at the next briefing. The only bright point, and it was only slightly bright, was the four-day break — I was going on my four-day break that morning. With a little luck, the pilots would have all forgotten the forecast by the time I returned.

At this stage in my forecasting career, I was still in the process of developing the mind set that I would spend no time worrying about the forecasts after my shift had ended. In time, I would be able to deeply internalize that philosophy, but at Oscoda in the early 1950s, it was not quite there. In fact, I did worry about that forecast. I replayed the briefing in my mind all the way as I drove along US 23 to the cottage on Cedar Lake, about six miles north of the base, that we had been lucky enough to purchase on land contract terms. (Many fine homes have since been built there, but at that time, ours was the only home on that lake with indoor plumbing!)

In keeping with the routine I had established for the end of the string of night shifts, I stayed up through the day, attempting to reorient that biological clock. I chopped some wood, fed it into the stone fireplace, lit a fire. It was good to feel the bite of the ax as it sank into the wood and later to hear those sharp cracks as sparks erupted from the fire.

Finally wound down and with tiredness rising from a 24-hour-period of being awake, I went to bed and dozed off into fitful sleep. Long periods of deep shut-eye were broken into by bolt-upright instances of extreme wakefulness. The night advanced. For some reason startled out of one of the deep-sleep episodes, I sat up, wide-eyed in the dark.

"Go back to sleep," Marie said, half awake. "It's only thunder."

"I thought maybe it was." I shuffled over toward the dresser, doing my best to avoid other furniture, and squinted at the luminous dial of our old reliable alarm clock.

"Is something wrong?" now Marie was wide awake.

"No, nothing's wrong," I said. "Just checking the time. It's four o'clock."

"And that's too early to be up." She slipped back into slumberland while I laid there awake — awake until, through the window, the sun began filtering through the jack pines, an old hackneyed expression running over and over in my mind: "How lucky can anyone get?"

At the end of the night shift following the four-day break which was followed by four day shifts, I made preparations for the briefing that would be held, as usual, in the alert hangar. It had been nine days since the 0400 thunderstorm. On this day the weather would be CAVU everywhere in the flying area and would remain so for several days.

The briefing was smooth. At the end, I asked for questions. There were none. The blond major, the briefing coordinator, stepped up to the lectern as I rolled the maps, preparing to leave.

"I didn't think you had a ghost of a chance of hitting that forecast," he said. "But there it was, at four o'clock on the dot — the loudest clap of thunder I ever heard."

There had been more than just a little luck in the timing. The thunder seemed unusually loud and captured attention because the Michigan north woods are very quiet, causing sound to project freely, but also because it occurred at the time of day when there was little activity, no competing sound. If the major could have slept through it, he'd have assumed the forecast was a bust. Nothing more would then be said about it. But I, of course, would sense lack of audience interest in reactions to my forecasts.

Turning toward the other pilots, the major repeated: "The loudest clap of thunder I ever heard." He could have dismissed it with a casual and privately acknowledged "Good job," but he wanted to be certain those fighter pilots all understood the significance of the forecast they had been given.

"Hey, it was no big deal," I lied, "only what had to happen — only according to the plan." I mentally noted that the major, a man who would take the time to make sure credit was given when he thought it was due, would be added to my Honorary Friends-for-Life list. Honorary only to preserve military decorum — he was a major and I was a tech sergeant. I never knew the major's name until, in June 1998 while playing in a golf tournament in St. Louis organized by the Air Weather Association, an organization of mostly retired AWS people, L. Dayton Blanchard, who retired as a lieutenant colonel and as a captain was one of the Wurtsmith forecasters, remembered he was John O'Diorn of the 63rd Fighter Interceptor Squadron.

The 0400 thunderstorm provided me with a credibility ticket that couldn't be beat. My forecasts were never again questioned at Oscoda and that was sort of phenomenal for the fighter pilots loved to make jokes of the forecasts and even of the forecasters themselves — some of whom they would nickname "foggy" or "stormy." I fully expected the moniker "thunder" to be attached to me, but my suddenly gained reputation appeared to be a bit too overwhelming for them and, also, since I was, at that time, the only enlisted forecaster at Oscoda, it might have been looked on as unseemly in that military setting. I had always liked the fighter pilots, a sort of happy-go-lucky bunch, and now I had a personal reason for it.

Top: The North American F-51 Mustang was a premier fighter of World War II. It had the range needed to escort B-17 bombers deep into Germany. *Photo by Robert Nordstrom.*
Bottom: The alert hangar at Wurtsmith Air Force Base, Oscoda, Michigan.

Top: Lenticular clouds at sunset.
Bottom: Pilots of the 63rd Fighter-Interceptor Squadron score a target on a wet day at Wurtsmith AFB, Oscoda, Michigan in 1951. F-86 fighters are in the background. *From "The Wurtsmith News," 23 July 1987.*

Chapter 19

Weather Gets in Your Blood

After two years at Oscoda, during which time the base had changed its name from Oscoda Air Force Base to "Wurtsmith Air Force Base" to honor a Michigan man, General Paul Wurtsmith, who had been killed when his B-25 crashed on a foggy day in North Carolina,[1] I decided to quit the Air Force. I had never thought of myself as a slow learner, but it was clear, nevertheless, that it had taken me much longer to arrive at that conclusion than had so many of my associates from basic training days and later. They had come up with the decision as if by instinct. I had to do it the hard way. I had to serve seven years before the reality dawned on me. But now I was doing it. I was leaving the service. Andy Anderson would be proud of me. He would only wonder why it had taken me so long.

Everyone leaving a base had to go through a base clearance routine. The chaps in the base administration office would provide you with a form neatly marked with boxes to be initialed at a long list of offices. This was done to make sure you were not leaving with some sort of unfinished business still to be attended to. When I left Selfridge, for example, I had to clear through the post office, commissary, the Red Cross, hospital, personnel services, auto registration, signal office, provost marshal, the personnel equipment section, air installations, library and gym. At each of those places, I had to wait until some person could find the time to deal with me. At that stage, there'd be a search through desk drawers and filing cabinets to see if, presumably, I had had a transaction of some kind with them which was still pending. And while I stood, transferring body weight from one foot to another, I'd be thinking that whenever a pilot walked into the weather station while *I was on duty*, he got instant attention from me. Most of the base clearance routine was a pure pain in the butt. At many of the offices, I'd be entering that domain for the very first time. So how could I possibly have had any business with them? And even if I did leave with, say, a book from the library that I forgot to return, my whereabouts would not be a secret. The Air Force would know where to find me because, wherever that was, it had sent me there! But I went through the drill with as much humor as I could muster. At the gym I had to interrupt a man busy shooting baskets who then had to search through his files to see, presumably, that I didn't have one of their basketballs checked out — how did you get a job like that? The last stop on my form was the Wurtsmith Air Force Base Hospital, a low-built complex of

---

[1] *The Wurstsmith News, 23 July 1987.*

wooden buildings connected by narrow, wooden, tunnel-like corridors where your footsteps would set off reverberating echoes. I had saved this stop for last because it would take longer. They would give me a physical before allowing me to become a civilian. Then they would initial the form.

The hospital buildings looked as if they could have been there ever since 1925 when the airfield was called Camp Skeel, after Captain Burt E. Skeel, CO of the 27th Pursuit Squadron at Selfridge, who was killed when his Curtis racing plane crashed at the Pulitzer-Schneider Race at Wilbur Wright Field (later Wright-Patterson AFB).[2] The buildings were probably there when the townspeople had gathered to watch Charles Lindbergh land his plane on the smooth ice of adjacent Van Ettan Lake. Lindbergh's visit was told to me one day by a town elder who came to the weather station to ask me about the loud explosion-like sound that had energized the residents of that then sleepy town of Oscoda. For all of them, apparently, it was a first encounter with the sonic boom produced by jet airplanes.

"What happens if I don't pass?" I asked. The doctor was a first lieutenant. This was in the days when doctors coming into the Air Force directly from civilian life skipped the grade of second lieutenant. Later, when that inducement still didn't seem to be enough to attract doctors, they would bring them in as captains. I could immediately sense that this lieutenant's total days in uniform were obviously fewer than the number I had spent in basic training, and perhaps even less than the number of days I had pulled KP. "If I fail the physical," I asked, "will I have to stay in the service forever?"

The lieutenant surprised me. He took my remark seriously. "I don't think so," he said, "but I'm not sure." Telling him I only meant that as a joke, I immediately put him at ease. We struck up a rapport. And in a matter of minutes I had passed the physical, as I knew I should have.

That done, I wasted no time in executing an about-face that placed me in the hospital corridor where I planned to quick-step toward that light at the end of the corridor, the outer door, the gateway to civilian life. I had not taken a dozen steps when the lieutenant called after me.

He walked up to me, looked around as if to ensure we were alone. "Say, sarge," he said, "before you leave, can I ask you something?"

"Sure. What is it?"

---

[2] *The Wurtsmith News.*

"Well, I wonder if you could tell me what it means when you're walking outside and someone does this?" The lieutenant was demonstrating, and not very well, *the hand salute!* It was only the great respect I have always held for persons who have the potential for working on my body at times when I may be anesthetized that kept me from roaring with laughter.

"It means," I said, "that you should do this," and I threw that lieutenant my snappiest salute.

As I drove out the main gate, I thought the Air Force, if it only knew it, might have done well if, in the manner of a queen knighting one of her subjects, it had sent someone to tap me on the shoulder and say: "You are now a commissioned officer." The great boost in pay and even greater potential for future pay advances might have been enough to convince me of the wisdom of staying in the service. Hell, they wouldn't even have had to teach me how to salute! Way back in basic training, the aptitude tests I had taken had placed me well above the score needed for attendance at officer candidate school. And from time to time, the thought arose that I probably should apply for OCS. But I was, to a certain extent, a captive of that rural Mid-west cultural upbringing in which it was considered to be in bad taste to blatantly strive to advance yourself, not for the sake of increasing your knowledge, which was perfectly acceptable, but for the sake of advancement alone. If you went to OCS as a first-three-grader NCO, you already knew all they could teach you about the military, you were only going there to be given permission to pass through the door into a higher social order, a repugnant pursuit to a farm boy from rural Michigan. It would be many years before I would understand the shortcomings of that kind of thinking.

I also held on to the thought that, some day, the military would grant professional commissions to weather forecasters. That had been the experience of Captain Julian Voss, one of my fellow forecasters. Voss told me that one day when he had driven us in his '29 Model A Ford out onto frozen Van Ettan Lake to introduce me to ice fishing. But Voss's professional commission had been granted during World War II, a time when many such appointments were possible. (One of Detroit's auto company executives, William S. Knudsen, had been brought in, I was told, as a four-star general.) Still, I thought the professional commissions might happen again; and in reconsidering the possibility of applying for OCS, I felt exactly the way the Detroit Tiger's great star, Al Kaline, had felt when they asked him why he never became a manager. "They never asked me...." he said, "(and) it got to the point where I figured that if I had to ask, it wouldn't have been worth it."[3]

---

[3] Cantor, *The Tigers of '68*, 163.

While I probably would have stayed in the service had I had a commission, my prime motivation in leaving was to relocate where I could seriously pursue a college degree. Our home near Selfridge, with three or four colleges within commuting distance, was such a place. No college courses had ever been offered in the Oscoda area.[4]

To its credit, the Air Weather Service did its best to show me why my decision to quit was not exceptionally wise. The colonel commanding our parent organization, the 12th Weather Squadron at Stewart Air Force Base, New York, sent me a letter designed to fill the gaps in my reasoning process.

The colonel made a very good start, trying to enlist the aid of Marie in the salutation and in the letter itself. Because it was addressed to her father's house, our home of record, I received the letter after I had made my irrevocable decision to leave the Air Force. But it would have made no difference. One of my unsaid gripes was that, as the only enlisted forecaster, there was a significant disparity in salary between myself and the other forecasters who then were all captains. I never doubted they deserved their rank, they were as nice a group as anyone would ever want to meet, but the fact remained they made a lot more money than I did and I had been assigned more responsibility. Because Captain Wolfe had made me Station Chief and NCOIC of the weather observers, in addition to my forecasting duties I was responsible for the observer's quarters and for their work. He particularly wanted to improve accuracy of the weather observations and asked me how to do it. "The answer," I said, "is easy. The observers are as good or better than any you'll ever find but are often too busy to double check their obs. Have the forecaster check all computations of each ob and initial each ob certifying to its accuracy." This might have decreased my popularity with my fellow forecasters. Being gentlemen, they did not show it and even made me think they enjoyed improving the obs. My simple suggestion apparently resulted in the detachment being named most outstanding in 12th Weather Squadron and 3rd Weather Group.[5] And I began, again, to believe I was being underpaid. The colonel's nice letter, reprinted below, could not change that feeling.

---

[4] That would change in later years to the extent that someone stationed at Wurstsmith could even earn a master's degree. Those newcomers to Wurtsmith as well as all others in the Air Force probably never realized how lucky they were to have those expanded educational opportunities.

[5] Letter dated 27 April 1966 from Colonel Donald J. Wolfe to WO Theodore L. Cogut, titled "Recommendation for Commission." In the letter, Col. Wolfe wrote: "Had Air Force policy allowed his obtaining a commission at the time (in 1953) I would have insisted on his submission of application. During his tour at my weather facility, then TSgt Cogut was made station chief and NCOIC of my observer section. This action was dictated by the fact that he was clearly my most outstanding NCO and most capable forecaster. Throughout his tour of assignment in these duties our station was consistently noted as the most outstanding unit in the 3rd Weather Group and 12th Weather Squadron. "

12th Weather Squadron
Stewart AFB, New York

6 February 1953

Sgt. and Mrs. Theodore Cogut
825 E. Kenneth Avenue
Royal Oak, Michigan

Dear Sergeant and Mrs. Cogut:

    Two months from now you both will be faced with the serious decision of whether or not Sergeant Cogut should continue his career in the Air Force. This decision is not only of vital interest to both of you, but is of equal importance to the 12th Weather Squadron and the Air Weather Service. Sergeant Cogut's high degree of efficiency and fine record in the past has been a matter of personal satisfaction to me.

    Perhaps I can be of assistance. Here are some of the things which are important factors for you, Sergeant Cogut:

    1. A permanent job with training and educational benefits already exists for you in the Air Force. You have completed seven year's service; your career is well established, and excellent opportunities exist for advancement. Although you may have a civilian job waiting for you, your absence of these past years may mean that you must undergo new training or indoctrination. It may be that training or education for a new position will be required before you can secure a job of your choice.

    2. The $10,000 free insurance policy. This would cost a person 25 years of age approximately $3,830 for a 30-year period with a civilian insurance company in term insurance, or approximately $20,00 per month in ordinary insurance.

    3. While the GI educational bill offers you a chance to go to school with ($160.00 per month for more than one dependent), a similar opportunity exists in the Air Force whereby you may attend USAF service schools, and receive your regular monthly pay. You can also enroll in a correspondence course or off-duty schooling under "Operation Bootstrap" and secure college credits and a degree. Perhaps attending college in civilian status seems attractive, but this $160.00 must cover rent, groceries, doctor and dental expenses, insurance, car expense, tuition, books, etc.

    4. You will be eligible for promotion to Master Sergeant and with a pay increase.

    In addition to the security and financial aspects, the opportunity of living in Europe, Japan, and other interesting places will undoubtedly appeal to you, Mrs. Cogut. Although now there is a waiting period for dependent travel to some countries, the Air Force is trying to reduce this separation period.

    Of interest to you both is the excellent Air Force retirement program. For example: A Master Sergeant can retire after 30 years' service with a monthly income of $229.32. It was interesting for me to learn that a similar retirement plan in civilian life would require a $50,670 annuity policy with Metropolitan Life Insurance Company.

Medical facilities including treatment and hospitalization at minimum cost are available for dependents.

In conclusion, I wish again to assure you that your service is appreciated, and you have my personal interest and best wishes for a successful future. If I can be of further assistance to you, please feel free to write me at any time.

        Sincerely,

        CHARLES A. BECKHAM
        Lt. Colonel, USAF

While it may have been advisable to appeal to the wife of the reluctant soldier in most cases, it was not so in ours. Marie had always left those decisions to me. If she had been like many other wives, it was more than likely that I would have long since been pushed into OCS. As for the economic analysis, that made no impression at all. Beginning with my basic training days there were always those around who would say, "you had to find a home in the Army (or Air Force) because you couldn't make it on the outside." My response: Who says I can't?

The good colonel had written: "you will be eligible for promotion to Master Sergeant," and "a Master Sergeant can retire after 30 year's service with a monthly income of $229.32." He couldn't have known it, but that point made even less of an impression. I knew that, had I stayed in, I would soon have been promoted to master sergeant, in fact, Captain Wolfe had told me so, but I was still too young to enjoy the prospect of spending 23 years in the same rank (master sergeant was then the top enlisted rank); and a retirement income of $229 a month did not seem like much reward for *30 years* in the military. In addition, and perhaps of most importance, I was still young enough and brash enough to think that, on the outside, I would soon become a captain of industry. I basked in the swell-headed notion that I had mastered the difficult science of weather forecasting. I had reached the top of that game. If I could do that, it was reasonable to think I could do the same in another field. But I did appreciate the colonel's letter. In my experience, it was unprecedented for a colonel to show such concern for a tech sergeant and his wife.

So we settled back into our home near Selfridge, and I quickly enrolled in a full-time college program. This was to lead to a degree in the humanities as a degree in meteorology, which many would think to be my first choice, would have been a ridiculous effort for I already had more education in that science than a university bachelor's program could provide. In any event, it was not an option. The University of Michigan at Ann Arbor, the closest institution offering meteorology, was beyond commuting range. It was not long, however, before I realized that my academic goal should have been the study of economics — my financial situation showing that I needed considerable education in that area.

Obtaining full-time employment with the US Lake Survey and later with the Engineering Department of the Wayne County Road Commission provided a practical education in cash flow matters. At the Lake Survey, I became familiar with the channel-depth requirements of the big freighters that plowed their ways through the Great Lakes and connecting waters. An interesting job, but it paid even less than my previous compensation as a tech sergeant. Somewhat painfully, I was being reminded of Colonel Beckham's advice regarding the economic realities of life on the outside. In the effort to stretch my paltry salary, to the somewhat disconcerted looks of our neighbors I plowed up our rather large front lawn that I had previously envisioned as a highly manicured golf-course-like frame for our home. On that piece of land, the once emerging showpiece of a suburban life style, I planted tomatoes to complement the snap beans I had planted on our adjacent lot. Marie, a top-notch example of the military housewife who somehow finds the way to survive on meager resources, made large quantities of tomato-bean soup which she canned in mason jars and stored on shelves in our utility room. That winter, with most of our money going out for furnace oil; we would literally live off her canning efforts. On Sundays, sister Ange and her husband Bob Newman, an executive with the Manufacturer's Bank in Detroit, would come out to ostensibly play pinochle and for Bob to tap my brain for college algebra answers as he worked homework problems for the college course he was taking. But that was just a ruse, they were really only visiting to bring a roast or some other cut of meat, providing us with the only meat dinner of the week.

The realization that I was not in an envious economic condition came vividly to the fore that same winter when I saw Marie cutting up my old blue Air Force overcoat to provide the material from which she would make coats for our daughters. What a talented lady! She decorated the collars of the tiny coats with pieces of red remnant cloth. The coats did not at all look like fragments of a military garment. They were attractive, stylish little blue coats. For buttons, Marie asked if it would be proper to use the metal buttons with the Air Force insignia. It would save a little more money, but we decided that would not be done. We would buy buttons, spending at least that much on her creations. As I watched her sew on the new buttons, I was reminded once again of how greatly I had been blessed.

Our situation improved dramatically when, on the advice of brother-in-law Bob Newman, I took the test for a surveying job in the Wayne County Road Commission and was lucky enough to place second behind an applicant who had a Ph.D. This was also an interesting job and my new salary then surpassed that of a tech sergeant. Among other assignments, I helped lay out the alignment for interstate highway I-95, the Edsel Ford Expressway, in the Detroit suburb of Harper Woods.

But to keep those jobs I had to cut my college program to half time, and my prime goal in leaving the Air Force had been the pursuit of a college degree. At the rate I was going, I would have retired from the Air Force (with 20 years) before I got that degree. And it went without saying that the job and school requirements left me with not much more time to spend with my family than when I was in Saudi Arabia and they were eight thousand miles away.

Starting out from our home in that northeastern suburb of Detroit, in the mornings while it was still dark, I drove 40 miles through the heart of that city to a western suburb where I met up with my survey crew. At the end of the day, I reversed the trip except that I stopped off at the engineering college (I had switched colleges to pursue engineering-type courses) about half way home where I waited for it to be late enough for the night classes to begin. After classes, I drove the remaining distance home and wished the Edsel Ford Expressway that I had been doing the survey work for was completed, that would have shortened the trip by a half hour. It would then be about 2300 hours, still time enough to squeeze in a few hours of homework. With that kind of schedule, I had little time to enjoy the childhood of my three lovely daughters — Leta, Willa and Pamela. Oh, I then had money enough, it seemed, partly because I was being paid much more in the surveying job, but also because I didn't have the time to go out and spend it. On the week days and on weekends, when not trying to catch up on my missing sleep, I'd be making engineering drawings for homework. I was slipping into a totally time-consuming routine, and the end was not in sight. Little by little I became aware of a missing part of my life.

The civilian jobs were interesting, and it wasn't as if there was not a weather connection. At the Lake Survey, precipitation amounts falling on the separate watersheds had to be quantified to arrive at the impact on the surface levels of the separate Great Lakes and, therefore, on the channel depths. At the engineering department, to find the property corner stakes in winter one had to pound and pound into the frozen earth with a pickax. With almost every swing of the pickax, sharp pieces of iced earth flew back to sting your face. This work was so slow it often seemed spring would arrive to melt the ground before we could uncover the stakes.

And in 1953, there was just plain interesting weather in southeastern Michigan. At about the time of the deadly tornadoes in Flint, through the picture window I had made for the bay window we saw ball lightning — the only time, either before or since, we've ever seen it. To say it was an eerie display would be one way of trying to describe it. Spherical in shape and closer in size to a softball than a basketball, it glowed with a soft blue luminescence. Behaving just as if it was only there to have fun, it bounced along a wire that ran between two utility poles. After going about half the distance between the two poles, it hopped down to the road directly in front of our house. Then it skipped along the road about a hundred feet and was gone. I wanted to study the meteorological data to see if there were clues as to why we would have seen it, but of course I had no data.

I thought of the corporal in my early weather station days at Westover who, when I asked why he had reenlisted, had said, "You'll understand when you've been in it for a while. Army weather gets in your blood." And indeed the corpuscles swimming around in this former weather forecaster's blood must have looked like those special meteorological symbols that, in one distant time of mission-urgent mind-sets, we had plotted so carefully on the weather maps.

That did it, I decided it was time to drive to Oscoda and visit Captain Don Wolfe, my former detachment commander. But here I have to confess it wasn't just the ball lightning that made me think of the trip. Coincidentally, Don had just sent an old Air Force photo of me. There I was, in the weather station, seated at the map table, looking as if some earth-shattering meteorological pronouncement was about to be made. Captain Wolfe had clipped this note to the picture: "Sure wish you were back in the service."

All the detachment commanders of my Air Weather Service career had been first-rate gentlemen. They were, without exception, the kind of persons one would want to have as next-door neighbors. This was certainly true of Don Wolfe. On more than one occasion, Don and his wife Lucy had the weather troops at their home on Lake Huron for a barbecue, a much-appreciated touch that we had not experienced in any other weather station. And Don had the sensitivity, it appeared, to know when to send that picture. Did he know it was about the time I'd be missing those 0700 fighter pilot briefings? With that simple gesture, a note clipped to a photograph, he'd get me started on a return to meteorology. To be sure, I didn't need much encouragement, but I probably never would have made the attempt without it.

Passing through the sleepy town of Oscoda, we took the left fork in the highway at the old, familiar, and homey green building, Tony Decker's restaurant and tavern (which, sadly, has since been replaced with a fast-food eatery), got a temporary pass from the air policeman at Wurtsmith's main gate, and drove up under the jack pines to the weather station.

As soon as I walked in the station, one of my old weather observers told me that 12th Weather Squadron at Stewart Air Force Base in New York was unaware that I had not reenlisted, and after I left Wurtsmith they published orders promoting me to master sergeant. Thus, I became, probably, the only civilian ever to be promoted to master sergeant. (Of course, it had no effect.) Don Wolfe asked me if I had decided to reenlist after all. I said I thought about doing it but there was a huge problem. I had been out of the service too long to be allowed to reenlist. Don thought about that for a minute or two and then said he knew of a way to do it that would get my old rank of tech sergeant back.

I should, he said, apply for enlistment in the reserves. He knew the colonel in Washington who handled the reserves and was sure I could return to the Air Weather Service by that route. I tried that and, at first, it looked as if Don, who I thought tended to look on everything as an opportunity, might have been a little optimistic.

The Air Force Reserve recruiting station, located in Detroit's federal building, was in the charge of a tech sergeant. "I can enlist you in the reserves," he said, "but since you have now been out of the service about a year and a half, the best rank I can get you is airman third class."

"Good-bye," I said, "it was nice meeting you." I was willing to take a cut in pay to be back with the fighter pilots, but I wasn't going to let my family starve.

"Wait a minute," came the reply. That tech sergeant, after all, probably had a quota to fill. Then, rising from his desk: "There's a chance I can get you a higher rank, but you'd have to apply for a grade determination which would be done at Reserve Headquarters in Washington."

"Let's do it," I said. This was what Don Wolfe probably had in mind when he said I could get my old rank back. But the recruiting sergeant wasn't thinking old rank.

"But I have to tell you," he said, "that the very best they'll do for you is airman first class (three stripes)."

"That won't be good enough," I said. "Think of my record. I've been a tech sergeant and was a weather forecaster for five years. But go ahead and submit the grade determination form." There was nothing to lose in making the attempt. I knew the colonel in Washington who was in charge of weather reservists would have a better appreciation of the value of my experience than would this tech sergeant. The grade determination form would show I had experience making forecasts for the Berlin Airlift, for the Middle East and for the fighters at Selfridge and Wurtsmith. But if it should turn out that only airman first class would be granted, I'd forget the whole thing, satisfied that I had made the attempt.

Weeks went by, then one day, I received a phone call from the recruiting sergeant. "I need you to come to my office," he said. "I have a letter from Reserve Headquarters in Washington about you and there's a big mistake in it."

The "mistake" happened to be that it was recommended that I be enlisted in the reserves as a tech sergeant, my old rank, just as Don Wolfe had expected. "They just couldn't do that," the sergeant said. Now, if I still wanted to enlist, he could do it but only if I were to accept the grade of airman first class.

He would have gotten away with it, and probably had with others. But years of being nice to people, an operating procedure that I had learned at the weather briefing counters, nice even to the few who didn't deserve to be treated nicely, suddenly evaporated. "Sergeant," I said, "what's the rank of the officer who signed this letter?" I was pointing at the signature block. All along he had been resenting the fact that a cotton-pickin' civilian was going to have the rank bestowed him that he had served so long to attain himself.

"Colonel," he said.

"And what do you do when a colonel tells you to do something?"

"Come with me," he grudgingly responded. "Maybe we can convince the Coast Guard lieutenant down the hall to swear you in as a tech sergeant."

With that, I was assigned to Selfridge Base Weather for my reserve duty, one of my favorite old outfits. Now, on weekends, I became one of the weekend warriors that I had sort of tolerated in the past. We now know the reserves are the military backbone, but back then we regarded the weekend warrior weather forecaster with amusement. They wanted to be helpful but didn't know how to be, and they usually got in your way when you were trying to analyze the meteorological situation. Few of them seemed to remember much about weather forecasting. A major who was a medical student at the University of Michigan was one notable exception. He was impressive as he went into the theoretical basis of the meteorological situation with you. When he received his medical degree, he went on active service as a lieutenant at the Fitzsimmons Hospital in Colorado. Like me, he took a cut in pay to return to active duty. But even to him, we would not entrust the forecasting decision. Now, however, things were different. Now, I was *one of them*. And it became very clear that a finer forecaster than a weekend warrior forecaster would be hard to find!

Remembering what we used to expect from reservist forecasters, I was looking forward to a couple of days rest, or at least of extremely light duty, on my reserve weekends. But once again, as when I had tried to get a daytime-only forecasting duty station while filling out my assignment wish list in Saudi Arabia, my best laid plans had gone agley. There were those at Selfridge who knew I had once been the active duty forecaster in that very weather station, and the word evidently had gotten around. I would, therefore, know of the local weather effects that were peculiar to Selfridge. Knowing how local terrain affected the weather at your weather station was most important and, incidentally, was a strong argument for keeping the forecaster at one station for a long period of time — for many years. But of course the assignments sections wouldn't like that scenario. Homesteading at a comfortable Stateside base couldn't be tolerated for how could they then fill the vacancies at places like Thule, Greenland or Dhahran, Saudi Arabia?

On the second day of my first weekend duty stint, the forecaster on duty nodded knowingly to Major Putnam, the detachment commander. "He's ready," he said, and I was presented with an isobar pencil and a pen with which to sign the pilot's weather clearance, the Form 175. There had been a few changes at Selfridge Base Weather since the days of my tour there some four years earlier. Of particular note was the acquisition of a state-of-the-art radar — the CPS-9 — which had replaced the old 50-mile-range AN/APQ-13 radar that had been salvaged from the World War II B-29 bombers. The CPS-9 had a vastly improved range, being reliable to 150 miles out, and a capability of providing a cross-sectional, or vertical, view of the water-laden clouds. By knowing their vertical extent, one had a means of judging the relative severity of the storms. This radar was better in depicting real-time storm movements than anything else then in the Detroit area. Word of the marvelous device reached the management of the Detroit Tigers baseball team which prompted phone calls regarding the prospects for raining out the ball games. When first called with that request, I balked, recalling the rule prohibiting forecasts for civilians that had begun at Westover, but was told in this case it was O.K. If memory serves, the calls had come from Spike Briggs, then the Tigers' owner. Of course, with the best radar in the area, it made perfect sense for us to provide the advice.

The rawin section had also upgraded its operation through acquisition of a GMD-1, a radio direction finder that replaced the old manually operated SCR-658, the "bedsprings." The GMD-1 was introduced to the military during the Korean War. An automatic-tracking system, it could record signals from the balloon-borne radiosonde from within the comfort of an office. No rawinsonde operator would regret the passing of the

SCR-658 which required that he sit outdoors in all kinds of weather. In 1956, not long after it had acquired this giant step forward in upper air data retrieval technology, Selfridge lost its upper-air sounding mission. Our GMD-1 was transferred to the US Weather Bureau at Flint, Michigan. I assumed this perhaps was a kind of replay of the Weather Bureau/Air Force meteorology competition that had reached a kind of high point with the snow forecasts at Westover. Thus ended an era in which the Selfridge forecaster had the benefit of upper air data that had been produced at his own station.

That reserve period of my military career only lasted about 18 months. At the end of that time, I was recalled to active duty, the last part of the plan that Don Wolfe had envisioned in my attempt to return to full-time forecasting. I then spent a half year forecasting once again at Selfridge, and then, shockingly, realized the impossible dream in an overseas assignment.

In this photo sent by his former CO, the author analyzes a surface map. On the left panel: a 300 millibar chart. To its right: a Skew T, Log P Diagram with the Selfridge AFB radiosonde plot. *Official Oscoda AFB photo, 1951.*

The GMD-1, introduced during the Korean War, was a radio direction finder that could track the radiosonde automatically, superseding the SCR-658 which required manual tracking. When employed at Stateside locations it was usually permanently mounted, but it could be dismantled and used in a portable configuration. This one was in the author's upper air sounding section in Vietnam. *Photo taken in Vietnam in 1966.*

CHAPTER 20

Shark Oil and Spider Webs

It was Washington's birthday, 1957, nighttime, and we were out over the Gulf Stream in a C-54 at 9,000 feet when I was reminded why it was important to brief pilots on turbulence. The kind of turbulence we were experiencing would officially be called "moderate" by weather forecasters, but merely classifying its severity was not to tell the whole story — it was the duration of the bumpiness that made it a matter of concern to almost all of the passengers — not an instantaneous stomach-above-the-top-of-the-cabin succeeded by torso-beneath-the-floor feeling which might be succeeded by only a half dozen or so similar thrills, but a continuous swaying-bumping-faltering-recovering search for that border of smooth air that always seemed to be just five more minutes of flying time away. Not one of our little group — not I nor Marie nor our three daughters who were six, eight and nine years old — succumbed to airsickness, a pleasant surprise for all of us. Perhaps it was the anticipation of the good things to come, for I had struck pay dirt in this overseas assignment. Someone at assignments must have felt I deserved nice treatment after that Saudi Arabia tour. This one was to be a three-year accompanied tour, meaning my family would go with me, and they were with me on this plane, and the place we were going to was the Island of Bermuda!

Once, over Middletown, Pennsylvania, during my hitchhiking with the AAF days, I had been in another C-54, also at night and we were in rough air, circling in the pilot's hope that the fog would lift and we could land. I was then a weather observer, not yet a forecaster, but even with my then limited knowledge of meteorological processes, I could have told the pilot he should immediately fly on to his alternate landing field, which he ultimately did. In the meantime, sitting in those canvas side-wall seats, nearly everyone began looking for the airsickness bags. I saved myself from that embarrassment by simply lying down on the floor, there was a lot of floor room between the side-wall seats. In that prone position with a duffel bag to cushion my head, the nausea cleared away.

This C-54, however, was fitted with two-abreast airline-style seats on each side of the aisle. Marie and I occupied two adjacent seats, daughters Leta and Pamela the next two. Daughter Willa sat in a seat several rows away. We were flying in a continental polar air mass that streamed out over the Atlantic after picking up its low humidity and cold air in its traverse over the North American continent. As that cold air mass met the convective currents rising off the warm Gulf Stream, the turbulence could not be ignored. I became concerned about Willa, sitting there away from our little family group. As a baby in those days before I left for Saudi Arabia, she often became frightened when she thought she might be left with a baby sitter, and I still thought of her as that sensitive soul. I went over to talk to her and made the acquaintance of the gentleman seated next to her who was the director of American International College's Bermuda extension division. While munching on a sandwich and apple that Willa had given him, he expressed appreciation for her thoughtfulness and said he had somehow missed being told at McGuire Air Force Base, the New Jersey base we had taken off from, that he had to order his lunch there or he'd have nothing to eat during the flight. We discussed the courses he would be providing at Bermuda, and he invited me to continue my interrupted college program with AIT's extension program, which I would do. I went back to my seat and looked over at my well-adjusted little eight-year-old and thought of how proud I was of her.

With an effective tailwind of about 15 miles per hour (13 knots to the crew and weathermen), we flew the 800 miles from McGuire in less than four hours and landed at about 2200 hours at Kindley Air Force Base, the USAF base on the British crown colony of Bermuda. The base had been named after Captain Field Kindley, an ace with 12 victories who served in the US Army Air Service in World War I. Prior to the creation of the independent Air Force, Kindley Air Force Base had been called Kindley Field. Thus, in a neat reversal of words, Kindley Field had been named after Field Kindley.

Kindley AFB was located near the eastern end of the 20-mile-long island. A US Navy Operating Base (NOB) was on the western end. On September 3, 1940, President Franklin Roosevelt transferred 50 destroyers of World War I vintage to Great Britain in exchange for the right to build bases on a 99-year lease in British possessions in the Caribbean and western Atlantic. British Prime Minister Winston Churchill desperately needed the destroyers to counter Nazi U-boat attacks on British shipping in the Atlantic. Bermuda was one of the British possessions involved in the agreement.

Just a little over a month from the time of our arrival, in late March 1957, I would have an opportunity to brief the pilots of President Eisenhower's plane. At the time when we were wearing the short-sleeved khaki-colored 505 uniforms, his pilots wore the more formal blues with long-sleeved light-blue shirts. In a uniform style that I would never see again, I observed that his pilots' shirts were tailored to accept cuff links decorated with either US Air Force or Presidential insignia. Entirely absent was the semblance of bravado sensed in some of our pilots. They listened attentively to the briefing, were pleasant and gave one the unmistakable impression of being long on flying experience. President Eisenhower was on Bermuda to meet with British Prime Minister Harold McMillan. It would be an important meeting, one that would reestablish the close relationship between the US and Britain that had been shaken the previous year when our two countries had been somewhat at odds over the Suez crisis. They would meet at the Mid-Ocean Golf Club, one of those sites that, because of what would happen there in the next year, would become permanently lodged in our memory.

I lost no time in locating Kindley's Base Weather. Weather and Base Ops, as usual, were located in the same building, a low one-story frame affair situated adjacent to the civilian airlines terminal. The terminal also was a one-story building of frame construction. A young Bermudan, a descendent, probably, of one of the black persons who had been shipwrecked on the island's reefs in the 17th century, ran a shoeshine stand at the terminal door. Inside the terminal, one could obtain sandwiches and drinks at the snack bar which was set up on the weather station side of the building. The men on duty in Base Ops and Weather found the snack bar location to be rather convenient and would sometimes spend a 10- or 15-minute break there. Both the weather station building and the terminal appeared to have predated the Roosevelt-Churchill agreement on the base.

To a weatherman, one aspect of the weather station was immediately striking. There were no windows in the forecasters' map analysis and briefing areas. Not one window. This, I thought, was the kind of weather station that would have gladdened the heart of the warrant officer forecaster at Westover who had been my mentor. "Never," he said, "ever look out the window after you've made your forecast."

If I were to always fully embrace that restriction, it was the warrant officer's belief that I would not be tempted to change a forecast through a chance glimpse through the station window of a darkening cloud or, with the opposite kind of forecast, a beautiful clear blue sky. My forecasts then would have to be based exclusively on the best meteorological data to be found on the weather charts and the good warrant officer, presumably, wouldn't have to shrink from saying he had taught me everything I knew. In this station, the forecaster had no such worries. As far as windows were concerned, he was to be kept in the dark.

Back at Westover in the years 1947 to 1949, I thought there were more maps being analyzed than I would ever see again, and there *were* a lot of maps, but Kindley's Base Weather operation in 1957 outdid even the Westover performance. To handle all the map analysis and briefing requirements, the forecasting staff was organized into shifts of three to four forecasters each. In addition to the standard surface, 850 millibar, 700 millibar, 500 millibar and 300 millibar charts, the analysis routine also included the delta H or height change chart of the 500 millibar analyses. The delta H charts had been introduced to me at the Specialized High Altitude Forecasting Course at Chanute in 1950, and I had then decided they were not of much, if any, value. I continued to hold that they were essentially a waste of time, time that could be better spent in concentrated study of the forecast problem. For a long time I made no impression on anyone who might be able to delete their analysis from the long list of forecasters' duties. (They would, however, be dropped before I left Kindley.) Similar to the mathematical front-moving formula we had practiced with in the Weather Forecaster Course in 1948, the delta H charts told you where you had been — not where you were going. We also created prognostic charts for the surface, 700 and 500 millibar levels. Kindley Base Weather was a first-class map factory.

Because there were many who wished to visit that island paradise, including what seemed to be an unusual number of headquarters inspectors (if you had a choice, would you rather inspect a base that was under a foot or two of snow?), our customers covered the waterfront. We provided weather and flight-level wind forecasts for MATS (Military Air Transport Service) aircraft, KC-97 air refueling tanker planes, Navy aircraft of various types including blimps, Air Force fighters, Air Weather Service WB-50 weather reconnaissance planes, RAF planes, RCAF planes, commercial airliners flown by KLM, Pan Am, TWA, British West Indies Airways, Avianca, British Overseas Airways Corporation (BOAC) and others. Their destinations covered the compass: from New York to Florida, Canada, the Caribbean, South America, the Azores, North Africa, Europe and even moving points at sea — aircraft carriers. We were responsible for providing weather support for every plane leaving Kindley — military or civilian — and for others overflying the island.

The mental picture that most Stateside folks have of Bermuda is that of a perpetually sunny isle where celebrities like Mark Twain had gone to enjoy the weather. As an advertisement of an old steamship company that probably plied the Bermuda waters in Mark Twain's time had put it, the "Bermuda Islands (were) a Convenient Picturesque and Salubrious WINTER RESORT."[1] While we were there, tourists from the US would flock to the island in winter, riding Mopeds even when the weather was rainy, as was not unusual in the winter. But the tourists didn't mind, the weather was still gorgeous when compared to Stateside.

The winter of 1957-58 was a period when the 11-year cycle of sunspot activity, a phenomenon known as far back as the 1840s, reached a

---

[1] Published by the Quebec and Gulf Ports Steamship Company, New York and Bermuda Division. Although usually thought of as only one island, Bermuda actually consists of a number of small islands, the principal ones being connected by bridges or causeways. Their entire extent is only about 20 miles.

record maximum. The event was accompanied by an unusual amount of squally weather over the North Atlantic, making it a most unfriendly place for planes or ships at sea. Weather reports transmitted from ships reported 40-foot waves and turbulence encountered by aircraft was almost routinely no less in severity than moderate and often severe. For weeks, the ocean to the east, west, north and south of Bermuda was covered with thunderstorms. Electrical discharges from the cumulonimbus and space radiation bursts from solar flares were interfering with radio transmissions. These communication interruptions often left us without current weather reports. Also absent were the coded local weather forecasts, called "TAFORS," made by the forecasters at those distant bases to which we were sending aircraft. Forecasters sending planes to Sidi Slimane, in Morocco, a base whose actual weather data and forecasts were sometimes missing even when communications to Europe and the US seemed adequate, were particularly vexed by what then amounted to a total data blackout.

One night, Ron Jackson, a friend on my forecasting team, who became one of the first senior master sergeants when that "supergrade" was established, was a Navy veteran of the Pearl Harbor bombing, and was from Riverside, California, informed me of an interesting coincidence. After many comparisons with the forecasts from his home base at Riverside, he discovered there was a strong correlation with the forecasts made by the forecasters at Sidi Slimane. When the Riverside forecaster was forecasting low stratus ceilings and fog, it was a good bet the Sidi Slimane forecaster would be doing the same.

It so happened that on a number of occasions we could receive the Riverside forecast but not the Sidi Slimane forecast. Because at those times the last actual weather observation from Sidi Slimane was often too old to be of any value, it was jokingly suggested that the Sidi Slimane forecast correlation with the Riverside forecast might be employed to get the Kindley forecaster out of his missing data jam. Of course, the Air Weather Service forecaster had had the scientific method drilled into him too deeply to ever seriously contemplate that expedient. But it does occur to me that we might have, in the Riverside-Sidi Slimane paradox, an opportunity for that graduate student who is looking for a new field of investigation. The student should be reminded, however, that it was not the actual weather nor the verification of the forecast that seemed to correlate — it was only the forecast.

In the wintertime, the great temperature difference between the frigid continental polar air that swept off the US and Canadian shores and the warmer waters of the North Atlantic was a perfect recipe for vigorous convective activity. Thunderstorms whose tops were unusually low — only in the 10,000 to 15,000-foot range — erupted everywhere over the routes to the west, northwest, northeast and east. This meant that the passengers and crew of those piston-engine aircraft flying to or from Bermuda were going to experience a variety of turbulence and of atmospheric phenomena that long would be remembered. An RAF lieutenant was one who would remember.

This RAF pilot had just flown through those energetic convective cells in a small twin-engine bomber on the long haul from the Azores. To take advantage of less severe head winds he dropped down to 6,000 feet. He found there would be a big problem with that altitude — the turbulence. But he had to stay there because of his fuel limitation. And then there was the electrical phenomenon. "The aircraft," he said, "literally glowed with St. Elmo's fire — the wings, propellers, the radio console, all had radiated an eerie light."

As he described that flight to me, it became clear that this pilot and his two man crew, the copilot and the radio operator who had been operating that glowing radio, had been stunned by their experience. To say that it was nerve-wracking wouldn't begin to describe it. They must have been relieved beyond measure when our tiny island came into view and probably offered prayers when their wheels touched terra firma.

I asked that RAF flyer, who was still ashen-faced from the ordeal as he talked, how he would have characterized the turbulence. "Would you say it was light, moderate or severe?" These were the only categories to which we were allowed to assign turbulence when we briefed or debriefed pilots. If, for example, a pilot said: "It was rougher'n a cob." We would have to pester him until he would agree it was either light, moderate or severe. I explained to this pilot that I had to fit his turbulence into one of the three categories. I already knew, of course, that it wouldn't have been light or moderate. But he would have nothing to do with that limitation.

"No," he said. "Not severe. The turbulence was utterly fantastic, just utterly fantastic!"

I debriefed the "leftenant" on the winds he had experienced and turned to attend to other chores — this was a map factory. But he did not leave. When I looked up from my map, he began to describe again that long night flight from Lajes to Kindley. As if fearing that I might think he was one who could too easily be impressed by the weather he had flown through, he repeated and then repeated again the details of the turbulence and of the St. Elmo's fire. When he finally left, I knew without the slightest doubt that I would not want to be in a plane decorated with St. Elmo's fire, and I also knew that "utterly fantastic" was a new and much higher category of turbulence than any I had known of before.

One of the interesting aspects of dealing with pilots and their planes over an extended period of time was that certain aviators seemed to sort of become identified with the kind of plane they flew. When you thought of a particular aircraft configuration you would also think of a specific pilot. And some pilots, it seemed, even came to physically resemble their planes. When I looked at a C-124, a four-engine piston-driven plane that was then the cargo workhorse of the Air Force, I thought of a certain pilot who was built like a professional football defensive lineman. He was wide in the shoulders like someone who might have spent a dozen years lifting weights, but also wide in the middle. Chunky, but not what we would call fat, he would in no way be taken for Steve Canyon, the handsome comic strip aviator. He had the thick hands of someone who had known heavy work before working the cockpit levers. He wore the folding overseas cap and a plain military jacket — no ribbons to pretty it up. This captain's plane was the C-124.

Now if you had seen a C-124 lift off a runway 50 times, you would still be amazed, at the 51st time, that that machine could fly. It just didn't look as if it could fly, and it was said to provide a somewhat uneven ride. Within our little forecasting circle, it had acquired the nickname "Old Shaky." One day, while debriefing that captain, the C-124 pilot, I asked if he would characterize the turbulence as light. I had a recent report from a Connie indicating that the turbulence was light over the route just flown by the C-124. "Hell," he said, "the C-124 has *built-in moderate* turbulence."

I always felt comfortable briefing that C-124 pilot. He really seemed to know his airplane, and I was sure he was a pilot who could be depended on to do the wise thing when the weather turned sour. It was therefore a bit of a surprise to me to have him say he wanted to take off for Lajes on a day when a hurricane was poised over that route. Over the past couple of days, I explained, all the other pilots who wanted to fly that route had decided to wait out the storm. He wanted to know if I had any pilot reports that would confirm the bad weather, and of course I had none. All the other aircraft that might have headed that way or might have come to us from Lajes were standing down. And our WB-50 Hurricane Hunters' plane of the 53rd Weather Reconnaissance Squadron was not flying into the storm that day. He interrupted me before I could finish my little speech.

Leaning over the map that I had tried to resume analyzing, he looked directly at me. "Listen," he said, "that hurricane's no problem. The airscrews on that baby (he demonstrated with his thick fingers the spinning motion of the 124's props) will bite into that wet air and pull us right through it." I didn't think the hurricane was as vigorous as it had been, there were some indications that it was in a weakening stage. And I didn't think I was dealing with a pilot who had a date with an angel (as in the Westover safety poster). So I signed his clearance. He took off for Lajes. Made it all right.

And the next time he saw me, he smiled as if he had a secret that he wouldn't tell. He didn't say anything about the hurricane episode. I assumed that Old Shaky had shaken him up a bit.

Hurricanes that might have been a threat to Bermuda itself were of course a major concern. When a hurricane was forecast to pass over the island, Kindley's planes were evacuated to the mainland, usually to Charleston, South Carolina or Dover, Delaware. This was an expensive reaction to a man's forecast and a source of stress for many of us.

For a number of years now, the public has been accustomed to satellite images of hurricanes obtained from the National Hurricane Center in Miami, and so it is rather common knowledge that it is more difficult to predict the track of a hurricane than that of an extra-tropical low pressure cell, the typical weather system of the temperate latitudes. It was even more difficult in the late 50s when satellite images were not part of the observational package. We could pick up the hurricane on our CPS-9 radar, but only when it had approached to within about 150 miles. By that time, the plans for plane evacuation had to have already been made.

One hurricane that approached close enough to be picked up on our radar vividly illustrated the difficulty of predicting a hurricane's track. This storm, like many in that area, moved up from the southeast. For several days, as it drifted in close to us, visibility was somewhat diminished by salt-spray haze and the surf on Bermuda's south shore began pounding with increasing frequency, both phenomena commonly seen as a hurricane approached. We were convinced this hurricane would at least brush Bermuda and began issuing forecasts advising everyone to prepare for damage. Island residents boarded windows to protect against flying objects and filled their bathtubs to provide water should a power outage disable the pumps that brought water up from their cisterns. (Each home had a cistern to store rain water that was collected on the white-washed, limestone roof. Privately collected water off the roof top constituted the "island water works.")

While these protections were in progress, in the southeast quadrant tell-tale spiral bands showed up on the weather radar. The bands radiated out from the storm's center which now was astraddle the radar's 100-mile marker (a bright circle on the radar screen's green background). Island wind gusts picked up to about 45 miles per hour. The storm then began to curve toward the northwest. This turning was expected, but we were surprised to see that the storm's center was staying on the 100-mile marker. Incredibly centering itself exactly on that 100-mile marker, it then began curving northward, then northeastward, eastward and southeastward. For someone who believed in the occult, it would have appeared that the storm was lost as it approached the island and, finding the 100-mile marker, began to follow it as one follows a highway route on a road map. Just when I began to think the path of this storm, which had been downgraded to a tropical storm, was going to trace a perfect circle around our island, a maneuver that would have been even more unbelievable, it shot out to the east toward the Azores. Memory now suggests that this could have been the same storm that ran across the path of a German naval cadet ship, sinking the ship with the tragic loss of those young sailors.

In our efforts to predict landfall of hurricanes we had competition from the shoeshine man who had his stand set up next door, outside the airlines terminal. For the future track of hurricanes and that BIG QUESTION — whether the storms would hit the island — we relied on our skill in using the steering winds of our upper air charts. The shoeshine man employed an entirely different method. When a ship or our Hurricane Hunter's WB-50 found a hurricane, or even if one was rumored to be out somewhere over the ocean, he would consult a vial of shark oil that many island residents had hanging on their walls. The shark oil vial even could be found, as I would later learn, at the Royal Bermuda Meteorological Office. If the shark oil was clear, the weather was to be good. If the oil was cloudy, bad weather was on its way. This was the Bermuda equivalent of the Weather Wizard that many forecasters had received as gag gifts. That one consisted of a witch and a Dutch boy and girl who were either within or outside a little house. Operated by a humidity-sensitive mechanism, the witch would be outside when the weather was to be bad, the boy and girl would be outside when the weather was to be good. I was never able to understand the basis for shark oil as a weather predictor, but I had no trouble with the shoeshine man's spider-anchoring-its-web-to-the-ground theory. Also according to him, if a spider was seen anchoring its web to the ground, a hurricane was on its way. That one made sense. If you were a spider and a hurricane was coming, it would have been entirely logical to anchor your web to the ground.

Information as to the intensity and location of hurricanes came to us from the weather recon WB-50s. After briefing the crews a number of times on flights into the center of a hurricane, I was surprised to learn that the intrepid Hurricane Hunter aviators did not seem to regard a flight into a hurricane as unusually dangerous. On their way in and out of the storms, which they usually penetrated at the 10,000 foot level, the Hurricane Hunters calculated wind speeds and wind directions and released dropsondes. An opposite of the radiosonde in its direction of travel, the dropsonde floated down on a parachute. Like the radiosonde, it was fitted with a radio transmitter and sent back the same kind of information — temperature, humidity and pressure.

When not flying the hurricane missions, the WB-50s flew a set course that extended far to the north of Bermuda to a point southeast of Nova Scotia and back. The information from the dropsondes they sent back and their calculated flight level winds, they usually flew near the 500-millibar level, was important to the forecaster at Bermuda and the forecasters at other bases.

One terrible day when one of the WB-50s was well north of Bermuda, the plane and its entire crew, including several of my good friends, vanished. We were told a Japanese ship had seen a flaming object fall into the ocean at about the time the WB-50 disappeared, but whether what they had seen was our weather plane was never established.

Years later, when the so-called Bermuda triangle became a subject of popular discussion, I thought of our weather plane and of two other unexplained aircraft disasters that had occurred while I was at Bermuda. One of these involved a Navy plane that was inbound from the Azores. It went down a considerable distance south of Bermuda. We guessed its navigation equipment was giving the crew erroneous readings or perhaps was inoperable, and they ran out of fuel while searching for our little island. In the third incident, a lieutenant who took off from Kindley in a century series fighter blew up on climb out. This plane would be the only loss of the many thousands I would clear and for a time it shook my belief that all pilots who got their weather clearance from me were the beneficiaries of the unaccountable good luck that seemed to attach to me.

Like the lieutenant of the Traverse City thunderstorm of some ten years earlier, this lieutenant had been in and out of the station a number of times, in fact, he had been waiting to take off for days while having maintenance done on his plane and in the meantime came into the station ostensibly for yet another check on the weather but really just to participate in conversation. It was clear he loved flying. And this would have been hard to explain to others, but I developed a rapport with this pilot quite unlike that that usually developed. Even though counted only in those few days, it was an acquaintanceship that seemed like those one would expect to last a lifetime. When it happened, the sky was clear and the visibility unrestricted; and I remembered thinking, as he prepared to leave, that a pilot who loved flying could have had no better opportunity for a takeoff into a gloriously blue sky.

Weather satellite technology and computer produced progs made their appearance while we were at Bermuda. These were received by FAX. Those early satellite cloud pictures, whose descendants would become almost indispensable to the forecaster in the years to come, were not yet of a quality that would make them very useful to the forecaster. The same was true of those early computer-produced progs. The progs, produced by the National Meteorological Center at Suitland, Maryland, did, however, hold out the promise of that great day coming for the weather forecaster — the day when he would be freed from the long, tedious hours of constructing prog charts. The map factories, like Kindley and Westover, might then become something of interest only to a historian.

Those first computer-produced progs were constructed on what meteorologists called the "barotropic model." Computers had not yet reached the stage where they could handle the complex mathematical formulae necessary to define the state of and changes to the atmosphere. They would be greatly improved in the years to come, but at that time, the best that could be done was to accept certain simplifications. As it was explained to us, among them was the assumption that wind directions would not change with height and that air of a different temperature would not be advected. Of course, every forecaster knew forecast maps constructed under such constraints would be somewhat disappointing.

The barotropic model would often predict a tail wind at the 8,000- to 10,000-foot level for the flight from Bermuda to Charleston, South Carolina when the actual wind was sometimes a head wind. Any pilot will tell you that is a serious forecast error, especially when you are dealing with long over-water flights. Reluctantly, we were forced to then disregard the barotropic forecast charts.

Speaking of wind forecasts, we were faced one day with a very disturbed navigator, a lieutenant, who marched into the weather briefing room holding high the wind forecast he had received at a Stateside base. I wanted to ask him about the weather he had just flown through, but he didn't want to talk about that. He wanted to talk about the wind forecast he had been given. Flustered, red faced, eyes popping, he shouted: "The winds are all wrong! You guys don't know what you're doing!"

I suspected that the forecaster at his departing station had given him winds off the barotropic prog, and I tried to engage him in a calm conversation about those winds, but he was too agitated to allow it. The forecaster in charge of our shift, a major, then stepped up to the briefing counter and asked : "How can the winds be wrong?"

Of course, the navigator was far too agitated to be influenced by the major's attempt to create normal dialogue and restore professional decorum. He waved the wind forecast above his head and repeated: "The winds are all wrong!"

Now I did not subscribe to the notion that as one ascended the rank ladder in the military, wisdom was also in the ascendancy. But the major's next remark made me think that it might well be true. He looked directly at

225

the navigator and said: "Lieutenant, winds can't be wrong. Lieutenant, **you can't gig God!**" And that proved to be the calming remark.

I do not want to give the impression that I considered the work of navigators to be trivial. Quite the contrary, I depended on their expertise on thousands of flights, including some where I was one of the aircraft's passengers. In fact, I admired their work, and had navigator friends, but theirs was a profession, like that of the weather forecaster, that sort of encouraged the telling of the humorous but mostly untrue incident. One such occasion occurred on the day a twin-engine plane, a T-29, flew into Bermuda from a Stateside base with a load of student navigators aboard. These students were probably all quite-new lieutenants. On that training mission, they were to compute in-flight winds based on their calculations of the plane's position as it flew to Bermuda. In addition to the normal side windows, there were a number of celestial domes, a row of windows along the top of the fuselage, that could be used to shoot the stars. Knowing there had been an abundance of navigating talent on that plane, I could hardly wait to debrief the pilot, for to tell the absolute truth, there were times when we suspected, probably entirely without foundation, that the navigator of a particular plane that had just landed might have been dozing when he should have been figuring his winds or perhaps had lost a figure or two in his computations. On this plane, there'd be a lot of corroborating calculations, so the winds would be quite infallible. When the T-29 pilot walked up to my forecasting counter, I asked him what winds he had. He responded: "I have no idea. I had 15 navigators on board, and they didn't know where we were!"

No group of pilots or navigators ever took their jobs more seriously than the pilots and navigators of BOAC (British Overseas Airways Corporation). They appeared to us to be more military-like in their work than the military itself. As one example, the first time I stood outdoors and saw the BOAC crew drive up to the station, two vans were used. The captain, all alone save for the driver, was in the first van. The rest of the crew rode in the second van. Although there might have been some other reason for it, it looked to me as if those traveling arrangements were intended to serve as a reminder of who was in charge.

BOAC flew the Boeing B-377 Stratocruiser, the commercial airliner version of our Air Force KC-97 Stratotanker, on weekly runs to London, some 3,800 miles from Bermuda. Powered by four Pratt & Whitney 3,500 hp engines, it had a cruising speed of 340 mph.

The first request for a wind forecast to London at their flight altitude of 17,000 or 19,000 feet, was made several days in advance of the flight. A navigator or flight planner would stop by the station, pick up the wind forecast that had been prepared for longitudinal zones, briefly discuss the winds in that courteous British way and return to the BOAC office which I believe was in Bermuda's city of Hamilton. They would then work up an overall wind factor for the flight — a wind factor of plus 27, for example, meant that the resultant wind over the entire flight was effectively a tail wind of 27 knots. Within the next day or two, the procedure would be repeated. Based now on the prog chart he had updated, the forecaster would work up a new set of winds for the same flight. The navigator or flight planner would pick up the winds, return to his office, and work up a new wind factor. This would all be done again in the immediate hours before the flight to London.

That triple wind forecast requirement was an unusual procedure by American standards. It was unquestionably thorough and that would have been a source of comfort for the passengers, had they known about it. But it tended to make the forecasters a bit nervous. We feared that the flight planner had loaded the aircraft to the very last knot of our forecasted winds — and what if we were wrong by a knot or two? However, they obviously had built in their safeguards, for I never heard of a BOAC B-377 that got into a low fuel predicament on that long, eleven-hour flight.

In the course of that three-year tour, there were one or two little single-engine planes that landed at Bermuda whose pilots also aspired to cross the Atlantic. These were prop-driven planes such as you would find at your typical small-town airport. When the tail winds were good, these intrepid aviators would leap off the island, heading east, ambitious to cross the Atlantic in their version of Lindbergh's famous flight (but of course, he had done it non-stop).

My most interesting flight clearance at Kindley could have easily been this one: A pilot placed his Form-175 on the forecaster's counter and said to me: "Fill in the weather and the winds."

I looked at the 175 and asked: "Where you going?"

"Can't tell you," came the reply.

A strange response, but you got used to all kinds at that forecasters' counter. I tried again. "What's your altitude going to be?"

"You don't need to know."

That last response triggered a thought. In the military, when the phrase "need to know" came up, you were probably dealing with secret information, and I remembered that Eddie McDaniel, my friend who was a crew chief on one of the KC-97 aircraft, had told me just a day or two before that the Air Force's most secret plane was then parked on the field.

This pilot was telegraphing exasperation that was born, evidently, out of my slowness in grasping the importance of his mission. "All you need is the winds," he said, "and I've got them right here." He handed me a slip of teletype paper. Three or four wind directions and speeds had been typed on the slip. There was nothing to tell what altitude the winds were meant to apply to nor what route he'd be taking out of Bermuda.

I stood there looking at the so-called "winds," trying to formulate yet another approach to narrow the information gap. He didn't have time for that. "Just put the weather on the clearance," he said, "I'm going to the snack bar and will be back for it in a few minutes."

Among the many thousands of weather clearances that I would complete, this was the only time I'd be faced with that kind of situation. The forecaster's unwritten but time-honored SOP (standard operating procedure) didn't cover a situation like this. We always knew, indeed we *had to know,* where the aviator intended to land, when he intended to land (his ETA) and how high he'd be flying.

And it was also important to know the kind of plane we were dealing with — this would give us an idea of its sensitivity to turbulence and its capacity for landing at other landing fields should the weather at the destination or the alternate airfield go sour. But I clearly saw I had all the information I was going to get from this particular pilot, and it left me in a quandary. How could the clearance form be handled?

I compared the winds he had given me to those on our 850- 700- 500- and 300-millibar charts. His winds bore no resemblance to any wind field on those charts. This pilot was either flying in his own non-parallel universe or his forecaster was the ground hog. As I considered the wisdom of entering his winds on the clearance, a bit of hurt professional pride welled up. Aside from the doubt as to their accuracy, I had never before entered winds on a clearance form that I had not developed myself or had at least verified against the charts.

Reluctantly, I took up my Form-175 pencil. In the sections for turbulence, icing and winds, I wrote: "Unknown at the unknown altitude and flight route." For destination weather: "Unknown at the unknown destination." And in the forecaster's signature block: "Unknown prognosticator."

I laid the 175 on the briefing counter and walked into the radar room to see if there were any cloud buildups in the area. If there were, I'd bring them to the attention of the enigmatic aviator, at which time, I was sure, I'd be upbraided for the way I handled the 175. The CPS-9 scope was clear.

On my return to the forecaster's counter, I saw the 175 was gone, taken, apparently, by an unknown pilot who was going to climb to an unknown altitude, fly off in an unknown direction and, at some unspecified time, land at an unknown destination.

If my weather station had windows, and of course this one had none, I would have looked out to see if I could identify the unknown plane. I thought of walking outdoors to look down the runway when the phone rang. The length of the ensuring conversation eliminated my chance of seeing the remarkable plane that most of you will have guessed the identity of.

The caller was Dr. Mackey who was head of the Royal Bermuda Meteorological Office. Dr. Mackey, a polite and unflappable scientist with an ever-present sense of humor, had had the misfortune of having his meteorological station burn down. To continue providing weather forecasts for Bermuda's citizens, he called us regularly to receive a description of the meteorological charts and a reading of our forecasts. The forecasts then were relayed by him to be broadcast on island radio and television, with, as I remember it, any additional wording that he might contribute. His interruptions were always pleasant diversions from the work at hand. On this day, I described the synoptic situation, explained my reasoning in preparing the forecast and, in deference to his long experience in Bermuda forecasting, asked, as I usually did, if my forecast seemed reasonable to him. Being a British gentleman, and perhaps not wanting to undermine my confidence, he said it did, as he always did. As I prepared to hang up the receiver, the good doctor added: "And, by the way, the shark oil also agrees with your forecast!"

Top: The C-124 Globemaster cargo plane. This one is on display at the Pima Air and Space Museum in Tucson, Arizona.
Bottom: Bermuda homes were built of limestone blocks sawn out of the island's rocky surface. Thinner slabs were used for the roofs as seen in this photo. Roofs were whitewashed and the rain collected on them was diverted to a cistern under or beside the house — that was the island's water supply system. *Photo by Marie Cogut.*

Top: One of General Eisenhower's Constellations is shown here on display at the US Air Force Museum, Wright-Patterson AFB, Dayton, Ohio.
Bottom: The Boeing B-50 Superfortress was an outgrowth of the famed B-29 and looked quite similar. It had significantly more powerful engines than the B-29 and a higher vertical fin. This one is on display at the Pima Air and Space Museum in Tucson, Arizona. A C-97 is the plane partly seen at right.

At the bottom of the pilot's clearance form there was a block to be filled out by the forecaster. This example was made out at Kindley AFB in 1959. For destination weather at Bermuda (call letters ZQUK) the forecaster has indicated 1,500 feet scattered (abbreviated to 15), 10,000 feet scattered abbreviated to 100), high overcast (the slant mark indicating 20,000 feet or higher) and 12 miles visibility. For the return to Bermuda, he is forecasting 1,000 feet broken, 10,000 feet overcast, 2 miles visibility with light rain showers (abbreviated RW-), and northeast landing winds of 15 miles per hour with gusts to 25 miles per hour. The teletype machine print hammers were manufactured with symbols for the three types of cloud conditions, the symbols used in this example. They were:

⓪ scattered, 1 through 5/10 sky is covered,

⓭ broken, 6/10 through 9/10 sky is covered,

⊕ overcast, 10/10 sky is covered

Chapter 21

We Found Paradise, Almost

During that three-year Bermuda tour, we lived in five different houses. For our first night on the island, our sponsor had found us lodgings at Harrington House which was located on Harrington Sound. The fact that Harrington House had a name and no house number we would find to be universal on the island — Bermuda had no house numbers.

On the second day, we moved to the top floor of an apartment, called "the penthouse," at a place called "Spanish Point." It was on the north shore of the island near limestone cliffs where graceful white birds called "long tails" swooped and swerved above the breakers, soared up along the high rock walls and merged with the equally white, cottony cumulus. The apartment itself was named, appropriately, "Oceanview." As the name implied, from our penthouse windows we had a great view of the Atlantic.

We had arrived on Washington's Birthday, in late February of 1957, and it was rather cold by a native Bermudan's standards, but for this family just arrived from the Michigan cold, it could only be described as balmy. By the time our second winter rolled around, however, we were acclimated and, like native Bermudans, we would think the weather was cold and were astounded to hear tourists marvel at the warmth.

Before reporting for duty with Kindley's Detachment 10, 9th Weather Group, I visited the Austin dealer in Hamilton, the capital of this British crown colony, to look at a tiny Austin A-35. I say it was tiny, but that was only by American standards. Actually, it was about the same size as all other cars on the island. This was the saloon model, a family car even though it seemed small to us, and more practical than the sporty red MG I really wanted. The dealer opened the bonnet (the hood) to show me an uncluttered engine that I could see would be a joy to work on. Changing the spark plugs and resetting the distributor points gap, something that I always did myself, would be a snap. The boot (trunk) too, he assured me, was large enough to carry groceries home from the commissary. We were convinced and bought it. As I started to drive it out of the showroom, in an accommodating touch that I would find typical of him, the dealer handed me a bottle of touch-up paint which exactly matched the car's gray color. (A few months later, when the car's starter failed without warranty coverage, he removed the starter, rewound it, replaced it in perfect condition, and would not take the payment that I was insisting he take). Thereafter, at every car washing, I took out that little bottle of paint and, with a tiny brush, covered any nicks that had appeared since the previous washing. We were going to take exceptionally good care of our first new car.

That '57 Austin A-35 seemed to be about half the size of the '55 Ford we were forced to leave in Michigan because it was too big to meet the vehicle size-limitation law, one of Bermuda's automobile laws that would seem strange to the newly arrived American resident. The only car then being produced in the US that was small enough to be allowed on the island was the bug-like Nash Metropolitan.

The little Austin proved to be sensible transportation on Bermuda where roads were narrow and the speed limit only 20 miles per hour! Since there was little likelihood that one could suffer a total loss of automobile on an island where no one could drive over 20 miles per hour, I toyed with the idea of either buying no collision insurance or buying reduced coverage. It was not, however, a completely loss-free environment. Since automobiles were rather new to the island, only having been allowed for about 10 years, there were a number of inexperienced drivers. Still circulating was the story of the first two automobiles that were brought to Bermuda in 1947. According to that account, the two horseless carriages, the only two on the island, suffered a head-on collision on Front Street, the main thoroughfare of Hamilton, the colony's capital city. With those thoughts in mind, I eventually purchased full coverage for the first year, telling myself I probably would reduce the coverage in each of the succeeding years.

Some American service families complained about the very low speed limit and also about life in general on Bermuda. To us the attitude was unexplainable. We were determined not to join in their grumbling. When they complained about the weather, one had to admit they had a bit of a point. You did feel cold in winter after your first winter and the summer temperatures seemed at times a trifle warm because of the island's high humidity. But the surrounding ocean so moderated the summer temperature that the mercury never rose above the 80s — there weren't many places Stateside that could boast of those kinds of summers. And the winter temperatures, after all, would not dip below the low 50s even on the coldest days.

While some pined for the familiar surroundings of a Stateside community, we were having the time of our lives! In that new little Austin, we drove around the island, through Blackwatch Pass, over the tiny bridge at Somerset, to the Royal Dockyards, to St. George's ducking stool and the oldest church in the Northern Hemisphere. At Easter time, we stood on the cliffs near Spanish point looking for whales — our Bermudan friends insisted that whales appeared there every Easter. They were not in the slightest disappointed to see that the whales forgot that appointment during the three years we were on the island. Like the Bermudans, we did not lose the faith, making our appearance at the cliffs each Easter. We took the girls to the Saturday's kids bingo at the NCO Club where Willa announced she was going to win the super jackpot, and *did.* The super jackpot was a cleanup prize of all the toys that had been left over from a number of bingo games. The toys covered an entire banquet-sized table, so many that we had to have friends help us bring them home in their car, ours overflowing with Willa's loot.

In Hamilton, we watched the Gombey dancers on Front Street and, at dockside, boarded the *Queen of Bermuda* to attend a bon voyage party for departing friends. One of the forecasters took me on a pleasure cruise out into Hamilton Harbor where his boat engine died and we were left stranded for a whole afternoon while passersby in other boats waved at us in good cheer, thinking our frantic motions were just our way of expressing joy at being lucky enough to be in the harbor on such a glorious day. Out on the ocean on another day, I tried deep sea fishing in the boat chartered by the weather detachment, and immediately got so seasick I was half hoping the detachment clerk would pound a hole in the bottom of the boat as he swung at, with a baseball bat, and mostly missed, a wildly thrashing shark that he had pulled into the boat. (His aim may have been compromised by the case of Heinekens that had been loaded aboard.) If he had punched a hole in the boat, I knew we'd either head for shore or drown, and I was too sick to think which would be the better choice. Nevertheless, when I recovered, I looked back on those boating incidents as grand experiences.

After the midnight shifts, Marie and the other wives would meet their off-duty husbands for golf at the St. George's Golf Club where servicemen got a break in green fees and where I learned the game (somewhat). On St. George's first hole, a blind dogleg right over a large hill, you'd usually look in the cup thinking you had a hole-in-one but instead would see you had lost your ball in the Atlantic, the "rough" on the back edge of the green. This was on Bermuda's eastern end where an old coast artillery emplacement, Fort St. Catherine, stood guard. There were countless visits to Bermuda's picturesque south shore — John Smith's Bay and Horseshoe Bay. Before ever thinking we might some day walk on this enchanting island, we had admired the travel posters of Bermuda which usually depicted fabled Horseshoe Bay and its pink beach made from an infinite number of sea shells pulverized by eons of pounding waves. Beyond the pink sand, awe-inspiring juxtapositioned patches of blue and turquoise waters beckoned. This had all once been but a dream. Now we were here, on those very beaches, the girls romping through the surf while we, seated at an easel, watched warily for the undertow and tried capturing in oils the enchantment. We *were* having the time of our lives!

And we even managed to upgrade our lodgings, moving to avoid the noise of children playing in the courtyard at Oceanview, not a good sleeping situation for a shift worker. We went to a single family home called Homeleigh Cottage in Warwick Parish. (The island was divided into parishes. From west to east these were: Sandys, Southampton, Warwick, Pembroke, Paget, Devonshire, Smiths, Hamilton, St. George's. Oceanview was in Pembroke Parish.) Then another move to a cottage closer to the airbase, in Smiths Parish, this one named Lion Rock, after a prominent limestone formation in front of the cottage that resembled a lion. With that last move, we knew we had arrived at the best accommodations possible. At Lion Rock, which was on Harrington Sound, the opportunities for swimming were great. There was just one little concern: the water was 12 feet deep at the very edge and the girls did not know how to swim. But I knew they would soon learn.

"I can give you three inner tubes," said the attendant at the base gas station. He had glanced toward the back seat of the car to see my three girls as I asked if he had a couple of old inner tubes the girls could use while swimming.

"Thanks," I said. "But I only want two. Two would be just fine."

He walked back into his garage and came back with three inner tubes. I repeated that I only wanted two. "But I have lots of them," he protested. Probably because he had more old inner tubes than he knew what to do with, he was urging me to take more off his hands. "You can have three, or even more. I can see you need three."

"We do need three, take three," a chorus of girls' voices from the back seat agreed. But I made him take one back. You see, I had a plan. First, the water immediately in front of our house at Lion Rock was, as I've said, 12 feet deep. If the girls were going into that water, they'd either be floating on inner tubes or actually swimming, and they did not yet know how to swim. But if they had only two inner tubes, with two of them leisurely floating, the one standing on the rock bank would get so flustered she'd take that calculated risk of jumping in and grabbing for a tube when the unwary floater drifted too close to the bank. In the process of thrashing through the water that girl would learn how to swim. And that's what happened. All three girls learned how to swim there and swim well. If you discount the sheer ingenuity of my plan, their swimming expertise came about without any real help from me. Of course, I loved my offspring too much to allow any one of them to jump in when not in my presence. But the plan was so well constructed that I never had to interfere with a serious rescue. At least that's the way I remember it.

So we went on with our lives on beautiful Bermuda where the charming pastel-painted houses all had white roofs. Where passion flowers grew wild on the hill behind our house, where a single two-inch-long yellowish-orange goldfish swam in a two-foot-wide glass-fronted tank a clever mason had built into the limestone wall beside our backyard walkway, and where deep-orange five-inch-long goldfish, with those attractive floppy fins, swam in the five-foot circular pool built beside the sideyard stairway. And we drove under a pink-flowered canopy of oleanders on our way to painting excursions or picnics on Horseshoe Bay.

Nowadays, it seems, it is a rare family that does not have at least two automobiles, but that was not the case in 1957. In fact, it was one of my pre-Bermuda dreams to advance myself economically to the point of becoming a two-vehicle family and I didn't then know when, if ever, it would happen. But one day on Bermuda I ran into a bargain I couldn't pass up. A sergeant in the weather detachment's maintenance section was rotating Stateside and offered me his 4-cycle German Fox motorcycle for $100. I bought it.

After driving it home, somewhat erratically (it was my first motorcycle), I became concerned with its harsh metallic sound and upon dismantling the engine found that the short drive-end of the crank was out of round. Of course, that was why I had acquired it at so low a price. There was no way to make a repair so I reassembled it as it was and made a mental note to always be absolutely sure there was plenty of oil in the crankcase. In all the time I had it, it continued to produce that metallic sound that would cause me to seriously throttle down as I approached my neighbors in order to not so conspicuously announce my arrival, but it never faltered and it provided a fuel efficiency of 120 miles per gallon of gas.

We were now a two-vehicle family. During good weather, I rode the motorcycle to work and Marie drove the Austin to take care of family errands. Seven-year-old Pamela was the first to risk a ride on the back of the bike. Willa and Leta followed in this experiment of motorcycling, but I got the impression they would rather be conveyed in the Austin. Marie shunned the bike completely. Little did she know that that was going to change and change dramatically.

One fine Sunday morning, our idyllic lifestyle came skidding to a halt. I had stepped out on our front porch to take in the morning breeze coming in off blue Harrington Sound. There were sailboats on the sound and long tails doing acrobatic turns above. They reflected themselves on the water surface as they dove toward and then swerved upward from the sail boats. My next-door neighbor, who had just finished 9-holes of golf, drove up to his parking spot, got out of his car and called up to me: "Looking for your car?"

Until he asked, I had not noticed the void in my car's parking spot. The Austin A-35 was gone. At a moment like that, of course, we tend to fear the worst. I remembered I had not long ago downgraded the Austin's insurance coverage because the car had passed its first birthday and its insurance was then up for renewal. It was no longer covered to the extent of a total loss. And why should it be? We were on an island where, for Heaven's sake, the speed limit was only 20 miles per hour. It was hard to imagine a total loss, even in a head-on collision. But what had happened to our beloved Austin?

"I think I saw it," my neighbor continued, "down in the water off the ninth hole at the Mid-Ocean Golf Club."

I thanked him and called the local constabulary. In a matter of only minutes a constable was at our house. He had driven up on a motorcycle rather than the black, fast Sunbeam Talbott that the constables usually were seen in. "Do you have transport?" he asked.

"I have a motorcycle."

"Climb on it and follow me."

One of the benefits of being a police officer that may be little appreciated by the general public is that when you want to go somewhere, you can drive as fast as you damn well please. This is probably true everywhere in the world. I had to really rev up the old bike to keep up with the constable. On this island where, in the 18 months of my residence I had never (with a few minor exceptions) driven at a speed over 20 miles per hour, I was easily exceeding twice that rate and still losing my leader as he went around the many curves. Typical of the island, some of the curves were blind, disappearing behind limestone walls from which the road had been carved. At other times, this quaint Bermuda thoroughfare would remind me of how lucky we were to be on this charming small corner of paradise, but now I could only think of the calamity that lay ahead.

As we sped along, I reflected that this incident had been foreshadowed with the theft of my bike some weeks earlier. On that occasion, I had also called the constables. After about a week with no word as to the bike's whereabouts, Eddie McDaniel, one of our close friends on the island and the crew chief on a KC-97 refueling plane,[1] came by to pick me up on his

---

[1] As was the case with many good friends we would meet in the service, we sadly lost touch with the McDaniels after we transferred to our next assignment. Many, many years later, in January of 1995, in one of those "stranger than fiction" occurrences, while waiting for a haircut at the Davis-Monthan

bike and the two of us scoured nearly half of the island's country lanes before we found it— parked, unhurt and out of gas, alongside a stone wall not far from John Smith's Bay. (I do not wish to leave the impression that Bermuda was a place rife with crime. Quite the opposite was true. In fact the theft of my bike and car were the only such instances that I ever heard of on that island. I was just the one unfortunate one.)

    The urgency with which the constable proceeded on ahead made me think the police department's performance was not going to be the same as it had been with the theft of the bike. A slight surge of optimism was surfacing. My little auto had apparently been stolen and may have been wrecked, and this good constable was speeding along in order to lay hands on the perpetrator(s) while the trail was still warm. The faster we went, the better I felt. I hoped the car was not badly hurt. *Badly hurt?* I'd been thinking of it as if it were a part of the family.

    It was then only a little over a year since the Mid-Ocean Golf Club had been the historically significant location of President Eisenhower's meeting with Prime Minister Harold McMillan. They couldn't have picked a more picturesque site. Situated on the south shore of the island, the course could boast of spectacular views, and the fairways were as well kept as the greens of some golf courses I would play on. If our car had been damaged beyond recovery, at least the thief had not killed it at some obscure duffer's course.

    The Austin's tracks were all over that otherwise perfectly manicured ninth fairway. During the night, the thief or thieves had been having a ball, driving it around and around that smooth stretch of grass. When their little game got boring, they perched the A-35 atop a high hill, aimed it at the ocean, and let it go. It came to rest upright in a pile of rocks, completely submerged save for the roof. I hurried down to look at it.

    From the water's edge, I selected the tops of some partially submerged rocks to be used as stepping stones and tiptoed on them to the driver's-side door. The water was clear. Inside the car, their black and white stripes contrasting sharply with the red leatherette upholstery, swam a school of sergeant majors. A small but attractive fish that is a favorite of curators of aquariums, the sergeant majors were swimming back and forth through the hoop made by the steering wheel. I left them to their sport and returned to the grassy bank to talk to the constable.

AFB barbershop in Tucson, Arizona I met another one of the crew chiefs of those KC-97 tankers. He told me he knew Eddie McDaniel and even remembered that the tail number of Eddie's plane had been 1339.

"Will you be taking fingerprints?" I thought I was asking a question that needn't be asked. And so I was.

"Oh no, the water will have made them unusable," answered the constable.

"But what about the top of the car? They may have placed their hands on the top. Couldn't you get prints there?"

"Other prints will be there, including your own. I'm afraid taking prints would be a waste of time."

I had had visions of the Queen's finest, the Scotland Yard kind of detective, for whom no challenge was too great, who, like the Royal Canadian mounted police, always got their man and for whom this job would be a piece of cake. But I was either dealing with a small town kind of police force who didn't have the know-how to tackle this problem or it was a problem that was beyond the capabilities of even a Scotland Yard. And yet, based on all that I knew of crime-solving which, admittedly, came from Hollywood movies, it seemed so simple a crime to solve.

Then, in a more cheerful vein, the constable offered: "But I know who did it." Now we were getting somewhere. Vengeance at last! Like a fanatic, I had personally changed the oil in that little car every 1,000 miles on the dot, had changed the plugs before they really needed changing, had carefully set the gap in the distributor points each time I changed the plugs, and had covered every tiny scratch it had sustained with the paint from that little touch-up bottle given to me by the Austin dealer.

"Then, you'll be arresting him or them?"

"I'm afraid we can't, the constable replied. "We know who did it but we can't prove it." And so ended the "attempt" to bring that criminal or criminals to justice. There was nothing more the constables could do. Now I'd have to break the news to the insurance company.

The insurance company provided me with yet another shock. They informed me that I'd have to decide, *before the car was pulled out of the water,* if I wanted to have a payoff (at the rate I had recently reduced the coverage to, which was about half the car's value), or have the car repaired.

That was not an easy decision. Unless the car was pulled out of the water, I wouldn't know if any significant damage had been done to the undercarriage when it hit those rocks at however many miles per hour it had accelerated to as it rolled down from the Mid-ocean's ninth fairway. If I knew it had been so damaged, I'd definitely want a payoff.

This seemed to be the time to consult with legal experts. I wondered if the insurance company was within its legal rights in requiring me to make that decision before the car was pulled out of the ocean. And I also wondered if there might be some legal problem connected with either choice, after all, most Americans knew practically nothing about Bermuda laws.

"Just try to get the best deal you can from the insurance company." That was the advice given me by the major at Kindley Air Force Base's legal office. The diploma that hung on his office wall indicated that the major had a law degree from an American college or university, its ornate lettering suggesting that it was a reputable institution of learning. That wouldn't, of course, necessarily indicate that he knew anything about Bermuda's laws, but the same could be said of any Air Force legal officer based at Kindley. The best that you could hope for was that the legal officer you were consulting had become familiar with the more important Bermuda laws, just as every forecaster had to become familiar with the local effects of every new base he went to. In regard to having to decide between the payoff and the repair before the car was pulled out of the water, I said that just didn't seem fair. But the major didn't see that I had any choice in the matter. I was expecting a different reaction, again, because it seemed so unfair.

"Will I be restricted from buying another car if I accept a payoff?" I don't know why I even asked the question except that some of the island's laws seemed quite unlike what we were accustomed to and Bermuda was still in the process of getting used to having automobiles in its midst. They had only been on the island for about 10 years. I was told, for example, that there was a law restricting you from purchasing another car in Bermuda if the new one you bought was less than five years old. I couldn't see where that law could apply in my case, but it must have been in the back of my mind when I asked the question. Years of dealing with meteorological situations where things were not always as they seemed had probably made me a bit paranoid.

The major, predictably, reacted as if I had asked the most ridiculous of questions. "You have nothing to worry about," he said. And he repeated: "Just try to get the best deal you can from the insurance company."

After that interview, I concluded that the wisest move was to accept the loss of approximately $400 (to us, at that time, still a fair amount of money) that the payoff would represent. I rationalized that the salt water would probably silently eat away at the car's vital parts and that one day, some months after the car had been repaired, with Marie driving, the steering gear would fail. So I took the payoff. Then it would be off to a dealer and the purchase of a new car. Maybe this time it would be the red MG I had wanted before buying the Austin. But the bag of surprises had not yet been emptied.

Within hours of cashing the insurance company's check, I received a phone call from the Bermuda Transport Control Board. I needed to know, the caller said, that under the Bermuda law governing the selling of a car purchased new, I would not be allowed to buy another one for a period of one year. I responded that I had not sold the car. It had been stolen and wrecked, and I had merely accepted a settlement from the insurance company. They indicated that they knew all that and that I nevertheless fell under the purview of the law they had just explained to me. There was nothing to be done about it. I could not buy another car on that island until my three year tour would be close to ending.

I was back at the base legal office within the hour. The major who had given me the advice quickly understood that that advice might have been better. Sympathy practically dripped off those office walls. I couldn't blame him very much for not understanding Bermuda laws, they were certainly different. But I did reflect that it would go quite bad for me should I display the same ignorance of the local meteorological effects in my job as Bermuda's weather forecaster. I would later conclude that he was new to the island and simply did not yet have time to know Bermuda's laws.

"What would you suggest I do now?" I asked. He didn't know. Upon reflection, a hardship transfer out of Bermuda was one thought, but I didn't like it. Perhaps more than any other American service family, *we really liked Bermuda.* But what were the alternatives?

We lived seven miles from the base, too far to walk on a routine basis. But of course I had the motorcycle. I could ride the bike to work easily enough in good weather, but what about bad weather? If you rode to the base on your motorcycle through a rain storm your Class-A uniform would be in such poor condition you wouldn't be allowed to work the shift. As mentioned in the previous chapter, we were wearing the khaki-colored short-sleeved 505 uniforms. Within a few years, the 505s would be replaced with the 1505s which were made of a synthetic material and would drape nicely even after exposure to rain, but the 505s were made of cotton. They would lose their starched shape with the slightest exposure to moisture. And there was also the small matter of groceries to haul and places to go as a family.

Like all the others I had known, my detachment commander at Kindley was a straight shooter. A lieutenant colonel and veteran of World War II during which he had been, as I recall, a fighter pilot, under his tenure the detachment was awarded the Air Force Outstanding Unit Award. He shared my appraisal of the situation — that the Bermuda law, as a result of the insurance company's behavior, was treating me most unfairly. He did much more to help than I would have expected him to do. He knew I didn't want to leave Bermuda, which seemed to be my only avenue of relief, and I got the impression that he also didn't want me to leave. Nevertheless, he suggested that I apply for a hardship transfer just to focus attention on my predicament. If the transfer should be offered, I would not be required to accept it. This was only to be a paper exercise. The application would be the vehicle for bringing the situation out into the open where those who might be able to do something about it could see how ridiculous it was. In a letter written mostly by the colonel, the problem was set forth in the request for curtailment of my overseas tour. This paragraph stated my dilemma:

> I wish to make it perfectly clear that this assignment, though not of my own choice, has been otherwise entirely satisfactory. This is particularly true for me since my last overseas tour was in Saudi Arabia where my family could not be with me.

The colonel wrote an attachment to my letter. To fortify our real hope of only using the hardship transfer application as a way of energizing someone to take action in restoring my right to an automobile while not really effecting a transfer, he outlined his forecaster manning situation. He explained that, within a year's time, losses due to tour expirations would constitute forty percent of the forecasting staff.[2] We decided that, as a first step, a copy of the letter be sent to the American Consulate.

From my days as a farm boy, I can recall a saying to the effect that trouble is not a lonesome traveler. And that was true in our case. The business of the car was bad enough, but it wasn't going to be the only negative we would have to face, for it was during the time of the car trouble that our landlady required her house at Lion Rock. Her daughter would be moving into it, and we had to move within 30 days. But where to?

Moving from one house to another seems a simple thing, but not so on that island in the late 1950s. As a rule, an empty house was almost unheard of. If you wanted to move from the one you were occupying and into another, you usually had to make arrangements for it many months in advance. The standard procedure was to begin looking around to see whose tour was about to end and you asked that person to put in a good word for you with his landlord. As soon as we found out we had to move, that was exactly what I did. The staff sergeant who was the dropsonde operator on a WB-50 weather recon ship was leaving for the states in a couple of weeks. He said he'd contact his landlord to arrange for our taking his house. But cruel fate intervened, for he was the dropsonde operator on that WB-50 that went down

---

[2] He was kind enough to also add: "T/Sgt Cogut is an outstanding weather forecaster and briefer and would be a serious loss to this detachment."

without a trace while flying its standard weather recon flight north of Bermuda. Since we knew of no other available houses, someone suggested I should contact the staff sergeant's wife, whom I had never met, and tell her of the arrangement I had made with her late husband, so we could have her house. There was no way on earth I was going to do that.

I dreaded the prospect of sending Marie and the girls back Stateside without me, there to stay for the remaining 18 months of my tour. But that seemed to be the way we were heading. At that point, Marie began walking up and down the Bermuda roads, stopping all she met to ask if they knew of a house for rent. One day, while walking up the rather steep hill on Harrington Hundreds Road, Marie found a miracle in the form of a Portuguese lady who was working in her yard. The Portuguese lady understood no English and Marie understood no Portuguese, so it was also a miracle that, by sign language, one managed to communicate the need for a house and the other managed to communicate the availability of one. Not long afterward, we moved into their house. Its name was "Seabreeze." One crisis was past, but the other lingered.

My application for hardship transfer drew a letter in response from Kindley's chief of administration, a major, who wrote: "sending this request to the American Consulate is not recommended." Instead, he recommended that I make an appeal to the Transport Control Board. He also suggested that I'd be well-advised to seek the counsel of a local barrister.

But before doing that, it was decided to bring my problem to Kindley's Base Commander, a full colonel. He had a good relationship with some of the British VIPs on the island and, if he brought the matter to their attention, something might be worked out informally. He brought the matter to their attention, and they were astonished at how the law was being applied in my case. But they knew of nothing that could be done.

By this time, several weeks had gone by. It was still summer. The weather was warm, and aside from getting slightly soaked in a passing isolated rain shower, the typical form of summer precipitation in Bermuda, riding a motorcycle to work in that season was not a problem — in fact, it was fun.

But a way of bringing groceries to the house still had to be found. I began carrying groceries in two saddle bags that I attached to the bike, but these would only hold about half of what we needed. After a few shaky trial runs, I discovered I could bring home nearly as much as when we had the car if I also carried groceries in a cardboard box which could be picked up from a stack of boxes that were usually near the commissary door. The box had to be just the right size to be cradled within my arms and the handle bars as it rested on the gas tank.

Not long after I had mastered balancing the box of groceries for the ride home, Marie surprised me. On my next grocery run, she climbed on the rear seat. This was the girl who had previously sworn she would never ride on that two-wheeled contraption. The impetus had to have been, simply, that it was tough to sit home and wait while I had all the fun of shopping. And she probably rationalized that if she didn't get on that darned bike, I'd keep buying the wrong things. After that she never hesitated to don her motorcycle helmet whenever there was any kind of shopping to be done or places to visit.

It might also have been true that she had noticed I was getting more and more expert as a bike driver. In fact, my confidence in riding that machine soon stretched into a feeling of over-confidence. One rainy day, while riding alone with my box of groceries balanced on the gas tank, the machine slipped out from under me on the wet uphill curve as I was making the turn from Harrington Sound Road to Harrington Hundreds Road. Fortunately, the bike was equipped with crash bars, and the bike and I began skidding along on the crash bars, up the steep hill on Harrington Hundreds. After an initial slight bump as it hit the blacktop, the box of groceries began sliding down the hill. Before it could come to a stop, an egg carton bounced out of it and onto the blacktop, joining the box in the downhill slide. Next to join the downhill run were five eggs that rolled out of their carton. They wobbled along beside the card board box and the nearly half-empty egg carton.

When the bike scraped to a stop, I ran down the hill in the hope of retrieving the escapees before any of my neighbors, who thought I knew how to ride that bike, could see what had happened. I passed the cardboard box and the egg carton, both had yielded to the forces of friction. I temporarily ignored two of the eggs that had rolled to a stop on the gravel shoulder and went in hot pursuit of the other three that were gaining speed on their downhill run. Positioning myself ahead of the lead escapee, I grabbed it, stuffed it in my pocket, and waited for the other two to roll into my cupped hands. Then I recovered the egg carton, picked up the two eggs at the side of the rode, noted that *not one of the five had been cracked,* and placed them all back in their carton. I returned the carton to the cardboard box, righted the bike, saw that, like my cargo, I had not even suffered a scratch, looked up and down the road, observed that I had not been observed, and drove the rest of the way home, relieved that there had been no witnesses to my embarrassment and yet regretting there'd be no one but me to tell of that incredible incident on a rainy day on that steep hill on Harrington Hundreds Road.

Thereafter, I approached all curves with considerably more caution, even when the pavement was dry. It was not long before all three girls, like their mother, would climb aboard the bike, one at a time, of course, for a ride to wherever we had to go. They obviously felt safe on it but there may have been a bit of a problem with the dignity of the conveyance.

One day, as I was transporting 11-year-old Leta to an after-class school activity, she tapped me on the shoulder when we approached within a half block of the school. "You can let me off here," she said. I stopped the bike at the curb. Leta climbed down off it, straightened her dress, unstrapped her helmet, handed it to me and said, "You can go now." She was not going to let classmates see her riding up on that noisy two-wheeled transport. Like any father proud of and wanting to comply with the wishes of his little girl, I killed the 4-cycle engine. With one foot on the curb, arms supported by the handle bars, I watched her walk away and marveled that, even at that early age, she knew exactly what she wanted to do when she grew up. She wanted to be a history teacher. (She would attain that goal and a number of others.) How amazing that seemed to this weather forecaster who didn't even know about his profession until he was an adult and then only fell into it in an accidental way. When she entered the school, I kicked the starter. The old Fox with the metallic growl roared to life.

Some of our dearest friends, Eddie and Bobby McDaniel and their two small boys, lived near Spanish Point, at Oceanview, the apartment house in which we had spent our first winter on the island. Their only means of transport had also been a motorcycle. While we still had the car and when they wanted to visit us, we'd pick them up for the visit to our home, about five miles away. Now, when we wanted to visit them, Eddie would arrive on his bike, and he on his bike and I on mine would each make two round trips, first taking two of the girls to the McDaniel home, where we would drop them off, and then returning for the third daughter and for Marie. At the end of our day of visiting, we'd repeat the transport arrangements.

About that time, word got back to me that the officials at the Bermuda Transport Control Board were indeed sorry the law was imposing a hardship on my family. They regretted that they were powerless in the matter. It was, simply, the law.

The major at the base legal office then contacted me. He said I should see a Bermuda barrister, Mr. Peter Smith, whose law office was in Hamilton. I went to see him, told him my sad story. He said he would see what he could do.

By this time, we were approaching the winter season when Bermuda's weather at times can be quite challenging for a bike rider. Like the winter before, there seemed to be an unusual amount of squally weather. To insulate and guard against the cold, windy rains, I wore a leather jacket over my uniform and a dull green rain suit as an outer garment. The rain suit consisted of two parts: a pair of bulky waterproof trousers, held up by suspenders, that were wide enough to be pulled up over my uniform trousers; and a waterproof pull-over with hood and a drawstring at the neck. With the hood over my head, I would tighten the drawstring, snap on a pair of goggles, and pull my crash helmet down tight over the hood. Each leg seemed to be as wide as my upper torso. Dressed that way, I knew I looked like the creature who had just crawled out of the black lagoon, but I could drive through any rain squall that I could see into, and some that I couldn't, keep my Class-A uniform dry and, except for minor wrinkling from the heavy press of the rain suit and leather jacket, and a drop or two of rain that managed to sneak by the rain suit, I could, when finally unwrapped, look quite militarily presentable. Once inside the weather station, while hurrying to minimize the jibes tossed at me by my weather associates, after all, I *did* look ridiculous, I would retreat to the privacy of the radar room, remove those outer garments and hang them to dry for the ride home at the end of shift. All that dressing for the mere ride to or from the weather station was a bit of a bother, but there was some satisfaction in knowing that, with all that bulky clothing, I probably couldn't so much as be scratched should I take a spill. Fortunately, the incident of the rolling eggs would be my only mishap with the bike.

As the weeks went by I was becoming somewhat well known on the island and not for any reason that I might have wished for. Dignitaries who heard of my predicament expressed dismay at the way the law had worked in my case. It wasn't supposed to work that way, but it had, and they were all very sorry to know it had. After giving him sufficient time to look into the matter, I paid the barrister in Hamilton another visit. He had indeed looked into my situation and saw, as the others had, that the law was not intended to work the way it had in my case, but it *was* the law. The only thing that he could see to do was to have a bill introduced on the floor of the colonial legislature. This would not change the law. Naming me specifically, the bill would merely provide me relief from that law. The barrister said he would try to find someone who might introduce such a bill.

All the other attempts to come to my aid had, in the aggregate, taken months. This one would take longer than the rest. In the end, I would never know the results of the private bill that was to be presented on the floor of the colonial legislature. I would reach the end of my three-year tour trying to be allowed to replace a car that had been stolen. Even now, these many years later, I recall the entire episode as one that can hardly be imagined.

But it might not have stopped there. As February of 1960 approached, marking the end of that three-year Bermuda sojourn, I requested an extension of my tour into the summer of 1960. As uncomfortable as was the transportation, we still, as a family, loved our time in that little spot of paradise. We had learned to cope without an automobile for what would become 18 months, we knew we could do it for a little longer. One impetus for the extension was the facility I was acquiring for painting the enchanting Bermuda scenery. I had fixed the bike with apparatus to carry my easel, canvas boards, and paints and had traveled to a number of off-the-beaten-path locations to paint. I knew that this kind of opportunity would not come again.

We still believed that to live on the island of Bermuda was a dream that could not be topped, and we did not yet want to be awakened. But in one more surprise development, my request for extension of tour was denied. As far as I knew, tour extensions had been approved routinely. The official explanation was that there would be a hump in forecaster rotations at the time I wished to extend to and that, therefore, the request could not be granted. Of course, that could have been the reason, but it was also true that my predicament and the inability of anyone to solve it had become somewhat of an embarrassment. Someone might have noticed that I had been spending a lot of time talking to the BOAC flight planners, and it may have been suspected that it wasn't always about the flight-level wind factor. I might have been overheard asking the BOAC representative about the possibility of obtaining a weather familiarization flight to London, an innocent conversation that could have precipitated the fear that I intended to raise the level of embarrassment by seeking an audience with the Queen.

Top: Willa (left) and Pamela pose on dad's NSU Fox motorcycle in front of our home at Lion Rock, Smiths Parish, Bermuda.
Bottom: A Boeing C-97 Stratofreighter cargo plane at Wheelus Field, Tripoli, Libya in 1951. This plane was also made as a refueling tanker, with designation KC-97, and as a civilian airliner with designation B-377. KC-97 tankers were stationed on Bermuda.

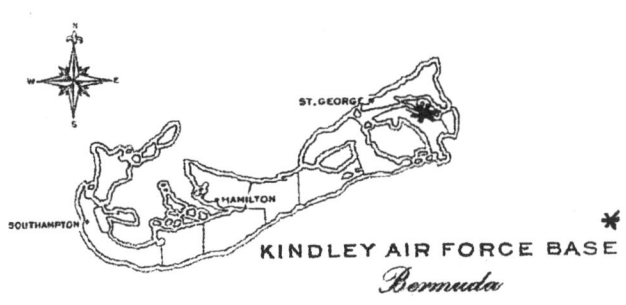

Top: Fishhook in shape, Bermuda was the remnant of an ancient volcano. Kindley Air Force Base was located on its far east side.
Bottom: A cruise ship has docked at St. George's Parish, Bermuda's eastern end. *Photo by Darrell Mach.*

Chapter 22

Peace Is Our Profession

In that February of 1960, we left our footprints on the pink Bermuda sands and climbed aboard a Boeing 707 at the Kindley air terminal, bound for Idlewild, the airport in New York that would later be renamed JFK. We could have departed the island by sea, sailing to New York on the cruise ship *Queen of Bermuda,* saying good bye to our friends and to that enchanting island with a bon voyage party in our state room; but Marie reminded me of my seasickness on the deep-sea fishing boat, and we chose instead to depart on the Boeing 707, the Pan American jet that had made the first passenger jet crossing of the Atlantic from New York to Paris about 18 months earlier.

The contrast with the C-54 in which we had traveled to Bermuda three years earlier was almost too striking to describe. There was the utter smoothness of the ride, and above all, the quietness. A Boeing ad would state: "The 707 cabin, the most spacious aloft, will be so quiet you'll be able to hear the ticking of a watch."[1] In the C-54, with its four reciprocating engines, it would have been a little difficult to hear the alarm of an alarm clock. And we would be cruising at almost 600 miles per hour as opposed to approximately 250 miles per hour in the C-54.

On our climb out to about 30,000 feet, Kindley's terminal building and base weather shrank as if seen through a rapidly collapsing telescope. The manmade shape of the land upon which a runway had been built from sand dredged up out of the ocean, which was somewhat difficult to appreciate while you were on the island, now became quite obvious. The 3,000-foot-long causeway which connected the airbase to Hamilton Parish, that road upon which you and your bike had on many occasions dodged sea spray when the weather was bad and the breakers crashed over the bordering sea wall, was now as thin as a fishing line. The climb out was vibration-less and seemingly effortless. No struggling for altitude as was the case in that first flight in the old AT-11. So this was why the F-86 pilots often sauntered into Wurtsmith's weather station whistling and looking as if they'd just won Las Vegas's biggest jackpot.

We had hardly leveled off when the stewardess came down the aisle with champagne for the adults and soft drinks for the children. That was the only time I've ever been offered free champagne on an aircraft, so I'm

---

[1] *The Jet Age,* Time-Life Books, reproduction of a 1958 Time magazine ad. Alexandria Virginia, Third Printing, 1986, p. 82.

assuming our flight was a celebration of the first jet airliner to fly the Bermuda to New York run. And it wasn't just one glass of champagne. Even before we had quite finished that glass, the stewardess, a pretty picture of constant motion, came back to refill our glasses. And there seemed to be no end to the number of refills. With all that attention, and all that champagne, the flight became so pleasant we didn't really care much if we'd ever land. But the speed of the 707 ensured we would land and land much quicker than we thought possible, the entire flight taking just about two hours.

As we touched down at Idlewild, we saw our first snowflakes in three years. The customs agents got us through their procedures in a short time, and didn't even wonder about the strange hollow-packaging of my oil paintings, so done as to keep the wet paint from smearing. (Knowing that I'd never again have the opportunity, during our last week on the island, in an almost frantic effort, I had completed more than a half dozen seascapes and landscapes in oil, a surprising rate that I'll almost surely never again achieve.) From Idlewild, we reverted to the civilian version of the C-54, a DC-4, for the flight to Detroit and a short leave with our families. Enroute, one of the DC-4s props had to be feathered, but we didn't care, there were three others, and we were going home.

After that failed 18-month struggle to be allowed to buy a car, we were looking forward with great anticipation to this vacation in the motor city. Of course, buying a car was going to be one of the first things we'd do there. There had been one upside to that forced 18-month period of motorcycle-driving. That German NSU Fox motorcycle might have had a somewhat annoying metallic growl but it was extremely efficient, getting 120 miles per gallon. Our savings in gas purchases coupled with the fact that we couldn't buy anything very big (and therefore expensive) because there'd be no way to carry it home, allowed us to accumulate a small nest egg that was, nonetheless, large enough to purchase, for the first time, a new car for cash.

We bought one of the first examples of a Chrysler Valiant, an automobile that in the next model year would be called a Plymouth Valiant, and made preparations to drive to Pease Air Force Base at Portsmouth, New Hampshire — my next assignment. We were staying with my sister Ange and her husband Bob Newman at their home in Highland Park, an interior suburb of Detroit. On the night before our planned leaving, before retiring I asked Ange to look out when she awoke the next morning, and if it were snowing to let us sleep in. After three years on balmy Bermuda, we didn't think we'd enjoy driving to New England on snowy highways.

I caught the weather program on television in which a low pressure system was shown in the lower Mississippi River area. It looked as if I could outrun its weather band in a drive across southern Canada if it was not yet snowing when we left. But one should remember that, in those early days of television, the weather presenter often was a pots and pans salesman who had been given the weather talking chore because he could tell jokes reasonably well. So it is possible I wasn't being provided with the information I really needed and perhaps what I was given was not really the true weather picture. TV weather presentations these days are much better, but the weather presenters still don't understand that in order to be really useful they should provide the viewer with a good picture of the 500-millibar chart. They should also show the National Meteorological Center's computer-produced progs which often are more reliable than our hand-made progs had been. This could give the viewer a chance to evaluate the prog chosen for the prediction. When sister Ange looked out that morning, the weather was clear. We enjoyed an early breakfast with her and brother-in-law Bob and left for New Hampshire before sunrise. About a half hour later, as we were crossing the Ambassador Bridge, the span over the Detroit River that connects Detroit with Windsor, Ontario, Canada, a few white flakes floated down in front of the car. We wondered if we should turn back, but a flurry, after all, was not unusual around the Great Lakes in winter. This was probably only a slight snow shower that had drifted off Lake Erie. About 75 miles into Canada, however, we suddenly understood we were driving into a blizzard.

After two forecasting tours at Selfridge and one at Wurtsmith, I should have known better than to travel along the northern side of one of the Great Lakes in late February on a day when the flow was generally from the south. The snow we were driving through was being generated over Lake Erie, well in advance of the weather band connected with the low pressure system that I had seen on television. But it might have been just as bad if I had taken the southern route, along the south shore of Lake Erie. On that route, I would be closer to the weather band ahead of that migratory weather system and could well run into it.

With forward visibility of only two to three car lengths we were only doing 20 or 30 miles per hour along that Queen's Highway, the Valiant's tire tracks the only ones in the snow save for an occasional semi-truck whose driver must have had an urgent load to deliver. Snow had piled up even higher on the shoulders, giving us no place to pull off the road, a problem that became sort of acute when the driver's side windshield wiper stopped. This was one of the very first Valiants built and in the hurry to get it in the showroom, a worker had forgotten to install the retainer clip on the wiper linkage. I pulled into the driveway of a shut-down gas station which had luckily been plowed, and reconnected the wiper linkage with a safety pin donated by one of the girls. Eventually, we passed through and beyond Buffalo and were on the southern side of the lakes. The lake-effect weather band was behind us, but now we were well within the low pressure system's weather band. At Troy, New York, still fighting the snowstorm, we found a motel and stayed the night. The next morning broke bright and clear. This was the area in which my mother had spent her girlhood. I wanted to do some sight-seeing, but we had a reporting time to meet and pushed on to Pease Air Force Base, Portsmouth, New Hampshire, arriving near nightfall.

Pease AFB had been named for Captain Harl Pease, Jr., a bomber pilot and Medal of Honor recipient, who was lost on a bombing mission near Rabaul, New Guinea in 1942.[2] It may have been the extra-sharp-looking Air Policeman (AP) or the imposing sign at the main gate with the Strategic Air Command (SAC) emblem, the mailed fist holding arrows, or the motto: "Peace is Our Profession" painted on it that made you realize you were now in a different kind of Air Force organization.

Under the leadership of Lieutenant General Curtis LeMay, the general with the ever-present cigar, SAC had come of age and in that era of the cold war and the nuclear standoff, SAC would represent the ultimate in military preparedness. It would stay that way under LeMay's successors. At a SAC base, as Pease was, you were always conscious of the fact that you were in a combat-ready outfit.

Pease was a B-47 bomber base. The B-47 was America's first multi-engine swept-wing bomber. It was powered by six GE J-35 engines, each having a thrust of 3,750 pounds. The sleek plane had a normal range of 4,000 miles. With air refueling support from KC-97 tanker planes, that were also based at Pease, it was common for the B-47s' three-man crews to leave Pease on missions that lasted 15 hours.

The AP at the main gate directed us to the Pease guest house, a large wooden structure with a number of bedrooms that became the temporary quarters of visitors such as we. Located at the far end of the base, near Great Bay, the house appeared to have once been the home of the owner of a large tract upon which the base had been built. One car-length into the guest house driveway and we were stopped — stuck for the first time on our long trip, in the deep snow that had been deposited the day before by the same storm we had fought through Canada and New York. No amount of rocking would budge the comparatively low-slung Valiant. As it would turn out, this was a most appropriate introduction to Pease, a base that would present me with some extremely interesting weather situations and among these would be a number of first-rate snowstorms.

---

[2] *The Official World War II Guide to the Army Air Forces*, a Lou Reda Book, "Battle Honors," Bonanza Books, New York, 1988, p. 322.

When the B-47 crews took off on those flights of 15-hour duration, they were understandably very concerned with the ceiling and visibility that was forecast for their return. It wasn't merely the obvious question of whether their plane could land, they wanted to know, as they would put it, if they should take a toothbrush (translation: pack a bag) which they would need should they have to land at an alternate field. Woe to the forecaster who had failed to advise them that they would be landing elsewhere.

On my first appearance at Pease Base Weather, call letters PSM, I was surprised to find I wasn't greeted with the remark that forecasting the weather in that place was the toughest job in the world; and after I had been there a month or two, I thought that if ever there was a base where that could have been said without justifiable contradiction, it was Pease Air Force Base, New Hampshire. On days when there was no possibility of snow, fog could frequently be counted on to fill the anxiety void. In addition to radiation fog, the kind that is common over land masses, Pease's visibility could be severely restricted by sea fog that would roll in off the Atlantic.

Nor was I told, as I had been told at Dhahran, that weather prediction would be easy. The forecasters at Pease seemed to be taking a balanced, cool view of their responsibilities. Formalized forecast studies seemed always to be in progress — not because these had been directed to be done, but rather, because a forecaster saw the need for a better approach to a particular forecasting problem. Master Sergeant Robert McGlew compiled these into the detachment's official forecast study.

The most important of these procedures was the one developed by Sergeant McGlew for forecasting snow depths. His method correlated snow depths with future positions of the Hatteras Wave which in turn was determined from an inflection point of the contours on the National Meteorological Center's (NMC) 500-millibar prog. The accuracy of that procedure was impressive and was also a sure indication of the great strides that had been made in the NMC's computerized 500-millibar forecasts in the short time since the introduction of the barotropic prog that we had tried to use in Bermuda.

The most unforgettable snowstorm of that tour at Pease occurred with a nor'easter on Christmas Eve and Christmas Day of 1961. The first snowflakes of that storm began falling on Christmas Eve morning, at 0155. It was recorded by the weather observer on his WBAN (Weather Bureau, Army and Navy) form along with the temperature of 21 degrees and the remark: "vsby dcrng rpdly." By 0800 hours, the snow was six and one-half inches deep and increasing. By 1100 hours it was 10 inches deep.

We lived on base and I was scheduled to work the Christmas Eve-to-Christmas midnight shift. Normally, it took me just five minutes to drive to the station. This time, however, anticipating some trouble with the snow, I started out an hour early. I shoveled out the drive and a short section of street. With a running start through the shoveled section, I thought the Valiant would power its way through the unshoveled street ahead. But that low-slung car bogged down before I went half a block. I considered returning the car to its carport and walking to the station, but the road behind was uphill and there was no traction to be had in reverse gear. So I shoveled a little, powered the car forward a little until it bogged down, shoveled a little, moved the car forward a little and continued in that way until, about an hour late, I reached the station. Senior Master Sergeant Benjamin, the forecaster I was relieving, had in the meantime called Marie to see where I was, and she informed him that I was really on my way. With the snow as deep as it was, the accumulation was then measured by the weather observer at 18.1 inches, he had nowhere to go and was not concerned.

Through most of that midnight shift, ceilings and visibilities were below recommended takeoff and landing minimums. So aside from keeping track of the weather conditions and forecasting the future of the storm, the observer and I had little to do. It was truly a night when only Santa would be flying.

Getting home when that shift ended at 0700 hours Christmas morning was also a bit of a task. By that time, the observer was recording a snow depth of 26 inches. That would be the maximum depth of the storm. I had to dig the car out from its parking place in front of the weather station — it had been buried by a snow plow — and then wait for the plow to clear the street leading to our quarters. As soon as the street was plowed, Marie shoveled out our drive. I arrived home about two hours late which is not a good thing when you have three children in the house and, waiting for your arrival, they have not yet opened their Christmas presents. There was near open rebellion when I suggested we should all have breakfast before opening the presents. So we opened the presents.

Each Christmas, I tried to include some gag gift for one of the girls. This Christmas I wrapped a plastic windshield ice scraper with pretty Christmas paper and put it under the tree for 12-year-old Willa. When she unwrapped it and I saw her pained expression, I quickly said it was only a joke but she could keep it anyway so long as she used it to scrape the ice off the windshield of the car. Being a very fair-minded and generous girl, she said I could keep it all the time. It is interesting to note that, of my three daughters, Willa now lives in the coldest part of the country.

Afterward, the three children bundled up to go out into that winter wonderland of a Christmas Day with 26 inches of snow on the level and drifts two to three times as deep. It was what I insisted they do. I told them that, for as long as they lived, and regardless of where they lived, it was almost surely the deepest snow they'd ever see on Christmas Day — an accurate prediction up to this day. Then we listened to Christmas music on that perfect weatherman's Christmas Day, a day when the weather had declared its preeminence, preventing the weatherman himself from reporting for duty at his appointed time.

That forecasting tour at Pease was my first association with bomber crews, the real heroes of the Cold War. Their dedication to mission gave life to the SAC motto: *Peace Is Our Profession.* And I found the bomber crews to be just as witty as the fighter pilots when the time for levity presented itself. This was even true of the base's ranking officer, Brigadier General Jack Catton.

Unlike many other generals of my acquaintance, General Catton was a frequent visitor to Base Weather, and he came alone, without the company of aides that one might see when other generals came by. It was clear that General Catton had a sincere desire to absorb the weather situation firsthand. At his visits, we developed a routine of sorts. After I told him of the coming weather changes, he invariably would respond: "I'll bet you a dollar on it."

"Sir," I'd reply, "I'd feel terribly guilty taking your money so easily."

Years later, after I had left the Air Force and was serving in the Army Artillery, I was pleased to learn that General Catton had gone on to bigger things, becoming, I believe, the four-star commander of MAC, the Military Airlift Command, which was also the parent organization of the Air Weather Service.[3] It felt good to know that someone who appeared to have a genuine interest in weather was advancing into higher levels of the military.

Regardless of the weather station you happened to be assigned to, over time the preponderance of daily forecasting and briefing requirements would tend to blend into each other, so that, in your memory, the daily routines would be remembered as a sort of unimpressive average: You reported for duty; signed the forecaster's log; were briefed by the departing forecaster; checked the current weather observation at your base and at the bases to which the aircraft would likely be flying; analyzed charts; formulated your forecasts for those bases; briefed the weather detachment commander;

---

[3] This military "airline" had its beginning in May 1941 as the Ferrying Command, was designated ATC (Air Transport Command) in 1942, MATS (Military Air Transport Service) in 1948, and MAC in 1966. *Airlift!*, Marcella Thum and Gladys Thum, Dodd, Mead & Company, New York, 1986.

answered the phones, hot lines and radios (at Pease, there were nine such devices that might be clamoring for your attention, and sometimes all at the same time); briefed the relieving forecaster; and so on, making it practically impossible to recall the precise events of any random forecasting day. At any base, however, there was always a shift or two that could not perfectly be erased from memory. At Pease, one of these was that record-breaking Christmas Day snowfall; the other involved a B-47 training mission. The latter is now as easy to recall as if it had occurred only last night.

In the customary manner, I reported for that shift shortly before 2300 hours. As soon as he saw me, the departing forecaster said a large number of B-47s would be taking off on a long nighttime special training mission at about 0200 hours. After flying a considerable distance, the bombers would enter a training leg where they would be scored for accuracy.

I immediately saw that the weather at Pease would be good — both for takeoff and for the bombers' return. The crews would not have to pack a toothbrush. But a cursory look at the surface chart showed me that the weather on the training leg could be questionable. I asked the departing forecaster what he had been briefing for that leg. He had not predicted bad weather, he said, but he was not sure the weather would be good, adding that it would bear watching (usually a code phrase for all heck could break loose).

When he left, I had the weather observer plot hourly surface weather observations that were being reported at weather stations in the states upstream of the training leg. From the plots of the 2300 observations I could pick out an instability line in the developmental stage in that upstream area. Somewhat similar to the 0400 thunderstorm at Oscoda, there were no thunderstorms being reported, just some towering cumulus clouds and a line of falling then rising pressures, a telltale indicator of the instability line. But unlike the Oscoda episode, there wouldn't be time to leisurely track the line's movement by observing reports from the succeeding hourly observations. Within the hour, the first of the pilots and navigators would be drifting in for a prebriefing look.

Cut off from being able to track the line's speed with the succeeding observations, I did the next best thing. I calculated its probable movement using a percentage of the 500-millibar wind speeds applied on the line of the 500-millibar wind directions. This projected the line to be directly over the training leg at the time of the bombers' arrival, somewhere between 0400 and 0500 hours.

The bombers might be intersecting the projected position of the line, but it could dissipate before their arrival. Would it dissipate? It was something to consider. They'd be there near the coldest time of the night, so there'd be no surface heating to add impetus to the buildups. But the line appeared to have been just beginning to form at 2300, already a late hour in terms of surface heating. The conclusion? There had to be other dynamics at work. So it would be, I decided, pure wishful thinking to believe the buildups would dissipate, and being in the formative stage so late in the day, it was logical to think there'd be further development.

With those thoughts, I jotted down the forecast for the training leg: generally 8/8ths towering cumulus with isolated embedded cumulonimbus (the thunderstorm cloud). Turbulence would be moderate to severe.

Within the hour, the early arrivals began working on their flight plans. I provided them with their flight-level wind forecasts and the enroute weather, including, of course, the weather for that training leg. As I had long ago learned to do, I provided that information in a calm, cool way, trying not to alarm the crews while not understating its seriousness.

At about 0130, the crews were lined up, waiting their turns at the weather briefing counter. I had briefed only two or three crews when their commander, either a lieutenant colonel or full bird colonel (I cannot recall which), burst on the scene. Without asking anything about the basis for the forecast, which was something I would have expected, he proceeded to dress me down. At that time, I had been a practicing aviation weather forecaster for about a dozen years. This was a new experience.

Looking back on it now, I think it would be accurate to say the colonel was in a nasty mood. What was I trying to do? Why was I forecasting thunderstorms? Were any being reported? When a colonel confronts a sergeant in that way, the sergeant is expected to confine his answers to "yes sirs" and "no sirs." To his last outburst, however, I elaborated a bit: "No sir," I said, "No thunderstorms have been reported. But that doesn't mean there aren't any — there's a lot of distance between reporting stations — and even if they're not there now, that doesn't mean they won't be there later." This only threw gasoline on the fire.

"You're just trying to scare my pilots!" Now he was furious. Then, turning his back to me to face the crews, for the first time in my career I actually heard a commander tell his pilots to ignore my forecast! To try to see this from his perspective, you need to remember that this was at the height of the Cold War. SAC bomber crews were the primary deterrent to nuclear war, and training missions such as this one were regarded to be of the highest importance. When someone in SAC said *Peace Is Our Profession,* they were implying that crews such as these were preserving the peace through their dedication and efficiency, as they surely were. And it should also be remembered that the colonel and his crews were being scored on the efficacy of their simulated bombing runs. One couldn't scrub a training mission on the basis that thunderstorms were forecast and then not have the thunderstorms materialize. The crews had to handle the complications of the weather and this colonel, like a baseball manager, was trying to provide encouragement. That might require, as I tried my best to rationalize it, that he disparage the efforts of the duty forecaster.

Military protocol notwithstanding, I walked around the briefing counter to face that colonel. I wanted to give him a chance to retrieve the situation. "Nothing could make me happier than to have the weather turn out clear," I said, "but all the information I have is to the contrary. I can't change that." The colonel glared at me and walked away. The crews looked up from their navigational maps, exchanged meaningful glances, returned to their route-plotting, and said nothing. Theirs was not to reason why.

I returned to my side of the counter. Drawing on that reserve of calm, an aviation forecaster's legacy developed from countless hours of sweating out troublesome meteorological situations, I resumed the briefings which were now being conducted without that trace of levity that usually characterized them. Everyone was now Mr. Sobersides. The forecast, as it had to be, was identical to that given to the earlier crews. As I briefed for thunderstorms, the crews nodded and said nothing. Normally, they'd have extended the conversation a bit to further discuss that potential problem. I couldn't tell if they believed me. Perhaps, because of the colonel's outburst, they doubted the forecast; or maybe they believed it and were feeling trapped because he would not, or could not, cancel the mission. After awhile, I sensed that the crews were quiet because they were feeling sorry for me. They could have been thinking I was going to be proved wrong and that I'd soon be history. As the briefings continued, I believe I appeared to be calm, but under that facade rested a somewhat troubled sergeant.

In my several tours as a forecaster, I had briefed countless colonels and a few generals. Without exception those were pleasant occasions. Had I become smug? Was I conducting myself as if I were some high-and-mighty forecaster intoxicated with whatever power a forecaster had? Could that kind of personality change have come about without me even sensing it was happening? Well, even if that were the case, there wasn't the slightest doubt the forecast had to stand as I had given it — no evidence supported an amendment. That had to be the bottom line.

Soon all B-47s on that mission were airborne. With their takeoffs, an abnormal quiet descended on the station. There was nothing left to do but sweat out their return. I walked into the teletype room where the sometimes annoying chatter of the machines now seemed reassuring, a familiar weather station sound to make one think things were normal after all. There were no messages from the weather reporting stations on the training leg. I thought about calling the weather center at nearby Westover AFB (my old base at the beginning of my forecasting career, at Chicopee Falls, Massachusetts). In a reorganization, they had been given oversight responsibility for Eighth Air Force missions flying out of Pease. But if I did, I'd only be making small talk. Actually, I had called Westover before I briefed any of the crews just to see if the forecaster there agreed with my appraisal of the weather on that training route. Acquiescing in my analysis, he offered no thoughts of his own. I hoped he wasn't agreeing with me as a matter of professional courtesy or to avoid comments in a matter where there didn't seem to be an absolute requirement to stick one's neck out.

Still without any confirming reports of thunderstorms but believing they should be there, now that I had time to reflect on it I became deeply concerned for the crews' safety. When they saw how it was, would they go on in a heroic attempt to complete the mission? To protect the crews, I should have raised a big fuss, risking court martial if necessary. On the other hand, not since my days as a raw recruit, and perhaps not even then, did the forecaster have the authority to stop a plane from taking off. He was only there to advise. Once or twice in the past, when I believed the weather to be only a little worse than in this case, I had gone so far as to write on the pilot's Form-175: "Flight not recommended." But the pilots then were not on a SAC mission. They were only boring holes in the sky, only trying to get enough hours to qualify for monthly flying pay. They were not the Free World's first line of defense as these bomber pilots, heroes all, surely were.

The results of the mission finally began coming in at the end of that graveyard shift. I have forgotten the exact details and indeed probably was not told of them in their entirety save for the fact that it was a bad run and, but for the skill of those crews, could have been much worse. As I recall, the first B-47 was bounced around on the training leg but managed to get through. Severe turbulence forced the following bombers to abort. The mission might not have been considered so much of a loss if one took into account the value of the training the crews received in contending with those particularly adverse flying conditions, but I am sure they would have preferred not to have had the experience. The definite plus was that all the B-47s had returned safely, and for that I was supremely grateful. As I ran over the events of that night, I felt a sense of vindication — the best use of the available evidence apparently had been made in formulating the forecast. It would have been much better if there had been an earlier indication of deteriorating weather. If that had been known before I came on shift perhaps the mission would have been scrubbed, and my confrontation with that colonel would not have occurred. But with the information available, that was beyond any forecaster's capability. The reality of day-to-day forecasting was that you played the cards you had.

Near the end of that shift, the major who was then our detachment commander came in for his morning briefing. He already knew of the disastrous mission and asked me to tell him everything that had led up to it. Shortly after, General Catton arrived and was briefed by the major. I didn't want to hear what they said and tried not to listen, busying myself with the ever-continuing map analyses. I truly felt sorry for the colonel. Many would have condemned his deportment that night. He could have been one of those senior officers that I had heard of, but had never run into, who simply didn't like being briefed by an enlisted forecaster. They didn't realize that the enlisted man had the most experience in forecasting. By the time the officer forecaster approached that level he was elevated to the administrative office, away from the weather briefing counter. I didn't know what pressures this colonel had faced, and my Christian upbringing surfaced in the hope, truly held, that the incident would not negatively affect him. From my perspective, he had looked far from his best. Memory now suggests that he departed from Pease not long after. I would hear nothing more of the incident.

This is the work copy of the WBAN for the last several observations during Christmas Eve night, 1961. The first line is a special observation required because the ceiling and visibility had significantly deteriorated. The second line is a routine record observation, required near the beginning of each hour. Its breakdown follows: the ob was taken at 2155 (09:55 p.m.); "W" indicates the ceiling is indefinite, "5" indicates the ceiling is at 500 feet, "X" indicates the sky is obscured, visibility is 3/4 mile with light snow (S-) and fog (F), pressure reduced to sea level is 993.9 millibars, temperature is 24F, dew point is 23F, wind is north-northeast (040 degrees) at 20mph with gusts to 26mph, altimeter setting is 29.31 inches, visibility on runway 34 is 3/4 mile, there is drifting snow and the runway is covered with ice and snow ("CIASOR").

Marie shovels out the driveway in front of our quarters at Pease Air Force Base, New Hampshire on Christmas Eve, 1961. The snow depth would reach 26 inches by Christmas Day.

Top: This Boeing 707 jet was fitted out as Air Force One and used by Presidents Kennedy and Johnson.
Bottom: B-47 bombers such as this one and their Strategic Air Command crews kept the peace during the Cold War.
Both aircraft are at the Pima Air and Space Museum, Tucson, Arizona.

Chapter 23

The Clime Center and the Terror of Cheverly

We were living in Cheverly, Maryland, a Washington, DC suburb, as a result of my transfer to the Air Force's Climatic Center. We called the place the "Clime Center," a name that would be changed to "ETAC," (Environmental Technical Applications Center) while we were stationed there. But even after the name change, we still called it the "Clime Center."

Back in early 1946, when I first entered the Air Weather Service, though basic climatology was taught in weather observing school, there was a tendency to sort of sneer at people who called themselves climatologists. To deal with climatology, you were primarily looking at what happened in the past; but in meteorology, more precisely as a weather forecasting meteorologist, you were tasked with determining what was going to happen. To the weather forecaster, it was easy to see which was the most challenging field. In the beginning, I completely shared the view that climatology was not worth the time one might spend on it, but as I became more and more involved with weather forecasting problems, I became convinced that significant improvement in weather forecasting could only come about through a more perfect marriage of the two disciplines. At the Clime Center, I would have the opportunity to realize more fully the value of climatology.

Transfer to the Clime Center came about a few months after I attended the Climatology Course at Chanute. I was there in the summer of 1961, and it was my fifth weather school. On our first class day, the captain who headed the course, who I then guessed had a degree in educational psychology, probably surprised every student and me in particular (more accurately, at least mildly shocked) when he said that he didn't know where we would be stationed upon graduation, with the exception, however, of myself. He said that I was to be transferred to Washington, DC. I took his remark with a grain of salt for he had sort of eroded his credibility with me when he said, on that same first day of class, that he could then predict what our individual class standings would be upon graduation (but he did not reveal the predictions), which reminded me of the unsuccessful serial-number guesser at Dhahran. Nevertheless, he turned out to be right about my transfer to Washington, and he might well have been right about being able to predict our ultimate class standings. He was in charge of computing those statistics.

Speaking of statistics, a favorite instructor in that course was Mr. Howard Crombie who taught that subject. This was before the day of hand-held computers, a time when statistical problems were laboriously worked out on our slide rules or with the help of boxy Marchand or Friden calculators that clunked with the entry of each number. Crombie's teaching technique was simple: in what seemed like a process that would never end, he had us work out the many statistical measures of what also seemed like a never-ending and terribly lengthy list of numbers. It was a tedious process. To inspire us as we set about working out the mean, mode, median, standard deviation, standard error, linear regression, the correlation coefficient and other statistical functions, Crombie liked to tell the story of the Pittsburgh Pirates pitcher who warmed up before each game by throwing a baseball made of lead. When it was time for that player to ascend the pitcher's mound and throw out the first pitch, he picked up what seemed to be a ridiculously light regulation baseball and threw fast balls that were too speedy to even be seen. Come to think of it, I think Howard Crombie probably was the one with the degree in educational psychology.

We also were given a number of projects, the most interesting, perhaps, was the analysis and description of the climate of certain African and South American countries. Upon seeing that my assignments were Rhodesia and Colombia, one of my classmates suggested that I'd be better off studying Vietnam for that was where he thought we'd all be going. No one took him seriously. While the US was supporting with financial aid and supplies the South Vietnam government of Ngo Dinh Diem in his battle with the Viet Cong, in light of the result in Korea, a country, like Vietnam, divided along an arbitrary parallel of latitude, it was ridiculous to think we'd ever go the extra step of sending organized combat forces into that quagmire. To be directly involved in a land war in Asia, everyone knew, would be a foolhardy enterprise indeed; and we went on with Africa and South America.

Attendance at that school sort of came about as a result of the construction of a number of temperature-prediction curves. Always acutely aware of the need to save some of the time the forecaster spent on routine matters so he could spend more time on the critical terminal and route forecasts, at Pease I toyed with the hourly temperature-prediction chore. Hour-by-hour temperature predictions that ran out to 34 hours was a requirement of the ground crews who filled the KC-97 tanker planes with fuel to be transferred to the bombers in air-refueling missions. Fuel expansions and contractions, which dictated the amount of fuel to be loaded, made the hourly temperature predictions a matter of great importance to the fueling crews. With my method, which required wind direction and cloud cover inputs, instead of agonizing over each of 34 future temperature values, the forecaster could simply read them in their entirety off an etched plastic curve overlain on a graph preprinted with hourly temperatures.

A new detachment commander, Lieutenant Colonel Melvin Cobb, had arrived at Pease Base Weather and, I assume, had noticed my interest in digging into Pease's weather records with the undying hope of finding the keys to the forecasts. (In addition to the temperature curves, I had also, after poring over stack after stack of old WBANs, arrived at an objective method of predicting the time of ceiling deterioration to IFR when rain was predicted). In any event, Colonel Cobb recommended that I attend the Climatology Course. Cobb was a fine detachment commander whose superior intelligence was aptly demonstrated the day he recommended me for promotion to master sergeant. As an example of his thoughtfulness, instead of having me receive the news by slow mail, he informed me of that promotion by telegram while I was at Chanute, attending the Climatology Course. I sewed onto my uniforms the stripes that I would have had in 1953, over eight years earlier, had I not elected then to leave the Air Force. I was looking forward to reporting to my new assignment in Washington, wearing my equally new stripes.

But upon reporting to the Clime Center, which was located in a Civil War hospital building about three blocks south of the Capitol, I learned that I could keep my uniforms in the closet. Civilian clothes were the standard attire in that office in which civilian meteorologists worked side-by-side with military meteorologists. In summer, I wore a short-sleeve white shirt and tie; in winter, I had the opportunity to wear the single-breasted tailor-made Harris tweed jacket I had purchased in Bermuda from a branch of Alexandre's of Oxford Street, London.

The Clime Center was the problem-solving apex of the Air Weather Service. From one day to the next, one could shift from calculating snow-loading figures for government buildings in Alaska to the probable number of hours that air conditioning would be required in an American embassy somewhere in the world. I was one of the environmental analysts assigned to the Center's Engineering Design Branch with the primary duty of providing statistical climatological answers to such engineering design problems. We also prepared climatological outlooks for President Kennedy's trips abroad.

Our lifestyle at Cheverly, a Maryland suburb of Washington located on the Baltimore-Washington Parkway, was that of a typical suburban family, including, even, the office carpool. Completely boxed in by traffic, me and my carpool buddies would speed down Kenilworth Avenue each morning, slowing down only to avoid a stalled car that was usually broken down on the bridge over the Anacostia River, the victim of a troll that obviously lived under it. In the evening, we reversed the procedure and were usually pleasantly surprised to see that the troll had quit operations for the day, apparently only being interested in slowing us down if he could make us late for work. And there were the malls which, after Bermuda and New Hampshire, had to have been a special treat for Marie and the girls.

In addition to being a great place to work under the command of Lieutenant Colonel George Moxon, the great advantage of being assigned to the clime center was the opening up of a wide world of educational opportunity. On different nights of the week, I found myself attending University of Maryland courses that I had begun on Bermuda. I attended classes at Bolling Air Force Base, Ft. Meade and at the university's campus itself in College Park. I found it quite pleasant to be parked in the shade of a tree at Prince George's Plaza, then a fairly new plaza, reading a homework assignment while Marie and the girls shopped. I think it is no exaggeration to say that I owe my bachelor's degree to my family's unswerving loyalty to the shopping mall.

I was fortunate, also, in that, if not at the shopping mall parking lots, I was able to study at home in the quiet surroundings of our community. Cheverly, with its picturesque hills, charmingly curved streets, pink-blossomed springtime trees, and ivy-laced brick homes shaded by large oaks, was the perfect example of a peaceful town. But all that changed when Bunky drifted into our yard.

A once white animal whose coat had been permanently dirtied through thousands of rummagings in town garbage cans, Bunky was a very large and very active tom cat. Every night, residents of our otherwise serene community went to bed to a chorus of shrieking capitulations of the many other cats that Bunky challenged.

Unusual was the night when townsfolk did not hear the screams of the unfortunate feline victims and rare was the morning that we and our neighbors did not awaken to see garbage cans thoroughly rifled and the contents strewn all about. When the debris was scattered hither and yon as if by a whirlwind, we knew Bunky had a paw in it. The complete annihilation of neighborhood tranquillity clearly was Bunky's prime ambition. In the world of catdom, Bunky was the top-ranking town tough.

As with most bullies, when Bunky's competitors backed off he only grew bolder. Soon, his banditry, which had been confined to the nighttime hours, also became noticeable in the daytime. One afternoon, Marie stepped out the back door to find Bunky on the brick patio in what seemed certain to be a struggle to the death with another cat, a stranger. She could see the stranger was not going to win. That other cat, though courageous, appeared to be about to feel Bunky's crushing jaws for the last time.

Reaching into the fray with waving arms, Marie tried to stop the slaughter. She succeeded in diverting Bunky's attention from his feline prey only to find he then switched opponents. With lightning swiftness, Bunky closed his jaws on Marie's right arm. Satisfied that he had made a deep enough impression, he then slunk out of the yard — defiant, snarling — and on the prowl for other victims. This happened on a day when I was at work at the Clime Center.

As I entered the house that evening after work, I immediately saw that Marie's arm was badly swollen. I rushed her to the Andrews Air Force Base hospital. After treating the bite, the doctor decided that, as a precaution against the possibility the animal had rabies, Marie would have to take a long string of horse-serum inoculations. The shots would be administered in the abdominal area. I had heard they would be painful.

"How many shots will she have to take?" I asked.

"One a day for 15 days," was the doctor's reply. "Unless you can capture the animal so we can tell if it is free of rabies."

That night I began work on a device with which to capture Bunky.

As we pursue the grand goals of a formal education, many of us fail to appreciate the less pretentious gems of knowledge that fall haphazardly on us in our everyday lives. We think those tidbits of information must not be very important, else they would come to our attention only after spending long tedious hours under library lights or equally long hours seeking enlightenment from some professor's lectures. Something like that flashed through my memory as I pondered the problem of capturing Bunky.

Back when I had been stationed at Selfridge, a weather observer, who reported for duty fresh from the North Carolina back country, described a device he had made that was enabling him to put meat on his family table while being paid only as an airman third class. The meat was rabbit and the device he made, which would capture the animal alive, he called a "rabbit box."

I now saw I could make use of his idea. Pleased with myself for having listened attentively on the day that airman had described his live trap, even though I then had little interest in it, I began to make one of my own. Instead of using an onion as bait, which he had used to capture a rabbit, I would use chicken. And I would make the box a little bigger, just big enough to accommodate Bunky's tail. I didn't want to hurt him, only to capture him so the doctors could see if he had rabies. Shortly after midnight of the day that Marie had been bitten, I had it built and deployed on the patio. As I inhaled the aroma of the nicely browned chicken which I tied to the trip lever inside the box, I knew that Bunky could not possibly ignore this trap.

The bedrooms in our house were all upstairs. This was the middle of summer, and being without air conditioning as were most houses at that time, we slept with the windows open. At about 0300 hours there was a noise on the patio. While still in my pajamas, I rushed downstairs to see what was the matter, and snapped on the patio light as the box flipped with a clatter. I literally jumped off the back porch and onto that brick patio, but before I could reach him, Bunky had the trap door ajar and had run into the backyard shadows.

The next day, while working on calculating percentage frequency of occurrence of wind directions with which to construct optimum runway orientations, under both VFR and IFR conditions, for a proposed Army airfield, the problem of capturing Bunky rested uneasily on my mind. I could hardly keep focused on the runway project and could give little thought to the discussions regarding the Buddhists in Vietnam who were setting themselves on fire as a way of protesting against the South Vietnam government. My personal problem was becoming the topic of discussion in those hallowed Clime Center halls.

Meteorologists, typically, seem to love to have problems to solve and are eager to contribute ideas that could lead to their solution, but most of my associates at the Clime Center were the kind of city boys who had never so much as entertained a thought about how to go about capturing an essentially wild animal alive. Eventually, Jim Gullion, a tech sergeant from Bluefield, Virginia, came up with an idea. It was entirely within Jim's character to be there when someone needed assistance. He was at our house that evening, and we began to construct his brain child. It was beautifully simple. Since Bunky practically lived in garbage cans, Jim said, we would turn our garbage can into a one-way visit for him.

Our project was delayed a little because it was time to take Marie back to the Andrews Air Force Base hospital for her second rabies shot. Even though this was only the second of a prescribed 15 shots, you could see in the way she walked back to the car that Marie had been slowed down. She had a difficult time bending the slight amount necessary to step into the vehicle.

That night, Jim and I worked feverishly on his invention. By about 11 p.m. we had it ready for trial. This is how we set it up: we installed a hinge on one side of the garbage can's lid and a latch on the other side. The lid was spring loaded and fixed with a trigger that was attached by wire to a freshly roasted chicken which we placed in the bottom of the can. As an invitation to Bunky, the lid was left open.

Satisfied that the terror of Cheverly, Maryland could not possibly escape this device, Jim went home and we went to bed. At about 0200 hours, we heard a bang that could have been the garbage can lid closing. I got my flashlight, walked downstairs, stepped out onto the patio and peeked around the side of the house where the can stood. The lid had been slammed shut. Slowly, I opened the can — but only an inch or two. In the beam of my flashlight I saw Tabby, the neighborhood's pet tiger cat. I let Tabby out. After tarrying a bit to rub her back along my legs, she went merrily on her way. Resetting the trigger, I went back to bed.

About an hour later, thinking I heard a noise, I retrieved the flashlight, walked downstairs, stepped out onto the patio, and peeked around the side of the house. The garbage can's lid was still open. I must have been dreaming when I thought I heard the noise, probably a nightmare about trying to capture a most wanted criminal. Back to bed.

Again, this time around 0400 hours, I thought I heard a noise. The lid was slammed shut. As I slowly opened the can, the friendly eyes of an angora stared back at me. I released it and went back upstairs to try to capture the few minutes of sleep that were left before the day's normal wake-up; but all I did was listen for the bang of the garbage can lid. Catching that tom cat wasn't going to be easy. But one of these times, I told myself, I will look in that can and the cat will surely be Bunky.

The next evening, after taking Marie for her third shot, we returned to find Jim installing a doorbell on the stair railing next to our bedroom door. In addition to being a weather forecaster, Jim had a natural aptitude for electronics. The bell was wired, Jim explained, to ring whenever the garbage can lid was closed. This he said, would allow us to sleep soundly knowing the bell would wake us when there was a need to check the can. And that was exactly how it worked. We were able to sleep between bell-ringings. But the bell rang a lot more than we wanted it to. I released four cats that night — not one of them looked even slightly like Bunky.

"Have you caught the cat?" the doctor asked as I took Marie for her fourth shot. The hospital people, who were normally sort of blasé about inoculations, were beginning to cringe themselves when they gave Marie the shot. Though she kept that stiff upper lip, we could all see the shots were taking their toll. The doctor's remark further energized my desire to catch Bunky, but it really wasn't as if I needed any more incentive.

The story was the same the next night. Our garbage can was being visited by every cat in the neighborhood. On that third night of the garbage-can trap, we began to see repeat performers. Tabby came out of the can and rubbed against my legs again. Two of the other cats found in the can also were visitors from previous nights, Cheverly's cats apparently had identified that garbage can as the "in-place" for their nighttime snacks. But more importantly, it was clear that Bunky, a clever and sinister cat personality, was on to us. There was nothing to do but build some other trap.

In the carpool the next morning I gave Jim the details of what would be the third generation trap. "We'll staple a chicken-wire dome to a six-foot-square wooden frame made of 2X6s. It's size and the 2X6s will make it too heavy for Bunky to lift."

"Sounds like it'll work," Jim said, "provided you place it on the brick patio floor. That way, Bunky won't be able to dig under it." And that was what we did. We suspended the whole bulky, heavy affair from a branch of the big oak that overhung the patio. As with the other traps, this one would release when the bait was taken. We had it built on the eve of Marie's fifth rabies shot. After that shot, the pain was so great she had to support herself with both hands on the wrought iron porch railing in the simple act of stepping up the three short front steps of our house.

That night, at about 0100 hours, there was a terrific crash. Down the stairs I scrambled. On with the patio light. I looked out through the kitchen door's window. There, on the patio floor, staring malevolently at me from within the domed chicken wire trap was Bunky! No one could imagine a bigger sense of relief than what I then felt.

Mindful that I still had to get out there and remove Bunky from his wire enclosure while avoiding his steel-trap jaws, I edged the door open. Slowly and quietly, I stepped onto the patio. Bunky fixed me with his eyes, hissed, snarled, and as I continued to slowly approach him, dashed to the far side of the cage, butted it with his wide head, dug into the brick patio floor with his front legs, lifted the entire enclosure enough to get part of his body under the 2X6 frame, and literally pulled himself out with his front legs. In a flash he ran into the night, leaving claw tracks in the bricks one-thirty-second of an inch deep.

In the evening of that day, Marie's footsteps were slow and deliberate, that ever-so-slight lifting of a foot required to accomplish the forward movement that we call a footstep, as I guided her from the car to the hospital for yet another shot, became a hopeless effort in pain minimization. In fact, the pain from those daily shots in the abdominal area just could not be minimized. In the evening of the next day, the hospital staff brought a much-needed wheelchair out to the car to move her to the inoculation room. And she was not the only basket case. I have no idea of what I was then doing at the Clime Center — different devices that might be employed to capture Bunky constantly interrupted any possible climatological thoughts.

Unable to think of any new trap that might work, in desperation I resorted to the first — the rabbit box that I had tried at the very beginning — the one that the weather observer at Selfridge had described for me. I reasoned that the time that had passed since I tried that one might be sufficient. Bunky might have forgotten it and would fall for it again. But it had to be so big and heavy that, when he entered it, Bunky could not possibly tip it. I made this Mark IV trap three times as big as that first effort, both to increase its weight and also to fool Bunky into thinking he had not seen anything like it before. And to give him an even more secure feeling, I made the top of chicken wire so that he could see the night sky as he entered it, helping to develop in him the thought that he was only exploring another of his nightly haunts.

It was already daylight of the morning after deployment of that new and improved box trap and time for me to leave for the Clime Center when, in an almost offhand way, I thought I should check the box. I had actually given up hope that Bunky could be there, but the trap door had sprung shut. He was in there!

Not being willing to take any more chances with him, with a hammer and eight-penny nails I secured the door of his temporary lodging and left for work. At the office, I called the hospital to tell them I had at long last captured the animal. They said they would send someone out to our house to get him. In about an hour, Marie called to say that two little old ladies from the humane society were in the back yard. They were holding a short length of thin grocery string and were looking through the wire top of the box saying: "kitty, kitty, kitty." They told Marie they were going to open the box, tie the string around Bunky's neck and walk him to their car. "You'd better wait while I call my husband," Marie said.

Those kindly old ladies planned to sweet talk and lead by a piece of grocery string a tom cat that could fight off all the other felines in the town of Cheverly, even if they all came at him at once and he had three paws tied behind his back! I told Marie to decline their offer with the explanation that we didn't think Bunky would take kindly to having a string tied around his neck.

That evening, I delivered Bunky, still safely in the box, to the hospital where I was told the terror of Cheverly did not have rabies. Marie did not accompany me, which was very good. Her pain by then was so great I didn't know how I could have gotten her down the three front steps of our house. Her shots were thus ended eight short of the scheduled 15. I am told that inoculation for rabies is no longer the same, and is now not much different from other inoculations. For that great medical advance we should all thank our lucky stars.

In the days before the great advances in electronic computers, the slide rule was the close companion of anyone having to deal with complex numerical problems. This picture was taken in the Climatology Course classroom. *Official Air Force photo of the author, Chanute AFB, Illinois, June 1961.*

Chapter 24

1st Air Cavalry

For pinpointing of accessible targets, the air was normally not so effective as artillery....
  General of the Army Dwight D. Eisenhower[1]

In the village, poinsettia bloom beside a mud-walled hut where a woman with betel-nut blackened teeth squints at a tattered cloth as she pumps the treadle of her sewing machine. In a dark corner, chopsticks ablur, a gray wispy-whiskered man, gaunt, streams rice up from a bowl — it is breakfast time. On their straw-covered roof, a rooster pecks, scratches, crows his morning alert. In the yard, thin like his parents, a boy drags a scraggly bow across a one-stringed poor cousin of a violin he calls a "goa." Bananas hang from a tree. The belly of a low-slung pig makes a slow track in red dirt. At the shop next door, a carpenter looks for his wooden plane. He thinks of taking one last stab at the rough boards of a coffin that rests on flimsy props.

Beyond the village, women in black pajamas, pant legs rolled up, shield themselves with conical straw hats from the already fiery sun and plant rice seedlings in a flooded paddy. A water buffalo, his plowing done, watches the planters while standing knee-deep in the murky water. Ahead, more women, dressed like those in the paddy, gather at a river bank to wash clothes. They share the stream with hump-backed Asian cattle who send slow eddies through the sluggish yellow current.

Farther on, fields of tapioca. Bamboo thickets in which a monkey swings from stalk to stalk and a golden pheasant flashes its red feathers, unwittingly giving its position away. Orchids bloom in the steaming jungle.

We look for snipers, signs of mines and swerve to miss a flock of ducks herded by a man with a stick just as one would herd cows in Michigan. Three Vietnamese soldiers in greenish uniforms lie side-by-side, perfectly aligned on the beach-like sandy shoulder of the road, arms and legs as straight as if standing at attention. All three are young, all three were shot during the night — all three are dead. Whose side were they on?

****

---

[1] General of the Army Dwight D. Eisenhower, *Crusade in Europe,* Doubleday & Company, Inc. Garden City, New York, 1948, p.323.

I had come to this place, the Central Highlands and Central Coastal Plain of the Republic of Vietnam, commonly known as South Vietnam, as a result of the domino theory. If you followed world events, you had to believe that theory: as the Soviet dictator Stalin had abundantly proven, when you let one country fall to the communists, others will follow. And I had arrived in Vietnam in a roundabout way. Air Weather Service alerted me for weather forecasting duty in Thailand but dropped the alert because I had entered the Air Force's Bootstrap Program, attending University of Maryland classes full-time at its College Park campus, like any day student. The long years of nighttime credits from U of MD or whatever other college I happened to be near — Wayne State, Detroit Tech, American International College, University of New Hampshire — had finally brought me to my last semester of the bachelor-degree program, the criterion for acceptance to Bootstrap.

A few months later, with my degree almost in hand, I met a selection board at Ft. Myer, Virginia after which the Army announced I was to be appointed a warrant officer in the Artillery. Two other Air Force sergeants, Jim Gullion, my old friend from the Clime Center, and a new friend, J.D. Witherington, who had been assigned to the Air Weather Service's Scientific Services at Andrews Air Force Base, had met the same board and were also appointed warrant officers. We were the first weather forecasters to make that unusual move from being sergeants in the Air Force to warrant officers in the Army Artillery. In about a year, other Air Force sergeants would follow us, presumably after the Army had decided they had not miscued in appointing the first three. At the time the three of us were accepted, there were only 40 warrant officer meteorologists in the entire Army. Jim and J.D. almost immediately received their orders transferring them to the Artillery, but I did not. A holdup of some kind was in process, apparently, at Air Weather Service Headquarters. I thought the delay might have originated at Air Force level. Having paid for my final semester at Maryland, there could have been a reluctance to let me go. It was only after a number of queries of a major at the Pentagon in charge of those things for the Army, and his query of the Air Force that I finally received my orders.

I purchased the green Army officer uniform from Lauterstein's in Alexandria, Virginia, pinned on the warrant officer's bars, and began to attend to those family matters that were always precipitated by an impending move in the military. We would sell the house in Cheverly, leave Leta with trusted next-door neighbors so that she could finish her last semester at Bladensburg High School and transfer the two younger daughters, Willa and Pamela, to a high school near my new duty station.

Jim, J.D. and I reported for duty at the Army's Artillery and Missile School at Ft. Sill, Oklahoma, located at Lawton, Oklahoma. On my first day there, I was told the three of us were to attend a formal function that evening at the Hotel Lawtonian in Lawton. Dress was to be the Army's formal uniform, that navy blue jacket with Civil War-style yellow-bordered epaulets of rank and the royal blue trousers. Having not yet had an opportunity to order the dress uniform, I bought the acceptable alternate, a civilian suit, at a Lawton clothing store and attended the function. We three former Air Force persons and our wives were introduced by Colonel Buntyn who had a degree in meteorology and was in charge of the Target Acquisition Department, under which the Meteorology Division was organized. Colonel Buntyn warmly welcomed us and could not resist referring to Jim, J.D. and I as "those blue suiters" (Air Force guys) at the introductions.

At Ft. Sill, we were also warmly welcomed by Major Edward Mollichelli, head of the Meteorology Division, who had a masters in meteorology from Florida State, and by the warrant officer cadre who had come up from the Army enlisted ranks. We were soon familiarized with the details of Artillery ballistic meteorology operations. This was followed by brief assignments as meteorology instructors and then with assignments to Vietnam. Of the warrant officer meteorologists at Ft. Sill, two of the former blue suiters, J.D. and I, were the first to be sent to Vietnam — J.D. with the First Infantry Division and I with the 1st Air Cavalry Division (that we would habitually refer to as "First Cav").

My Vietnam assignment was to the 3rd of the 18th Artillery. Newly reorganized as an 8-inch howitzer battalion with a 175-millimeter gun battery and a later augmented 155-millimeter howitzer battery,[2] the 18th traced its heritage to Gettysburg. The first of the heavy artillery to arrive in Vietnam, 3rd of the 18th was attached to First Cav. The 1st Air Cavalry Division had deployed to Vietnam about a month earlier, following President Lyndon Johnson's July 28, 1965 televised announcement that he would be sending division-size forces to Vietnam. First Cav had been organized at Ft. Benning as the 11th Air Assault Division by Major General Harry W.O. Kinnard Jr.[3] This deployment of division-size units would mark a major turning point in the war. First Cav would be the first of the air mobile divisions. In Vietnam, it would test the feasibility of air assault by helicopter. Our heavy artillery augmented the division's 105- and 155-millimeter howitzer firepower.

---

[2] A gun typically has a shallower trajectory and longer range than a howitzer which, in turn, has a less pronounced vertical path and much longer range than a mortar.

[3] *US News and & World Report,* "The Warrior Class," July 5, 1999, pp. 29-30, an article about Kinnard's West Point class of 1939. During World War II, on D-Day, Kinnard, a 29-year-old colonel, had jumped with the 101st Airborne Division behind German lines. Later, during the Battle of the Bulge, when Maj. General Anthony McAuliffe said "Nuts" to the news that the Germans were coming to demand that his surrounded Americans at Bastogne surrender, Kinnard suggested that McAuliffe also use that as his later-to-be-famous official response to the Germans.

My mission: to weld together for duty with First Cav the battalion's 13-man ballistic meteorology section that the Army called a "Metro Section" and that I was trying to rename "Met Section" ("Metro as a short name for meteorology made no sense to me). In Vietnam, we would provide artillery trajectory corrections to compensate for changing meteorological conditions. The atmosphere affects the flight of an artillery round in much the same way it affects the flight of an aircraft. When compared to a no-wind condition, winds can either lengthen, shorten, or change the directional path of projectiles fired by artillery weapons. Similarly, when compared to a standard air density, changes in air density will either lengthen or shorten the distance the projectile will travel.

We were forming up with 3rd of the 18th Artillery at Ft. Lewis, Washington where the staff sergeant in charge of the Met Section briefed me on the situation. The men of the section had come from various Army posts scattered across the country. Our GMD-1, the equipment heart of the section, came from the 4th Infantry Division at Lewis and was being packed for shipment. I would have preferred that the men train as a team before leaving for Vietnam, but that was not possible. The sergeant convinced me things would go smoothly. In the coming months, I would find the men performing as meteorological professionals — his confidence was entirely justified.

We boarded buses for the trip south to Oakland, California and the Oakland Army Terminal where we boarded the *General Gordon,* a troopship. We'd be sailing for Vietnam, but first, days of waiting at dockside, peering out portholes to view the mob protesting our existence. Someone jokingly suggested that we nab the ringleaders, stuff them in duffel bags and take them with us to Vietnam. But a wiser one countered: "Would you want to have any of *them* guarding your perimeter?" And that, as the utterly deplorable parade of protesters and draft dodgers increased through those Vietnam years, was the best defense I could make on their behalf. With the opportunity for reflection that would come with the calming of the post-Vietnam era, I would see we were much better off having the protesters burn their draft cards, or go to Canada or Sweden or even Moscow than to be unlucky enough to have had them nearby when lives would depend on the fidelity of others.

Loaded with 3,900 men, 21,000 dozens of eggs, 3,000 cases of milk, 35 tons of potatoes, 150 tons of meat, 300 cases of lettuce, and 4,000 gallons of ice cream, we sailed under the Golden Gate. Haunting strains of "I Left My Heart in San Francisco" played on the Gordon's loudspeakers.

Remembering my experience with deep sea fishing off Bermuda, it would come as no surprise that I would be seasick from the time we left port to the time we would anchor off Vietnam. That would be nearly an entire month. For most of that voyage, I would be in the cabin shared with other warrants and a captain, an irrepressible personality and instigator, usually, of the poker game. They were all flyers. From the vantage point of my cot, I watched the aviators' nearly continuous poker game. I would be there because the only relief from nausea came while lying on my bunk, flat on my back, in a perfectly prone position. For that reason, I never entered the game. But also, even though Dad had been an avid poker player, I never learned the game. After entering the service, with a wife and later, children, to support, I had been motivated to not squander any of my pay.

At that stage in the Vietnam conflict, most of us thought the war wasn't much of a war at all, and the oft-repeated joke went like this: "It isn't much of a war, but it's the only war we have." I would then have considered it unlikely that the happy-go-lucky officers in that poker game could become casualties. After arriving in-country, I lost track of all but two of them. About half way through my tour, I learned that one of the warrants had been very badly shot up but was still alive. It is interesting that what I remember most about him was his detailed explanation of what he was going to do when he returned from the war. He would be going into the chicken-raising business, explaining that the fowl at different stages of their growth would be housed within successive elevated levels of the chicken house. I never really understood how that operation would work, but I hoped he would return to begin it. The captain, I learned, had been killed while trying to flush out some Viet Cong from their spider holes. As a flyer, that was not his job but it was entirely within his character to do it. He apparently couldn't resist the thought of temporarily trading the air war for the ground war. I never knew what happened to the other warrants in that cabin, but the war probably also ended badly for some of them for the warrant officer rank would become the rank with the highest percentage of casualties.

On those rare occasions that my stomach was in a quiescent mood, I liked to go to the ship's rail and watch the flying fishes while, at the same time, I looked for someone who might provide information that could be used when we reached the jungles. From veterans of the Korean War, I had been advised that, given the normal supply situation in a combat zone, I should give considerable attention to the matter of keeping my ballistic meteorology section adequately supplied with expendables such as radiosondes, balloons, parachutes, calcium hydride which was used in generating the balloon-inflating gas and replacement parts for the GMD-1, the radio direction finder that tracked the radiosonde.

One day when the Pacific was not too swishy, I was standing at the rail and met the warrant who would be in charge of obtaining the battalion's supplies. I asked him what could be done should I find myself in the situation of being short of one of the critical components needed to accomplish my mission. "We can query the supply depot," he said, "but they'll probably answer by saying 'sorry 'bout that.' Sorry 'bout that," he explained, "was the new cliché born of the Vietnam War." To have said "no sweat," the more positive response, I took to then be passé. I got the impression, later more than adequately confirmed, that Vietnam was not going to be a war that would generate optimistic sayings.

Several times when the Pacific was being cooperative, I found my maintenance technician, Specialist 6 McCurley, standing at the rail. This was the man who would keep the GMD-1, which we had obtained from the 4th Infantry Division at Ft. Lewis, and its ancillary equipment maintained. He was from the South and had at one time been an Army cook. I asked how difficult it would be to keep the GMD-1 running. He said it would not be a problem and, continuing the conversation, asked: "Chief, do you know how many pounds of flour goes into making pancakes for a hundred men?"

Of course, I hadn't the slightest idea. The fact that my maintenance man's background was in cooking did give me some concern as I thought of the complex electronics he'd be responsible for. But he was completely confident, and he tried his best to put me at ease. After we began to bounce that GMD-1 over the makeshift roads of Vietnam's Central Highlands and Central Coastal Plain, I would remember his confidence and would find that I couldn't have been luckier in a maintenance man.

We stopped one afternoon at Okinawa's White Beach and then, on the 20th day of our voyage, anchored off Qui Nhon, a city about half way up the South Vietnam coast. The ship's newsletter carried the following announcement: BE SURE YOU CRUSH OUT CIGARETTES BEFORE THROWING THEM OVERBOARD...VERY SOON SMALL FLAMMABLE VIETNAMESE BOATS MAY BE ALONGSIDE....

The little boats, true enough, were soon alongside, carrying fishermen wearing conical straw hats and black pajamas. After waiting several days to unload, during which we watched tracers arc across the night skies, we climbed into landing craft that took us ashore.

Reminiscent of a John Wayne movie, on one fine forenoon we splashed through the last few feet of the South China Sea and walked up on a beautiful sandy beach. With our arrival, and the arrival of other units to the south, including the First Infantry Division, we would boost the American presence in Vietnam to 125,000.

The first order of business on the beach was the issuing of ammunition. I received four clips of 7.62 millimeter ammo for my M-14 rifle, the personal weapon carried by the battalion. This was October of 1965, too early in the war for the familiar and much lighter M-16 which would later become the trademark rifle of the Vietnam era even though the M-14 was superior for long range accuracy. (When it did appear for issue, we called the M-16 the "Mattel toy" because it looked as if it was something purchased at Toys R Us.) There were 20 rounds in each clip. This we carried in two pouches attached to a webbed belt upon which also was attached a first-aid pouch and canteen. That done, we climbed aboard trucks and headed west.

In the afternoon, we camped alongside Highway 19, the east-west road that connected Qui Nhon to the Central Highlands and Pleiku. Eleven years earlier along that road, a French regiment called Group Mobile 100 suffered heavy casualties when they were ambushed by the Viet Minh. The Viet Minh were the antagonists faced by the French — our opponents would be the Viet Cong (VC) and the North Vietnamese Army (NVA).

It was a muggy day made worse by our heavy fatigues. The new light-weight jungle fatigues, like the M-16 rifle, which would commonly be seen in the media photos of that war, were not yet being issued. Those heavy fatigue uniforms, as well as the heavy M-14 rifle, would be with us throughout that tour.

Toward evening, we pulled into a rare flat area at the side of the road where road machinery, a bulldozer and grader, long unused, were parked. As we were setting up camp, a Vietnamese boy, about 12 years old, wandered by and our S-2, a major, who understood French, attempted to speak to him in that language. French seemed the logical second choice since none of us spoke Vietnamese. He asked the boy if there were VC in the area; but the boy did not understand the question, did not know the answer, or did not want to tell.

Just before dark, the battalion CO, a Lt. Colonel and veteran of World War II, sent word that all officers were to report to a tent that had been set up as a very temporary headquarters. We found him engaged in conversation with a slightly built soldier, who appeared to be a junior officer in the Army of Vietnam, an "ARVN," and an American captain, probably a green beret, who had accompanied the ARVN and understood Vietnamese. The Vietnamese soldier, as I interpreted the conversation, seemed to have requested that an attack be made on a certain target. He pointed at a spot on a map and fervently repeated: "VC, all VC!"

Being our first day in-country and then on our way to join up with the First Cav, and with our artillery still to be brought up, we were in no position to take action. And as I recall, the conversation also turned on the need, in any event, to obtain clearance before action could have been taken. What was most likely a VC target would elude confrontation with us that day.

That episode was our first exposure to the difficulty of fighting the Vietnam War. As believable as was the ARVN's word, you couldn't be absolutely sure his target was really VC. And if they were friendly Vietnamese, and, heaven forbid, you hit them, you wouldn't be winning the "hearts and minds of the Vietnamese people," a goal that had to be paramount if we were going to prevail against the VC and the NVA.

Afterward, in the dimming light, we dug foxholes, blew up our air mattresses and laid them beside the holes. Mosquitoes buzzed. This was malaria country. We dug into duffel bags, pulled out netting, propped it on sticks stuck into the sand, crept under it.

As it became truly dark, I remembered the Golden Delicious apple I had saved from the ship's mess and pulled it out of its duffel-bag storage — a last bit of evidence of the fantastic shipboard meals that would often be recalled while we "feasted" on C-rations in the months to come. I held that Golden Delicious up before me, rolled it in my hand in the starlight, partly seeing and partly imagining its soft yellow luster. Like a would-be juggler, I tossed the apple several times, enjoying again its solid mass each time my hand closed around it. As the starlight faded under a drifting altocumulus cloud, I missed the catch and the apple rolled down into the foxhole. I picked it up, brushed the sand off, watched the altocumulus deck move off to expose a bright white satellite, probably a weather bird, picking its way along a curved path high above the stratosphere, through the star field. I ate the apple. An hour later, small arms fire erupted nearby, telegraphing news of an engagement between militia forces and the VC. Still later, a First Cav UH-1 *Iroquois* that we would call the "Huey," whop-whopped far above our foxholes, staying high enough to reduce the chances of ground-fire hits.

In the morning, we convoyed to An Khe, through An Khe Pass. The pass was in the last stage of being secured by First Cav troopers. An Khe, located in the Central Highlands, was First Cav's base camp. It had begun to occupy that site when it arrived in August. Now, in late October, it was about to participate in the first full-scale combat action between the NVA and the United States Army. These hard-fought battles of the Pleiku campaign, principally the Ia Drang Valley, were also the first major infantry attacks by helicopters in all history.

The Division had its own ballistic meteorology section, a normal component of Division Artillery. During the Battle of the Ia Drang Valley, the warrant in charge of that section sent a visual team, a PIBAL team, to the battle zone to provide on-site ballistic meteorology support to the howitzers. The team came under fire. One of the ballistic meteorologists was injured and their theodolite holed in several places by NVA rifle or machine gun fire.

Unlike the fully capable electronic atmospheric sounding equipment with the radio-direction-finding GMD-1, a visual (PIBAL) team could only provide ballistic data when the sky was clear enough to allow sighting of the balloon through the theodolite's telescope. The speed and direction of the upper winds could then be calculated from the sighting angles. As for that equally important other variable, the air density, with a visual team the density aloft was calculated from tables based on assumed changes with height with the starting point being the density as actually measured at the earth's surface. To rely on assumed changes with height could not, of course, be as accurate as to use the temperature, humidity and pressure values measured with the radiosonde and then to transform those data into density values. The electronic method obviously was the preferred method.

The Division Artillery meteorology warrant whose visual team had been in the Ia Drang encounter with the NVA, came to see me. Unlike mine, his entire career had been with the Army. Back when I was in the Air Force, making forecasts in complete safety in Saudi Arabia, he had been in the Korean War, had been captured by the Chinese and had suffered severe frostbite injuries. He wanted to discuss the deployment of our two sections. One of our sections, he said, should be kept there at the An Khe base camp to support H&I (Harass and Interdict) fires, fires that would discourage the enemy from approaching the camp, and the other should be the mobile section, the one to go out with the Cav's combat elements. He then explained why his section should be the one to be kept at An Khe. In the effort to achieve air mobility, his section had been stripped of its heaviest equipment. All his vehicles — the two 2 1/2 ton trucks, which we called "deuce-and-a-halfs" (in our grammatically incorrect but preferred manner of speaking), the plotting van built on a deuce-and-a-half chassis, the section chief's three quarter ton truck, the trailer upon which the dismantled GMD-1 was to be transported and even the water trailer — had been taken away. In their place was an odd looking vehicle called a mule, a small platform on wheels that was a little longer than a golf cart which could only be used to move things around within the section's compound.

I asked how it would have been expected to keep him supplied if he were out on an operation. In the plan, he said, when he was out supporting a combat operation, his GMD-1 and supplies, including the large amount of water needed for production of the balloon-inflation gas, would be delivered to him by helicopter; but he feared that, for various reasons, primarily the likely perception that there would be other missions of higher priority, the delivery situation would be unreliable. Ironically, the effort to make his section more mobile by removing his vehicles, had in effect made it quite immobile. He wanted to know if I had any objections to being the section that would go out with the Cav's combat elements.

I had no objection, and if I had planned it, it couldn't have gone better. For one thing, given the nature of the Division Artillery warrant's Korean War service and the fact that his men had taken fire at the Ia Drang, I thought it to be no less than fair that they should now stay in base camp. Also, in the short time that I had to get acquainted with my 13 men, I knew most of them would have preferred to do what they had come to Vietnam for. They could not be kept very happy grinding out meteorological data in base camp only in support of H&I fires while the real action was going on out there far beyond the camp's perimeter.

Within a week or two, our equipment arrived, and we selected a small clearing between trees, found a small flat spot to park our plotting van, installed our GMD-1 on another flat spot, set up the balloon inflation shelter on yet another, connected the radiosonde recorder that would print out the data retrieved from the radiosonde, filled our 400-gallon water trailer at the combat engineer's water point, set up the balloon inflation apparatus and began making test runs of meteorological messages. At about the same time, the battalion's 8-inch howitzers and the 175-millimeter guns were emplaced. These would have a much longer range than the 105- and 155-millimeter howitzers that the Division had been using. (The diameters of 105-, 155- and 175-millimeter shells were, respectively, approximately 4-, 6-, and 7 inches.) We had only been operating about a week when the battalion commander stopped me on my way to the officers' mess tent to proudly state that the battalion had scored a first round hit on an enemy bunker complex while using the meteorological data supplied by my section. That was important, as it was about the only chance we'd have to hit the enemy decisively from base camp. Now that he knew how far our eight-inchers and 175s could reach, he'd take care to assemble out of their range, and we'd have to move those guns out of camp to hit his concentrations. Back at Ft. Lewis, I had some concerns about this meteorology team, thrown together from all parts of the US and just in time for boarding ship. They had not trained together and had not seen our equipment until it arrived at An Khe. This was their first chance to show what they could do. I couldn't have been prouder of them.

Soon, we were on the road, making up the rear of a First Cav convoy, usually just behind the ammunition carriers. Not a good place to be in the event of an attack, but *c'est la guerre*. In any event, we were not attacked. Once, we camped on a hill completely covered with small woody plants having clusters of flowers with tiny purple, yellow and white petals. Just a few months earlier, I paid $10 for only one such lantana shrub that I began growing in my backyard at Lawton. Here, there were millions of them, and no one was the least bit interested in them.

On those hills and at other locations, we installed our equipment, released balloons to each of which, with tough waxed string, we had tied radiosondes, and computed ballistic data for the howitzers and guns that accompanied us. Then, needed elsewhere, we dismantled the equipment, stowed it carefully to give it as much protection as possible on the rough roads and trails, and tagged along behind whichever battery needed our data. If it were B or C battery, we'd be with the 8-inch howitzers, if it were D battery we'd be with the 175-millimeter guns, those with the barrels that seemed to be as long as telephone poles. Later in the tour, a provisional 155-millimeter howitzer would be attached to us as air-mobile E battery, making 3/18 unique in the Artillery. Unlike the 8-inch and 175 millimeter, the 155 millimeter artillery piece would be light enough to be lifted by the Cav's helicopters. The Division then had more than four hundred helicopters, including the UH-1, OH-13, and CH-47, as well as four CH-54's[4]. I was told that all of the CH-54's then in existence, the huge choppers we called the "Flying Cranes," were at An Khe.

Bumping along, we think the Artillery song — *Over hill, Over dale, We will hit the dusty trail* —had been written expressly for us. Except that — there was very little dust. Mud there was. And rain. But very little dust. We were entering the six-month period, extending from November or December to March or April, of the northeasterly monsoon which was the generally wet monsoon in this part of the country. The rest of the year would come under the influence of the southwesterly monsoon which could also be wet at times.

One day we camped on a hill with very good radio reception. While trying to find a tune by Doris Day or Connie Francis we twirled the dial of our AM radio to Hanoi Hannah's frequency. Hannah informed us of a massive electrical power failure in New York. She told us the First Cavalry Division was about to suffer a humiliating defeat. A couple of weeks later, when our hometown newspapers caught up with us, we were quite surprised to learn she had been right about the New York brownout. But she would be as wrong as she could be about the fortunes of First Cav.

We participated in an operation that was to clear the enemy from the remaining contested area alongside Highway 19, on the route to Qui Nhon. One of our camp sites was on the grounds of a deserted and bullet-ridden Buddhist temple. It was near Thanksgiving Day. Incessant rains of the cold season monsoon had turned the temple yard into a sticky quagmire. It was raining when we arrived and would continue for several days. With the choice real estate taken by B Battery's eight-inch howitzers, we had to settle for a small spot of slightly elevated land, somewhat drier than the mud all around it. Small mounds bumped up out of the landscape, making it difficult to find a flat spot upon which to erect the equipment and personal tents. Most of the

---

[4] Col. Kenneth D. Mertel, *Year of the Horse — Vietnam,* Bantam Books, New York, New York, 1990, pp.1,26.

men placed their folding canvas cots within the dry balloon inflation shelter. I slept in the bed of my 3/4 ton truck which had been parked atop one of the mounds.

The uneven terrain wasn't the only thing less than ideal. We were set up under the barrels of B Battery's 8-inchers that were maintaining an almost constant fire. The ear-piercing concussion from those outgoing rounds jarred our delicate radiosonde recorder, the instrument that made traces of the signals transmitted to us by the radiosonde. With the firing of each 8-inch round, the recorder's pen did a jig, producing a squiggly line where a straight line was meant to be.

Thanksgiving arrived and we were still on the temple grounds, wearing ponchos to ward off the rains of the cold season monsoon. The battalion mess sergeant, who had set up his mess within the walls of the bullet-pocked temple, told my troops we were having turkey for our Thanksgiving dinner. I didn't believe it and my skepticism seemed to be confirmed as I stepped up for my dinner and watched the mess sergeant slice off a half-inch section of cylinder-shaped meat that had obviously been molded into that shape by spending considerable time in a can. It looked as if it were a large-can version of spam. This was a bit of a disappointment, especially since I had decided not to walk over to the chow line from the Met van in the steady rain but changed my mind when I was told turkey was to be served. The mess sergeant looked strangely pleased with himself, one might even have described him as beaming.

When I got my serving, I said, "I thought the troops were to have turkey today."

"Chief, that *is* turkey," the proud mess sergeant replied.

I tasted it. It *was* turkey. "And it's darned good, too!" I said. That mess sergeant had produced a minor miracle. And what a great change from our routine C-rations.

After we had been at that site for about a week, the small mounds all around us started to make sense. It began to look as if we were camped atop grave sites. Several days later, B Battery began to march order on short notice — the war was moving to another location. We packed our equipment as fast as we could and made our way down the road, barely within sight of the main column. As I looked back, I was quite sure we had indeed camped on a cemetery, and I felt remorse at having violated the sanctity of the site. I had been sleeping directly above a grave. I hoped no friendly Vietnamese had observed us.

Top: An 8-inch howitzer on Operation Black Horse north of Qui Nhon. The object in background appears to be a grave monument. February 1966.
Bottom: The author at the GP-medium tent, the "officers' quarters" while in base camp. An Khe, South Vietnam, April 1966.

Troops of the ARVN 22nd Division move out on patrol along a road on the coastal plain north of Qui Nhon. March 1966.

Chapter 25

300 NVA at 1,000 Meters...
or 1,000 NVA at 300 Meters

And we just cannot now dishonor our word, or abandon our commitment, or leave those who believed us and who trusted us to the terror and repression and murder that would follow. This then, our fellow Americans, is why we are in Vietnam. President Lyndon Baines Johnson, television address to the nation, July 1965.

Not long after we first arrived at An Khe, we were gathered together under some tall An Khe trees for a talk from Major General Harry W.O. Kinnard Jr., Division Commanding General. He welcomed us to the Division, mentioning, among other things, that we would be entitled to wear the famous First Cav shoulder patch — the black horse-head silhouette and black slash on a yellow background. In Vietnam, however, that easily spotted insignia was modified. Troopers wearing the patch substituted a greenish color for the yellow, this blended well with the fatigue uniform.

Although I frequently reminded the battery commanders that we needed time to dismantle and carefully pack the equipment, they routinely gave me far too little notice to march order. With their tracked howitzers and guns they could move out to the next firing site almost as soon as the move-out order was received, and they probably thought I could do the same. In addition to the two deuce-and-a-halfs and the plotting van with its ANGRC-19 AM radio, I had the 3/4 ton vehicle I rode in which contained our PRC-25 FM radio, and two trailers. We packed the dismantled GMD-1, including its 8-foot dish, in one trailer. The other trailer was a 400-gallon water tank on wheels, a vital piece of equipment, for without water we couldn't initiate the chemical reaction with calcium hydride that produced our inflation gas (nor could we drink). Of course, I knew the battery commanders had no control of the rapidly changing tactical situation that dictated the hurried moves.

In spite of the short notices, my men would always get the equipment packed in time to at least see the tail end of the battery on the road ahead — but there were times when we came uncomfortably close to missing them. It wasn't wise to be out there alone. I was a firm believer in the old adage — there is safety in numbers.

Proud feelings welled up as I observed the performance of that Met Section. I tried to remember the human relations lessons presented at the Air Force's NCO Academy at Orlando that I had attended while temporarily away (TDY) from my duties in Washington, but in one thing I was completely merciless. Every time we reached a new campsite where it was likely we'd be more than a day or two, every man not working on setting up equipment or working on data for the batteries, was to be digging a foxhole. They did that and even, sometimes admittedly at my insistence, wore their helmets while on the road. Once, when we were negotiating the hairpin curves through An Khe Pass, a place where a sniper could leisurely take aim at you from the bluffs immediately above your truck, I had to say: "Look, I know you're not afraid of being shot (probably not true), but just think of how silly you'll feel if a VC wiped you out by rolling a rock down on your head." And the helmet would then appear in its proper place, on that soldier's head.

When passing through known contested country, we kept our M-60 machine gun at the ready in the open bed of one of the deuce-and-a-halfs. We also had a rocket launcher and of course each of us had the M-14 rifle. Frequently on the move, we would occupy many different sites. From a defensive perspective, the preferred site was a hill, preferably a hill that was clear of trees or brush which could give cover to the enemy. If there happened to be such cover, we used our pioneer tools to clear it away. While camped one night on one of the several hills we would occupy, automatic weapons fire erupted a short distance away and pink tracers arcing high into the night began a slow migration toward our site. "What are we supposed to do if the tracers come right at us?" a young trooper asked.

"Dive for the foxhole," I said. "That's why you've been making Swiss cheese of this country with your shovel." And as an afterthought: "Don't fire unless you've been told to." In what had developed in a sort of unplanned way, our very best defense was to avoid exposing our position and to move so frequently that the VC or NVA would have a tough time planning an attack on us.

As Christmas approached, we were back at An Khe, this time to hear a talk from visiting Army Chief of Staff General Harold K. Johnson. General Johnson, like General Kinnard, was a much respected soldier. Johnson was a survivor of World War II's infamous Bataan death march. He made a point of saying, in a tone of great sincerity, that the President was thinking of us every day. And I wondered if, even at that early date, with casualty reports coming to him, President Lyndon Johnson was having misgivings about our presence in Vietnam. The first large-scale contact with NVA forces had occurred in the last weeks of October and the first part of November with the battles fought in the Ia Drang Valley area where the NVA goal was to cut South Vietnam into two parts. First Cav killed that NVA plan, but we lost

305 brave men[1] in the campaign and among them were troopers of the First Cav's 1st Battalion, 7th Cavalry, the descendant of George Armstrong Custer's ill-fated unit at the Little Bighorn (NVA killed were estimated at about twice that number). Shortly after, I paused one day in silence at the white cross that was planted on a small hillside at the 7th Cav's An Khe base camp. A passing trooper threw me a smart salute and shouted "Garry Owen," the title of Custer's 7th Cav song. A spirited "Garry Owen" became a part of the 7th Cav salute in Vietnam. It is interesting to note that the significant NVA defeat in the Ia Drang is all but forgotten, but the Tet "offensive" of 1968 is remembered by many and remembered as an NVA military victory, which it was not.

During that Christmas season, the great patriot, Bob Hope, put on a show at our An Khe base camp. Being on an operation somewhere along Highway 19, we missed him. On Christmas Eve night, we stood out there in deep mud and pitch darkness, driving rain cascading off helmets, down noses, down ponchos, in boots. A waterlogged soldier gave a password — "Magic Night."

It was inevitable that at some locations the Vietnamese would gather around to observe our activities. No one knew which of these might be the VC. At one of the sites, my men tried to shoo them away but to no avail. Finally, putting on a stern face, I walked up, waved at them and then toward their village, and turned away. When I looked back, they were gone. "What happened," I asked Lieutenant Gray (who was from Idaho), the battalion communications officer, who had come up to our location from battalion.

"It's your mustache," Gray answered. (Due to a water shortage on that operation, we had been told to stop shaving — thus the mustache.) I thought that was probably the reason, for it was commonly believed Vietnamese held men with whiskers in rather high regard. But at other locations, the locals were not so easily impressed and simply refused to leave. I could not bring myself to be nasty toward them so we took the opposite tack. Instead of trying to run them off, we demonstrated for them, just as if we were addressing visiting VIPs (generals, usually) at Ft. Sill, how we produced the balloon-inflation gas and how we released the balloons. I wanted them to carry back to their village the intelligence that the GMD-1's 8-foot dish was not some sort of radar that was an important piece of battlefield equipment but merely something that a small group of crazy Americans used while playing with balloons. And perhaps of equal importance, the gleeful reaction of the children to the ascending balloon clearly indicated that we were doing

---

[1] Lt. General Harold G. Moore, USA (Ret.) and Joseph L. Galloway, *We Were Soldiers Once ...and Young, Ia Drang: the Battle that Changed the War in Vietnam*, Random House, New York, 1992 p. 346.

our part in winning the hearts and minds of the Vietnamese people. (Unless we could do that, there'd be little point to the war.)

Some of the operations in which we participated were: Clean House I, II, and III, Flying Tiger in support of the ROK (Republic of Korea) Capital Division, Black Horse, White Wing/Masher and Crazy Horse. I also sent a visual team (PIBAL team) to support an operation at Tuy Hoa, on the coast south of Qui Nhon. We were setting up our equipment at many sites. For White Wing/Masher alone we occupied six sites. While the other in-country sections essentially remained static, because of the great mobility of First Cav we would become known as the most mobile section in Vietnam — a blessing as well as a concern. A blessing, because if you moved a lot it was more difficult for the VC or NVA to get a bead on you, a concern because conventional wisdom had it that the GMD-1 would fail if bounced around a lot. Well-built, with circuit components housed in drawers that could be pulled out to be worked on, the GMD-1 pre-dated the era of solid-state circuitry and jostling of its vacuum tubes was not a good thing. But my very able equipment technician, he who had begun his Army career as a cook, was performing maintenance miracles. He took justifiable pride in the fact that we were always able to respond to the call to move out.

White Wing/Masher, the big operation, took place in phases. It began in late January of that year of the horse on the Chinese calendar — 1966. Hanoi Hannah should have checked her calendar before she said the First Cav was in for a fall — that was not going to happen to the famous horse division.

Our first emplacement was not far from Qui Nhon. We had barely set up our equipment when orders were changed, we were told to move out. As usual, we packed up as quickly as possible and, just in time to see the battery's vehicles up ahead, we took our position at the rear of the rolling convoy. Word would get back to me that only hours or perhaps minutes later, our very temporary campsite was raked with mortar fire. Good timing was a good thing to be blessed with.

At another temporary encampment, during the phase called Black Horse, the firing battery commander had taken a chopper back to battalion, to the An Khe base camp, and I, being the highest rank on site and also the battalion information officer (an additional duty), was called to talk to a visiting reporter from a well-known east-coast US newspaper.

"You know," the reporter said, "this doesn't seem as bad as it does to the people in the States." Coming from a war correspondent who presumably could get to where fighting was going on, that remark was quite a surprise. This war was bad enough for the soldiers going Stateside in body bags.

"How bad it is," I replied, "will depend on exactly where you are and when." Then I explained that the CH-47 Chinook helicopter that was hovering above the adjacent hill, which, from my vantage point, looked to be densely vegetated and seemed likely to provide good cover for the NVA or VC, was preparing to drop troopers to the ground via a 60-foot rope ladder. And I offered: "I may be able to get permission to have you join them on the rope ladder." That could have been the reporter's opportunity to see something that looked bad, but he politely declined. I probably couldn't have received the necessary authorization anyway.

By that time in the war, for those of us in-country the protesters at home were no longer a source of jokes. Instead of simply believing that war was bad, to which we would all agree, I thought they were being energized by those who were trying to avoid participating in it or were themselves trying to do so. In effect, they were just anti-US-military. To be sure, they called themselves anti-*war*, but the war had been ongoing before we arrived, and the protesters did not then protest it. Back at the Oakland Army Terminal, while we were waiting to sail for this sad country, we could joke about them because we were certain that the good solid US citizens would never let them have their way. But it was different now, the protesters seemed to be gaining, and that contact with the press made me wonder if the media either knowingly or unknowingly were sabotaging our efforts. For one thing, it didn't seem that the war was being presented as an invasion of one country into another, as the Cav's contact with the NVA at Ia Drang Valley surely proved. Instead, from my admittedly limited view, it seemed to be pictured as a mere uprising within the country, a civil insurrection that we should have no part of. The obvious result of the protests was to give encouragement to the North Vietnamese invaders, and that meant the war would be prolonged, more Americans would go home in body bags and more Vietnamese, from the North as well as the South, would also die violent deaths. The protesters might have been too young and naive to think of that, they could have only been following what they perceived to be the popular thing to do, a mere fad, but those who remembered World War II and Korea would have known the results of those actions. *They* would insist on supporting us *if* they were given the facts. There is propaganda in all wars. The story told to us of the attractive daughter of a village chief who had a "V" cut out of her lip by the VC as a lesson to her father who supported the South Vietnam government (and us) might also have been propaganda. But the presence of the North Vietnamese Army, the NVA, in South Vietnam, was real, and the story of the chief's daughter became more believable as your time in-country went on.

After Black Horse, we left for an encampment beside ornate cemetery monuments. Impressive termite hills about seven feet tall and ten feet wide, tapered to four feet at the top. This was a very sandy area, probably the site of an ancient sea bed. For the first time since that first day in-country, to dig a foxhole was a joy, a venture down memory lane to when you were a child and, on an outing with your family, you dug holes on the beach. By that time, due to the ending of enlistments and the lack of replacements, I was down from my original 13 men to about seven. We were the only ones there, a situation that concerned me. It was decided we could support the battery from that position, but we were only there a short time when a runner in a jeep came by to tell us the battery could not pick up our messages on their radio and we were to march order and move up to their position. I sort of hated to leave that site, a place called "Phu Cat," where the digging was so easy. Two years later, on my second Vietnam tour, I stopped there again, after the Air Force had occupied the site. Where there once had been only our foxholes and small tents, there was now a grand Air Force installation with buildings and a runway. I could not fail to notice that, incredibly, there were *sidewalks* at Phu Cat. And I also remembered how good the living had been while I was in the Air Force.

The route to that new site consisted of a trail that ran for a good part of the way on the top of a berm separating two rice paddies. Memory suggests that we traveled some 15 or 20 miles along roads that were only trails. Our little Met Section was then the entire convoy and that did not seem a very comfortable situation to be in. As a precautionary measure, we had one man set up in the bed of our last deuce-and-a-half with our 7.62 millimeter machine gun. When we were about half way there, a Huey gunship swooped toward us and hovered close. The pilot was giving us the once-over. Looking directly into the barrels of his machine guns, I could appreciate how the VC and NVA must have felt in that situation. We could not contact him by radio. Finally convinced that we were not an NVA element in captured American vehicles, he flew back up the road. We followed our very welcome escort, arrived at our new site just before dusk, released a radiosonde, gave the ballistic data to the firing direction people, settled in for the night and were off again the next morning.

From there, we headed for Bong Son. When we reached that village, we took a sandy road out, at first following the An Lao River, picture-postcard scenery. The stream meandered between sandy banks. Water buffalo wandered contentedly in the cool water. At the river's edge, a waterwheel stood. Made entirely of bamboo, it was large — about 20 feet in diameter and about 4 feet wide. As the wheel turned, a great number of scoops, made of large-diameter bamboo (the kind the VC would use as 60 mm mortar tubes), three in a group and spaced at regular intervals along its outer edge, lifted the river water and deposited it into a bamboo pipe that ran into a ditch, then to a rice paddy — a beautiful example of native ingenuity. On the next day, while checking on a communications problem with Major O., Division Artillery Communications Officer, I made a sketch of it. A middle-aged Vietnamese man walked up, pointed at the wheel and said:

"Number One" (in their ranking system, things went from good to bad on a one to ten scale). The major asked, "You make?" and by gestures got the question across. With a broad, broad smile accentuated with a vigorous nodding of his head, he made us understand he had indeed made it. To which we both loudly replied: "Number One!"

We drove under a large white banner that had been stretched above the road, anchored to a building on one side and a road-side tree on the other. Recalling the large lettering now purely from memory, it read:[2]

WELCOME TO THE AMERICAN SOLDIERS WHO COME TO SAVE US FROM THE COMMUNIST INVADER

Shortly after passing under that banner, we encountered a stream of refugees that reminded me of similar refugees in the newsreels when Hitler's panzers swept through France in 1940. Sweeping toward us, carrying their most prized possessions, they flooded the road, slowing our progress to a crawl — carts and bicycles loaded with sway-backed pigs in baskets, chickens in cages, bags of rice, furniture, a man with a mattress on his back, children carrying smaller children. And they were only a part of the exodus. Many more, warned by our leaflets of coming battles, were airlifted out by Chinook. Over half the valley's population had been airlifted to safety.[3]

Up ahead, First Cav was preparing to challenge the North Vietnamese invaders — three regiments of the NVA 3rd Division. The battles around Bong Son, similar to those in the Ia Drang Valley, would confirm, if there was then any remaining doubt, that this was not merely an internal conflict. On the contrary, the North Vietnamese were conducting a significant military invasion of their southern neighbor. It was just the kind of behavior that Americans had always instinctively opposed. I was sure that when the news of the Bong Son battles reached home, the protesters would see the light and understand that protests, that precious American right we were fighting to preserve, should not be undertaken when it would put US troops in harm's way. Incredibly, I could not have been more wrong!

When you selected the site for your meteorology section, you were like the last guest to be seated at the dinner table. The choice real estate had in nearly all cases been taken. Our principal operating site of the Bong Son campaign, however, was the exception. It was as nearly ideal as anyone could hope for. We had set up on a small north-south plateau about 20 meters wide and about 45 meters long. The general landscape sloped away toward the

---

[2] If the wording was not exactly as shown, the essence of the message was certainly the same.

[3] *Yearbook*, First Cavalry Division.

south. This was good because the monsoon at that time of year brought winds from the north to northeast, effectively lengthening the plateau for balloon launchings. At that site, we had several mornings of restricted visibility due to fog. Missing my days of forecasting in the Air Force, I began to study our meteorological data, in particular the vertical movement of what we called "the subsidence inversion" and the moisture levels, to see if I could come up with some single-station analysis technique, similar to what we did on the Saudi Arabian desert. In this case, instead of forecasting sand storms, I'd try to forecast for fog or rain. One morning, while enjoying a cup of coffee with a new friend, a lieutenant from New York City, who had volunteered for Vietnam from the Korean DMZ to be where he believed he could be of greater service and, as he said, closer to the action, I began to describe the great influx of moisture that had appeared on the radiosonde sounding.

"What does it mean," he asked. "Rain?"

"Yes," I said, "and there's enough moisture for significant rain." Because I didn't have all the tools of the forecaster's trade at my disposal, such as a regional 500-millibar chart, the "forecast" was only meant as a sort of casual remark between friends that might, however, be useful to him in his artillery spotting that day. Unknown to me, however, our battalion commander, who had set up a temporary command post at Bong Son, had walked up behind us.

"What's that about the weather?" the colonel asked. Glumly, I then had to repeat my unofficial prediction. Given my forecasting background, I knew that offhand remark, just chit-chat between friends, would be taken by the colonel as a true forecast. It was almost a replay of that 0400 thunderstorm forecast at Oscoda some 15 years before, and an 18-year-old warning from Mr. Nobel, the instructor at Chanute Air Force Base's weather forecaster school, ran through my mind. Nobel was deadly serious when he said: "Never make an unofficial forecast." Heavy rainfall would have a serious impact on the conduct of the battles that day, but it was not my place to make forecasts, that was a responsibility of an Air Force forecaster such as I had once been. A forecaster at An Khe or Saigon, or wherever the AWS man having responsibility for the forecasts happened to be, could have been consulted on the day's weather. If he was contacted, I'd have no idea if my prediction agreed with his. Being a thorough man, I felt sure the colonel made plans in accordance with my forecast, and I went back to the plotting van, busying myself with another look at the sounding. The moisture was there all right, but did it necessarily have to produce heavy rain? Some time later, while it was still morning, light rain began. As the hours went on its intensity steadily increased, reaching a climax in the late afternoon. A heavy downpour, the heaviest of our entire time on the operation, literally soaked the Bong Son area. The mess tent, where we had our morning coffee-time discussion about the coming rain, was flooded and all but swept away.

We stayed at that location for several weeks. At night, the C-47 called "Puff the Magic Dragon" sprayed out curve after slow pink curve of tracers from its mini-guns, producing a prolonged grinding sound like that of your car when you inadvertently try to engage the starter with the motor running. Between such displays, we took note of the surprisingly long time it took parachute flares to finally hit the brightly lit landscape. In the mornings, I looked forward to those now regular chats with the lieutenant from New York City who, after that "forecast," thought I was possessed of an infallible talent. After our coffee, I would return to the Met van and the data we worked up to assist the batteries in laying artillery rounds on the NVA, while he would walk off into the brush. On one of those mornings, in a remark I found similar to that of the war correspondent's comment during Black Horse, he told me there wasn't much going on. A strange thing to say. We had been pumping out artillery rounds at so furious a pace that I was convinced smoke from the howitzer barrels had been dense enough to become the nuclei upon which ground fog had formed.

"Keep your head down anyway," I said, and I walked back to the GMD. As I peered through the GMD's orienting telescope, a tiny pebble, a fourth the size of a pea, hit my face. Where did that come from? There was no one around, the men were in the van working up a ballistic message for the guns. Then another of the same size bounced off my cheek. Looking downward after the third missile hit me, I spotted the culprit. A wasp was digging a hole in the gravel. But this wee digger of foxholes was making little progress, for as quickly as he shot a pebble skyward, another rolled into the hole to take its place. A determined fellow, he was still digging as I left.

In less than a week the lieutenant from New York City, my coffee-drinking companion and forecasting enthusiast, was dead. Shot through the forehead. Another name in a list that would reach an incredible 58,000 and a very personal reminder of the terrible price some have paid to preserve America's freedom. If there remained any doubts that we were in a real war, White Wing/Masher would forever dispel them. During White Wing/Masher, the First Cavalry Division itself accounted for 1,342 enemy killed in action (KIA) while other Free World forces, the US 3rd Marine Amphibious Force, the ARVN 22nd Division and the Republic of Korea Capital Division, accounted for an additional 808 enemy KIA.[4] Without the slightest doubt, we were winning the war, the only variable being how long it would take the North Vietnamese to quit. It seemed that happy event could not be far off.

---

[4] *Yearbook.*

After 43 days of being out on operations leading up to and including White Wing/Masher, we returned to base camp, to An Khe. A fortunate lull in our sector, giving us time to repair the GMD-1. Conventional wisdom had been right about the harm to be done to it through frequent march-ordering and moves over bumpy roads and trails. But true to his expectations and my hopes, Specialist McCurley had it repaired in time for the next battle. While he worked on repairs, one of my very able team chiefs, Sergeant Albiston, and I paid a visit to the supply depot at Qui Nhon where, instead of the routine Vietnam response, "sorry 'bout that," a helpful captain led us to boxes of supplies with the 1660 meteorological-item identification number. This excursion was one that we had been told to never do. But I was desperate. I had been routinely ordering replenishments of expendables and repair parts but their delivery was an agonizing exercise in often fruitless waiting. When circumstances got particularly desperate for certain items such as radiosondes, balloons or calcium hydride, I'd be advised by the supply warrant or his sergeant to revert to the "red ball" requisition — the highest priority of requisition that probably took its name from the World War II "Red Ball Express": the trucks that sped across Western Europe, carrying vital material for Eisenhower's armies. I would delay resorting to the red ball requisition until the situation became desperate because it would work to defeat the normal resupply process in which, by ordering on a consistent basis with a non-priority requisition, a demand would be built up that could more easily be anticipated by supply channels up the line. But my mission came first, so I picked up the supplies at the Qui Nhon depot. But "sorry 'bout that" was the common response I got to my request for replacement ballistic meteorologists. Due to expiration of their terms of service, I had been losing men beginning at the second or third month in-country. And, in addition, during the Bong Son operation, we discovered that one of my men was only 16 years old. We had him in a chopper and on his way home in record time.

Then there was Operation Crazy Horse. Crazy Horse began when a First Cav patrol made contact with an enemy patrol. Both sides then brought in reinforcements and the contact widened. This occurred in a valley not very far from and east of An Khe. It was the Song Con Valley. The valley was also called the Vinh Thanh Valley because there were three villages named Vinh Thanh within it. We Americans gave it yet another name. We called it "Happy Valley." Soon after the patrols made contact I was called into the battalion bunker for a briefing. Within a few hours, my 3/18th Artillery Met Section was loaded and moved out as part of a Cav convoy. These Happy Valley battles took place in the period 16 May to 10 June 1966. Shortly after entering the valley we came upon a yellowish grassy plain. The kind of place where, in the States, you'd expect to see antelope. Large canopied trees dotted the plain, and 18-inch-long knobby-skinned lizards climbed in the trees. It was a nice dry place lit by a warming sun. But we didn't stay there long. We moved farther up Happy Valley, beyond the three Vinh Thanh villages where the road deteriorated to a mere two-rut trail. Near the high end of the valley we made camp. Then the weather turned to what it frequently was when we were on operations. It rained. Not the kind that comes in strong and is shortly gone, this was heavy and steady — what we called "four dot rain," after its symbol as plotted on the weather maps. It seemed to have no end. We had dug our foxholes before the rain got serious. Now the clay around them was gumbo. Dirty, coffee-colored water filled the foxholes. Those hydrometeors (the high-faluting meteorologist's name for rain), striking the rain-filled foxholes, exploded as large bubbles that ran out of the top of the holes along rapidly developing muddy erosion channels.

We were just finishing a ballistic computation which we would broadcast over our PRC-25 FM radio to our companions who were farther up the valley, where they were fighting it out with the NVA Yellow Star Division, when a runner rushed up. There were, he said, either 300 NVA at 1,000 meters or 1,000 NVA at 300 meters — he couldn't remember which. I pondered the preparations that could be made to meet that threat. An absurd thought arose: I always hated clichés, and now I especially did not want anyone to feel compelled to say, "they had fought to the last man." When you considered the firepower that six men could bring to bear with their personal weapons, the M-14 rifle, and our one M-60 machine gun, it made little difference if the NVA were 300 or 1,000 strong; and the distance of 1,000 meters (six-tenths of a mile), though better than 300 meters, was still much too close. Our trucks were sinking in the mud, and we probably couldn't move them. Perhaps sensing my thoughts, one of the younger troopers asked: "Chief, will we have to jump in those foxholes with all that muddy water?"

"Just watch me," I said, "and do what I do. But you'll need to look closely because all you're going to see above the water is my M-14 and my eyeballs." But the NVA, it happily turned out, either didn't know we were there or thought we were not important. Perhaps we had simply lucked out through interception by a First Cav patrol.[5] In Happy Valley, just as in the other important engagements, First Cav was proving the NVA could not win the war militarily. Once again, First Cav had zapped NVA's General Giap. His attempts to split South Vietnam at its mid-section were dismal failures.

I was near the end of my tour. As I looked toward the arrival of my replacement, I was grateful that we had been able to help the South Vietnamese fight off their attacker. Before I could be tapped for another tour in this beautiful but extremely sad country, I was sure the Vietnamese would be enjoying a much-deserved peace.

---

[5] See the Happy Valley operation described in detail in *Battles in the Monsoon,* by the late Brigadier General S.L.A. Marshall who had been foreign correspondent for the *Detroit News*. *Battles in the Monsoon,* The Battery Press, P.O. Box 3107, Uptown Station, Nashville, TN 37219.

Top: The Vietnamese people were ingenious in the methods they devised for moving water into the rice paddies. Here, two boys, shaded from the hot sun, pedal a bicycle-like device that drives a chain with attached water scoops.
Bottom: The 3/18 Met Section at Phu Cat. A large termite mound stands to the left and in front of the GMD-1. Barely visible to the right, one of the ballistic meteorologists stands waist-deep in the foxhole he is digging.

Artillery erupts on the enemy's position during White Wing/Masher while below it an 8-inch howitzer is towed by a truck. The Met Section water trailer is in the foreground. In the bottom photo, our battalion's ambulance is on the ridge line and a Bell UH-1B "Huey" gunship hovers near the Met Section's van (below the cumulus) and trucks. *February 1966.*

Two photos taken during White Wing/Masher in February 1966 provide a dramatic contrast in helicopters. Our battalion's small OH-13 (top photo) and the Division's huge CH-54 "Flying Crane." The OH-13 was holed by small arms or machine gun fire but was still flyable. Met Section trucks and van (extreme left) are in the foreground.

Chapter 26

A Well-traveled Cat

One day while I was still in the Air Force and assigned to the Clime Center, our eleventh-grade daughter, Leta, brought a surprise to our home. We were then living in the Washington DC suburb of Cheverly, Maryland where the terror of Cheverly, Bunky the tomcat, also lived (chapter 23). But this was some time after our run-in with Bunky, and that was good because we needed some time to forget about cats. You see, Leta's surprise was a kitten.

When she came home from school that day, Leta approached Marie and I in a somewhat hesitant manner. She probably was not sure we would be ready for another introduction to a cat. Like her two sisters, Leta was an intelligent girl. She knew how to handle this one. Instead of saying anything, she held out her cupped hands within which lay a soft black and white ball of fur. I remember that even at that very tender age, the ball of fur knew how to soften you with its eyes.

Saying she had found the homeless creature in the hallway at Bladensburg High School, and looking somewhat as if we should consider ourselves uncaring parents if we refused, Leta wanted to know: "Can we keep it?"

With our frequent moves in the service, it seemed unwise to keep pets. We had never stayed anywhere longer than three years and not because we didn't want to. We thought we had an opportunity to stay longer than three years during the Bermuda tour, but mine became the only request for extension that we knew of that was denied.

So how did we answer daughter Leta? We said: "We'll keep him for awhile." Translation: "The little furry ball has just found himself a lifetime of prized vittles and comfy surroundings."

Leta was on the staff of the school paper, *The Scroll,* and the kitten had been wandering just outside the room in which the paper was published which explained why the kitten had to be named "Scroll." A teacher who was a member of the humane society, I suspect, had discovered a good way of finding homes for orphan pets. Just turn the little animal loose in the halls — some schoolgirl will take it home.

Because we didn't want him to become a neighborhood pest or get lost, Scroll was a cat who was never permitted to be foot-loose when he roamed the great outdoors. When outside the house, he was kept on a leash. The one-time waif quickly became in nearly all respects a member of the family and surely was a not unimportant source of comfort for Marie during my Vietnam tour.

After returning from Vietnam, I was reassigned to instructor duty at Ft. Sill. By that time, the 40 meteorological warrant officers that the Army had when we three former Air Force forecasters joined them, had just about doubled in number. But the number of meteorology sections in Vietnam had also increased, meaning that the demand for warrants in Vietnam had not diminished.

I had been enjoying that second tour at Ft. Sill, in particular the challenge of developing the Meteorology Elective for the Artillery Officers' Career Course, but the chances of staying at Sill for very long did not seem good. The encouragement given the North Vietnamese by the anti-US military protesters had injected great uncertainty into the prospects for peace. It seemed prudent to be preparing ourselves for a second Vietnam tour.

Two of our daughters, Leta and Willa, were then in college and the youngest, Pamela, was attending Eisenhower High School in Lawton. I realized that it wouldn't be long before the girls would be gone from our home, wherever that happened to be, and I wanted all of us to have something quite definite to remember of that period of our lives. I suggested a family camping trip to the West Coast and, after explaining how it could be done, won family approval.

I rented a pop-up camping trailer from the family business of one of Pamela's closest friends, and we left for the Pacific. At the end of the first day, we were at Carlsbad Caverns where we camped for the night.

Around midnight, another family group drove up and parked alongside. Scroll, a very curious cat, sneaked out through the canvas flap of the trailer top, no doubt to see what the new neighbors were up to. Apparently satisfied with what he had seen, he continued on into the strange new surroundings. After we arose, the neighbors said they had seen him atop the nearby picnic table when they parked. But he wasn't anywhere near the picnic table anymore, and they didn't know where he had gone.

Five very concerned family members then fanned out, looking under every bush, in every ravine, atop and under every car and trailer at the Carlsbad Caverns campsite. But Scroll was nowhere to be found.

I didn't want to waste all of our precious vacation looking for a cat so I proposed that we take a tour of the caverns, which had, in any case, been our original plan. This proposal met with firm objections until I assured everyone that that particular cat, after all, was an intelligent cat. Scroll had merely gone out, I continued, for a bit of a stroll. He would certainly be at the trailer waiting for us when we returned from the caverns and might even be a bit perturbed that he had not been invited to go along. The fact that he was an intelligent cat was something that the others could agree with. And they remembered the times, after we returned from having left him alone for a day of shopping, when he would assume a defiant stance in the kitchen and stare at us as we entered the house. And just to be sure we got the message, would nip at Marie's heels — not to hurt her — but to let her know how he felt. Somewhat reluctantly, they all decided we could temporarily call off his search and tour the caverns.

As we emerged from the caverns, at the precise moment near 1700 hours that the rangers said it would happen, the cavern's bat population, a literal cloud of thousands, flew out for its nightly foray. That kind of precision in forecasting was something that ordinarily would have been noted and would have evoked some comment from my family group. Over the years, they became accustomed to hearing remarks about precision forecasting and would somewhat routinely join that kind of discussion. But it didn't happen this time. They just sat there on the bleachers, glum, trying to be polite listeners to the ranger's presentation, thinking only of the lost kitty.

After the bat fly-by, we returned to the trailer and resumed the search for Scroll within our campsite and throughout the surrounding area. It became clear, even to me, that that cat was seriously lost. A not-so-minor nightmare surfaced — I could see myself spending my entire leave at this campground. I might even have to wire Ft. Sill for an extension of leave — and as I explained the reason for it, I'd have to say it was *to search for a cat.*

The next morning, we widened the search area. Still no cat. Thinking now that every possibility had to be investigated, no matter how unlikely, I walked into the campsite's convenience store.

"Can I help you find something?" the clerk, a young lady, asked. I don't know why, but I was then in the cat-food aisle.

"Yes," I said, "I'm looking for a lost cat."

She looked up and down the aisle. "It's not here," she said. "But have you searched in the dairy case area where the milk's kept?" Then, sensing that I was serious, added: "Why don't you try the old man who lives in the small house up the road? He likes cats. Maybe he has it."

This seemed like a pure shot in the dark. But as soon as I relayed the young lady's suggestion to Marie and the girls, they were off in a flash for the old man's house with me behind them, trying to keep up. And there, inside the old man's screened porch and tied to a chair leg was our wayward cat. His host explained that Scroll had come to the porch lured by a mouse within, and when the door was opened, had jumped inside and immediately nailed the rodent. Having quite a mouse problem, the old man would have kept Scroll. But he saw that he was dealing with a cat that had important family connections and untied him. Four members of our family group immediately grabbed the wanderer. There was great rejoicing in camp that night.

The overwhelming sense of relief felt when I saw I would not be spending my entire leave in a campground near the great bat cave at Carlsbad made the rest of the trip, at that moment, seem anti-climactic. Sensing that he now could enjoy special privileges, as we continued the trip Scroll assumed a relaxing position, stretching out to his entire length on the car's back seat and crowding the girls into uncomfortable positions as he did so.

We enjoyed Scroll's attempts to chase waves at Carmel, California. Being unfamiliar with maritime ways, he repeatedly followed the receding waves just a little too far and then had to take a frightening great leap backward as the new breaker came rolling in. Having walked in the surf at Carmel, incidentally, made Scroll a well-traveled cat indeed. He had been to the ocean when we lived in Maryland and now could say, if he were so inclined, that he had dipped his paws in both the Atlantic and the Pacific.

At the Grand Canyon, Scroll amused tourists who apparently had never seen a cat on a leash, a cat who was chasing a lizard down a canyon trail with a girl hurrying along behind, keeping a firm grip on the leash. It was Pamela. Several times she kept Scroll from becoming a Grand Canyon statistic as he tried to disappear over the edge in pursuit of a chipmunk. I was very proud of Pam's firm grip on that cat. The thought of having to search for him far down those steep canyon walls was definitely unnerving.

While homeward bound we stopped at a sandy spot in a campground near Altus, Oklahoma where Scroll again became the center of attention. Ever since the Carlsbad incident, he was kept tied even when we were all settled in for the night in the camp trailer. Several times that night he scrambled around within the trailer as far as his leash would allow him to go, banging into pans and making all sorts of other rambunctious noises. He was surprisingly agitated which I took to be an expression of frustration at not being able to roam around in the night as he had at Carlsbad. But at daybreak, as we opened the canvas flap, we saw what his problem had been. There, just at the trailer's entrance, we saw the trail of two snakes in the sand. The two had approached our trailer's entrance flap where it looked as if they had been standing on their tails in an effort to gain entrance. A close look at the trails revealed the unmistakable imprints of rattles. At that moment, Scroll's family status was elevated. He was no longer a mere accomplished mouser. From that time on he was known as hero cat.

Hero cat, aka Scroll, in a frequent relaxation mode atop the Hi-Fi set.

With so many places to lose him, we kept hero cat on a tight leash when we toured the Grand Canyon. He was under such tight wraps, in the camping trailer in this case, that he didn't even appear in this photo.

Chapter 27

XXIVth Corps Artillery

War is cruelty, and you cannot refine it....You might as well appeal against the thunderstorm as against these terrible hardships of war.   *General William T. Sherman*[1]

"I'm surprised you're here," the major said. He was the S-2, the officer in charge of the intelligence section of XXIV Corps Artillery. XXIV Corps Artillery as well as XXIV Corps Headquarters were located at Phu Bai, near the old imperial capital of Hue where the old-time emperors had reigned. The XXIV Corps operational area covered all of northern South Vietnam from a point north of Danang. "We had decided," the major continued, "that we didn't want a warrant officer."

Now that was a *big* surprise. I wondered why XXIV Corps Artillery had decided it wouldn't fill the meteorological officer slot on its Met Quality Control Team. As far as I knew, they needed someone in that job. I would have asked the warrant who was rotating out of that slot what the deal was, but he had departed before I arrived. And that, too, was a bit of a surprise. During my first Vietnam tour, unless he were a combat casualty, the Met officer didn't leave before his replacement arrived. I was told later that he was not a casualty. He seemed to have been lucky in leaving early. Perhaps they just didn't see the need for filling the slot. They'd been getting along without a warrant. Maybe now they wondered why they even needed one.

I knew that if I had been assigned to the job I would have been exercising technical supervision over two warrants in the corps areas who outranked me. Even though I had received my chief warrant officer commission as a CWO-2 back at Ft. Sill, I would still have been outranked by a CWO-3 and by a CWO-2 who had date-of-rank over me. That was a possible reason for the cancellation of this assignment, but if that *were* the reason, the major surely would have said so. Besides, amongst the warrant officers themselves, a distinction between the four grades of warrant officer never arose. That was not the reason.

Most of the Army Met warrants were dedicated to meteorology, but there were a few among them who would look for what might be greener

---

[1] General William T. Sherman, *Sherman, Memoirs of W.T. Sherman,* Lierary Classics of the United States, Inc., New York, p. 601.

pastures in other jobs such as mess officer or, in overseas assignments, to have charge of the officers' club or PX. And I thought commanders sometimes might have unintentionally facilitated the defection by giving out those other jobs as additional duties. I speculated something like that might have been the case with my predecessor at Phu Bai, but later concluded that also was unlikely for, while I was there, I was not approached to assume duties outside my field. At An Khe, in recognition of my degree in English and history, I was made Battalion Information Officer. That extra duty never got in the way of my primary ballistic meteorology job.

As Chief of the Meteorological Quality Control Team, the Met warrant's job at Corps Artillery was to maintain quality control of the data produced by the various Met Sections within the Corps' area of operations and to assist the sections with their technical questions. I would learn that the MQCT team at Phu Bai was composed of a sergeant and specialist. They did their best but were struggling, needing a warrant to help set policy and interface with the Divisions and Battalions in the Corps AO, letting them know that ballistic meteorology was an important function. This part of Vietnam, the Vietnamese I Corps area, was a hot combat area. Surely, if they needed a Corps Artillery Met warrant anywhere he was needed in this Corps.

I had laid my personnel records on the major's desk at about the same time that he said they didn't want a warrant officer. He began to idly leaf through them Then, saying the mess hall was open for the evening meal, he told me to go to chow while he looked at my records. As I walked to the corrugated steel building that served as the mess hall, the thought arose that I had come a heck of a long way, *all the way from Ft. Sill, Oklahoma,* just to find I had no job and that was the outcome even after the Army had sent me to that job, taking all the trouble to divert me from the one I had been alerted for! I didn't expect a red carpet, but I *did* think some sort of welcoming would have been in order. Because I believed that if you tempted the Fates they'd not likely disappoint you, I was not a volunteer for this tour nor had I been a volunteer for the first one. Nevertheless, I supported the American presence in Vietnam wholeheartedly and thought this was one place where I might be able to make a not insignificant contribution to that effort.

At the mess hall, spaghetti, one of my favorite foods, was being served. As I approached the cook at the chow line I asked him to place the spaghetti sauce in a separate compartment of the mess tray. That was my Army spaghetti routine. Learned over many years of standing in chow lines, that procedure would keep the spaghetti itself palatable in the event the cook should botch the formula for the sauce. But I needn't have been so cautious that time. In what I might have taken as a good omen regarding this tour, the sauce this time was excellent. I took my time with the spaghetti and the sauce, enjoying it, wishing to delay returning to S-2 where I'd learn of my fate. Ah well, I rationalized, I never liked the thought of being a headquarters troop anyway. Once, while a tech sergeant in the Air Weather Service, I turned down an inspector's assignment at squadron headquarters where, incidentally, the troops seemed to have promotion priority, because I didn't like being that far from the action in meteorology, that far from Base Weather and the flight line. I could live with going to a remote fire base that would be much like the ones of my last tour. Of course, we'd probably have C-rations there. No spaghetti.

The circumstances of my supposed assignment to this most northern corps area had not been routine. About a month earlier, back at the Artillery and Missile School at Ft. Sill, I had been alerted for Xuan Loc, a small encampment about 35 miles east of Saigon. After the chartered Boeing 707 touched down at Saigon's Tan Son Nhut Airfield, and as I was searching for ground transportation to Xuan Loc, I was handed a message that I was to report to XXIV Corps at Phu Bai. That was a most welcome and unexpected change. At Xuan Loc, I would not have been out on operations as I had been with the Cav. Instead, I'd be confined to that one location, grinding out ballistic meteorology messages day-after-day and would continue doing that month-after-month at that same place until the entire 12-month tour expired. I didn't know what duty at Phu Bai would be like, but any change at least held out the chance of lucking into something more suited to my mobile style.

With the anticipation of that new assignment rising, the 450-mile hop north to Hue went quickly. We made a smooth landing at the Hue airfield, immediately adjacent to Phu Bai. In a few minutes, I was bouncing along in a jeep with the words PHU BAI IS ALL RIGHT boldly printed in large white letters below the windshield. Someone in the motor pool had a sense of humor.

Now, upon leaving the mess hall, I stopped at the chow line to tell the mess sergeant the spaghetti sauce was especially good. His surprised and pleased reaction suggested he hadn't heard that remark in some time. It was exceptionally good for a combat zone, and he certainly deserved that compliment. Of course, I had an ulterior motive, I wanted him to guard the sauce recipe just in case I might, after all, be assigned to Phu Bai. Back at S-2, I encountered my third major Phu Bai surprise. The major said he had changed his mind — he would keep me. I would never know the reason for the change. There was something in my records that made him think I wouldn't be a total loss. Possibly, it was the fact that I had served with First Cav on my previous tour. First Cav was now operating in the XXIV Corps area. The major might have thought I could have been of some help to them in their relationship with that proud 1st Air Cavalry Division. Also, I could have been one of the few officers of the headquarters staff who had a previous

Vietnam tour. That kind of experience would have been thought desirable. In spite of that puzzling introduction to XXIV Corps Artillery S-2, I would enjoy working for the major and for his able assistant, also a major. They would let me run things my way and would offer sensible advice when I sought it. Vietnam had changed considerably in the two years between my tours. There were now a half million men in-country as compared to the 125,000 at the beginning of my first tour. At the base camps, instead of living in tents as we had in 1965-66, we were housed in "hootches," plywood huts with screened sides and corrugated steel roofs. The big advantage of the hootch was that it provided an effective barrier to the rats who would try to join you in your sleeping bag when you lived in a tent. At Phu Bai, I lived in a hootch with several captains and a couple of lieutenants.

"If you were with First Cav on your first tour," one of the lieutenants asked, "why don't you have medals?" It *was* true and true of nearly everyone in my 3rd of the 18th artillery battalion. No medals. It was said the battalion S-3, a major, had just not gotten around to making the recommendations. But I was about to respond with something jocular, like this: to get medals, you did something worthwhile and a meteorologist *always* was doing that, so he was usually ignored — or, and this was the most likely reason, my 3/18 Met Section operated so far out in the boondocks giving on-site battlefield support, for every First Cav operation of my time in-country, that battalion knew very little of where we were or what we were doing and probably nothing about how we were performing our mission — when one of the captains, (Captain Bengtsen of Wyoming) interjected: "He doesn't need medals, he wears the First Cav patch." I couldn't have said it better.

There were seven 13-man meteorological sections in the Corps AO. Six of these, the Army sections, were headed by a warrant officer at the outset. At one of those sections, the warrant was replaced with a lieutenant when the warrant rotated Stateside. The lieutenant had had more meteorological training than the usual line officer received and therefore was given the meteorological job. The seventh section, a US Marine Corps section, was headed by an on-site sergeant with overall responsibility exercised by a major at the Corps' Dong Ha headquarters unit. My job: with two enlisted ballistic meteorologists, to assist the seven sections as they supported their artillery units, assure the data they provided the batteries were accurate, set up their radiosonde schedules, and select or concur in their proposed emplacements when they moved. This Met Quality Control Team (MQCT) was the brainchild, I believe, of Colonel Buntyn who was head of Ft. Sill's Target Acquisition Department. It had been implemented in Vietnam between my two tours with an MQCT at Field Force II, Long Binh, and another at Field Force I, Nha Trang. A corps-level unit, an attempt had been made to set up an MQCT at Phu Bai but was, as we've seen, unsuccessful.

As soon as I knew I had not come all the way from Ft. Sill for no reason at all, I took a hop to Nha Trang to visit an old friend from Ft. Sill days, CWO Albert Brown, the black good-humored Chief of the MQCT at Field Force I. Albert Brown briefed me on his work. He had established a schedule for the radiosonde flights in his sector that I decided to adopt at XXIV Corps. At the end of the day, he found me an upper room for the night in the building he lived in, a nicely decorated two story masonry building, probably built by the French. That evening, he took me down to a stage show being held on the first floor. A Korean band, just arrived from Korea, played American tunes with verve. Their energy level reached the crescendo stage as they rendered "The Green, Green Grass of Home." We would just as soon not have had that reminder. Then a female dancer danced the finale — a risqué routine that was lurid enough to cause me to blush, and I was sure Albert also would have if he could have. Afraid that very much of that kind of life could make one forget meteorology, in the morning I left for Phu Bai.

I traveled extensively to get to those seven sections within the corps area, moving about by jeep or helicopter. The helicopter was sometimes the UH-1 Huey gun ship which could carry about a half dozen men, or the Artillery Commanding General's OH-6A Cayuse LOH (light observation helicopter) or Loach, a small chopper in which I was usually the only passenger. It was fun to land at an artillery unit's operating location in the general's chopper. Every time I dropped out of the Loach at those locations I was given the red carpet treatment. When it became clear that this visitor was only a warrant officer, that warm red carpet treatment cooled a bit but not by much for it was assumed that a warrant who could fly as a solo passenger in a general's chopper might not be one who should be treated too nonchalantly.

The Americans with the most hazardous occupations in Vietnam were infantrymen who were dropped into firefights by helicopter and the chopper pilots, usually warrant officers, and their gunners. But in that war without a front line, you could be killed just as dead when the VC or NVA shelled your compound. At places like Dong Ha, Camp Carroll, The Rockpile, Firebase Vandegrift and points north, the shelling might be by 122-millimeter tube artillery or by 122-millimeter rocket. At Phu Bai, at that time, it was exclusively by 122-millimeter rocket.

When that Soviet 122-millimeter rocket was coming directly at you, it made an ear-shattering noise similar to that of a jet fighter that streaks by just above the crowd at an air show. When it was coming in your general direction but not directly at you, it had the whirring sound that you had heard at a Fourth of July fireworks show. The missile was about five feet long, about five inches in diameter and, including its 14-pound explosive charge, weighed about 100 pounds. It reached an apogee of about 10,000 feet and had a normal range of about seven miles. Spoilers could be used to reduce that range, and atmospheric effects could shorten or lengthen range and change the directional point of impact.

As the 122-millimeter rockets landed in the soft sand at Phu Bai, they blasted out craters about 15 to 25 feet wide. With the explosion, the rocket's casing would fly apart along diamond-shaped fracture lines. With razor-sharp edges and hot enough at detonation to burn flesh, some of those diamond-shaped fragments were small enough to fit in the palm of your hand. Others were coil shaped, sometimes about a half foot to a foot long, about an inch wide and about 1/16-inch thick — projectiles designed to take off a man's limbs or his head. And there were other nondescript pieces that flew out from the explosion, deadly chunks of the rocket propulsion system as well as large segments of the rocket casing that did not fly apart with the detonation, even non-threatening airborne sand that sifted down on you, reminding you that your luck, at least for that moment, was still holding.

The first 122-millimeter rocket salvo fired while I was at Phu Bai included at least one missile with the jet-plane roar, meaning it was on-line with my body. This occurred at 0535 in the morning of March 22, 1969 (there are certain events in everyone's life that become indelibly connected with the time of occurrence). As we hugged the ground, counting explosions, I thought of Willie and Joe in my favorite Bill Mauldin cartoon, the one where Joe says: "I can't git no lower, Willie. Me buttons is in th' way." There were 13 explosions. One of them blasted out its crater on line with and just 40 feet short of my jungle boots. If the NVA or VC unit sending us those presents had intercepted and used in their trajectory calculation a ballistic message transmitted by one of our Met Sections, I was thankful they had miscalculated the range effect.

Though I often credited luck, at least in part, for the success of my long-past weather forecasting experiences in the Air Force, I would never openly admit to being superstitious. But with that 13-rocket salvo, in which no one had been injured although hootches had been damaged, I found myself recalling other occasions in which the number 13 had played a part. I recalled that my meteorology section in that first Vietnam tour had originally consisted of 13 men, and that I had been sent home at the end of that tour on the 13th of the month. I could soon bring to mind several other number-13 events that had turned out, if not good, at least not bad. As a recruit on that long-ago troop shipment by train from Oklahoma City to Long Island, I had 13 men to care for. On my first Vietnam tour, our chopper was an OH-13. And even though decades have now passed, that number still triggers the thought that luck will not be bad.

There were eight rockets in the next volley. That was at 0005 on March 28. One evening a rocket completely severed a 12-inch diameter utility pole. Its remaining 10-foot-high stump stood about 25 feet from the van that was my office. Shrapnel had cut holes in the side of the van. This occurred at about the time I normally would have been starting out the van door to walk back to my hootch. I had developed a pattern of returning to the van after the evening meal for a few extra hours on projects I was involved in, but that evening, for an unaccountable reason, I skipped the routine.

On another night, when the shelling stopped and while still trying to get my buttons as close to the ground as possible, I reached out and touched hot shrapnel that had lost its last bit of kinetic energy less than an arm's length away. The rocket had exploded at nearly exactly my distance but its impact point was left of my line. If they were trying to get me, I was pleased they didn't know how to apply the wind direction effect.

It was commonly said in the Army that you didn't understand the problem if those around you were screaming and shouting while you remained calm. The exploding rockets generated some hysterics, pleadings suggesting the Almighty was hard of hearing. From a nearby foxhole, a cry in the dark: "Oh my God! My God! They're going to kill me!" He was shouting for the rest of us who were not brave enough to ask for help. Faster even than the rockets, thoughts flew through caverns of the mind. Where were the protesters when you needed them? Why weren't they protesting the government that supplied the enemy with 122 mm rockets? You felt numb. The future had stopped and nothing you could do would get it moving again.

In the brief moments between explosions, you made silent counts of how many had so far fallen and made predictions of how many more were yet to be fired. After a number of such shellings, you became inured and told yourself that if they were going to get you, you wouldn't be giving them the satisfaction of appearing scared, even though you were. You were scared and, at the same time, angry. Where in hell was the counter battery? Where was our artillery? Why weren't we shooting back at them? And of course that attempt was made, but it probably wouldn't be effective in getting the enemy gunners for they had devised clever ways to be away from the rocket firing site at the time the rockets were fired. They would sometimes trigger the firing with a leaking pail of water. When the water had leaked out to a certain point, the pail lifted, firing the rocket hours after the gunners had left.

Recalling the experiences of my first Vietnam tour, I continued to believe your chances of going unscathed were improved if you kept moving. That undoubtedly was one reason why I found it appropriate to be frequently away, attending to my duties in various other locations within the corps area. In a jeep or chopper, chances were pretty fair that you'd be shot at sooner or later, but at least you'd be a moving target and could shoot back. Of course, my routine of frequent visits to the Met Sections made me available to them should they be in need of assistance that I might be able to provide, and it also let their commanders know that corps was interested in the performance of the sections. That was an important function. One of the newer warrants, a WO, who was assigned to one of the Divisions and had come from the Signal Corps, the outfit within which meteorology first became organized at the governmental level in the US, in 1870, particularly needed my help in extracting his men from duties such as filling sand bags and setting up concertina wire on the perimeter. On at least two occasions I managed to return his men to their ballistic Met duties. They had, of course, filled their own sand bags for protecting the GMD-1 and for personal protection.

Sometimes, the radiosonde runs taken by the sections were interesting from a purely meteorological point of view. One day at Dong Ha, which was close to the border with North Vietnam, the Met Section released a radiosonde directly into the center of Typhoon Bess. This typhoon had flooded the entire coastal area of XXIV Corps, turning the ditches into rivers, the rivers into lakes and making the rice paddies one with the Gulf of Tonkin. Given any opportunity, no meteorologist would pass up the chance to see a radiosonde sounding taken at a time like that. I left for Dong Ha.

That Met Section at Dong Ha, as well as the one at Camp Carroll which was some distance to the west, was so close to North Vietnam that shelling from across the border by Soviet-made howitzers, 122-millimeter tube artillery, was common. At Phu Bai we only had to contend with the Soviet 122-millimeter rockets, which were bad enough but didn't give you the same sense that someone was taking aim *directly* at your body. To give themselves some protection from NVA artillery, the section at Dong Ha performed its wind plotting and ballistic calculations inside a bunker that I believe was built by CWO Chuck Miller who rotated Stateside in the previous year. The warrant in charge when I arrived was CWO Bruce Layne, a friend from my days at Ft. Sill. In addition to the usual sand-bagged sides, the roof, which at other sites would be covered with logs and sand bags, at this weather station was covered with steel plate topped with about a foot of concrete. When you are concentrating on analyzing charts for density values and plotting ballistic winds, you do not have time to watch for incoming artillery. I thought that bunker at Dong Ha was stout enough to withstand mortar hits, but I was not so sure it could ward off a Soviet 122-millimeter tube artillery round. I hoped it would never have to stand that test.

Bruce Layne showed me the sounding they had taken in Typhoon Bess and it was impressive. It had been plotted on the Artillery's ML-574 chart that is used to calculate air density aloft. I replotted their data on the Air Force's Skew T, Log P Diagram, the chart that gives one an indication of the energy available in the atmosphere. The plot followed precisely what meteorologists call the saturated adiabat, an unusual occurrence. The energy available for free convection, and therefore the opportunity for rain, was practically unlimited.

Layne's section had also been keeping a record of temperature and precipitation that began in 1967 and confirmed what we suspected: it was hot along that Gulf of Tonkin coast and sometimes quite wet. Their records showed the highest temperature of the period was recorded at 110 degrees on 14 May 1969, and the greatest monthly precipitation was 16.25 inches, occurring in the month of November 1967.

Afterward, I stayed overnight, taking an empty cot in Layne's hootch, and we enjoyed popcorn that Bruce's wife had sent him along with a popcorn popper. What a treat! I hadn't had popcorn since I left home. We ate the popcorn and made small talk while I tried to keep my eyes off the walls of the hootch which had been decorated by NVA tube artillery shrapnel. Imagine, if you will, a plywood wall with thin slits cut into it that are about a half foot to a foot long. These slashes appear throughout the wall, from bottom to top, and their orientations are entirely random — an uncomfortable example of what shrapnel from 122-millimeter tube artillery does.

Savoring the flavor and the aroma of the popcorn in that incongruous setting, in that hootch up there near the DMZ, the only popcorn that you had ever seen in Vietnam, you could imagine you had been transported Stateside. Perhaps you were at a theater with your family or at the Ft. Sill O-Club. You could hold that homespun thought only until your eyes focused again on the walls that were so haphazardly embellished by knife-edged NVA shrapnel.

While traveling in the same area one day with CWO J.D. Witherington, whose base was near Quang Tri and who had met the warrant selection board with me at Ft. Myer, we went on to Camp Carroll. Like Dong Ha, Camp Carroll was within range of NVA tube artillery fired from across the DMZ. The met warrant at Carroll, CWO Porter, had constructed a substantial bunker for his weather station, similar to the one at Dong Ha. The camp had been shelled by tube artillery just minutes before we arrived and a communications specialist had been killed. One of the rounds landed about 150 feet from the weather bunker and squarely on a 3/4 ton truck, flattening its bed as if it had been placed under a huge press. A van parked close to it was smoking from the explosion. A cool and calm CWO Porter told me they had sort of deduced when the NVA would fire its artillery and had scheduled their radiosonde releases between those times. I hoped they would be as good in that calculation as they were with their ballistic calculations which were routinely on the money.

On another day, while traveling in the same general area in a jeep with CWO Chuck Wiggins, we came upon three Vietnamese standing at the edge of a rice paddy — a middle-aged man, a slightly younger woman and a teen-age girl. The woman frantically waved at us. Upon stopping, we saw the man's face was covered with blood. By gestures, the woman made us understand that the man, who must have been her husband and the father of the girl, had been hoeing when he struck a mine which exploded. She was pointing at the roadside paddy while making hoeing gestures and then threw up both arms to mimic the explosion. At first, I thought it also possible that he might have been trying to plant a mine in that dirt road, a mine meant for us, but could see no evidence of the road surface having been disturbed.

While Chuck took a clean towel from his bag and pressed it against the wounds to slow the loss of blood, I crowded them into the jeep. As rapidly as the jeep would go, we took them to the next camp, a Seabee's installation and to the camp doctor who dressed the wounds and told us they would take the patient by chopper to a hospital ship standing by in the Gulf of Tonkin. I asked him what the prospects were. The bleeding had sufficiently stopped, he said, but the mine had permanently taken the farmer's eyesight. Like any farmer in the States, he had only been caring for his crop. I assumed it was a Vietnamese mine for we would not likely have had any reason for planting one in that open rice field. He could have been opposing the VC and that was their way of taking care of him. We left for our destinations, the First Cav bases LZ Nancy and Camp Evans, where Chuck's section was to move to, occupying the well-protected underground site built by another friend from Ft. Sill, CWO Tom Shannon. On the way, this thought repeatedly arose: How could that Vietnamese family survive?

The 3rd Marine Division was about to launch an operation to clear the area in the vicinity of Firebase Vandegrift which was fairly close to the DMZ, not far from Khe Sanh. They needed ballistic meteorology support. I was preparing to move one of the Army sections to that location when I met Brigadier General Allan Pixton, Commanding General of Artillery, outside the S-2's office. At his introduction to me upon his arrival at Phu Bai, he had surprised me by saying: "I've heard of you." From the way he said it, I assumed what he had heard was not bad. He had most likely been briefed at Ft. Sill by Lt. Colonel Gilbert Polk who was in charge of the Meteorology Division, having replaced Major Mollichelli who went on to bigger things. While our wives visited in the house, Colonel Polk and CWO Witherington and I had often played 8-ball in my garage near Ft. Sill. Those gentlemen may be surprised to now learn that, as a good host, when I had distinguished guests at my pool table, such as they were, without being too obvious I did my best to hold back so that they could win. Colonel Polk's memory of that string of 8-ball triumphs could easily have led him to think of me in a friendly way and could explain why he would have said nice things about me, if he did.

General Pixton asked me why we weren't sending the Marine section to support that Marine operation. I explained that the Marines, specifically the 12th Marines, had a GMD-1, but it was inoperable, and as a further complication, they had no maintenance technician; however, I said, we could try to make them operational. (That the 12th Marines' GMD-1 would not operate had not previously been a problem because there had been no need for additional meteorological support at their location, it being handled by one of our nearby Army sections.) My last remark was founded more on hope than realistic anticipation. Not knowing what parts were needed and, even if known, if they could be found in-country, the possibility of getting the Marine section running in time for the kickoff seemed quite remote. Nevertheless, in that time-honored military response, I told the general we would do our best.

I left for the 12th Marines base at Dong Ha with SFC Forrest W. Adams, my right-hand man on the Met Quality Control Team. Sgt. Adams, like me a former Air Force person, was an excellent electronic technician. He examined the Marines' GMD-1. Just as I had feared, it needed a part that was nowhere to be found in XXIV Corps.

I discussed the situation with the major in charge of the Marine section and repeated what I had promised the general — saying we'd do all we could to make his section operational. Not wanting to seem discouraging, I left out of the conversation the fact that the prospects for success looked dim. He replied they would do everything they could to assist us in making their section operational.

Back at Phu Bai, I spent at least half a day working our phone lines trying to get through to my counterparts at Field Force I at Nha Trang and Field Force II at Long Binh. Much to my surprise, the part was available at Long Binh, in the Saigon environs. In a matter of minutes I was aboard an Army plane, a U-21, that small but fast twin-engine craft of the type favored by corporate executives. It cruised at 240 mph. We flew direct to Tan Son Nhut, and I got ground transportation to Long Binh where another old friend from Ft. Sill, CWO Given, handed me the priceless GMD-1 part. CWO Given was then near the 30-year mark in service. Making small talk, I asked why he had stayed in so long. "I didn't sign on as Christmas help," he said, which seemed fair enough. With the part in hand, I regretfully turned down a steak dinner offer at the Tan Son Nhut O-Club. Steak dinners were one of the advantages of being at or near Saigon, a bustling city of somewhat modern buildings, bicycles, Mopeds, two-wheeled carts drawn by a pony or ox, jeeps, three-wheeled pedaled taxis, rickshaws, small French automobiles and even an occasional attractive girl strolling beneath a gaily colored parasol — a world that had to be unimaginable to the jungle- and rice-paddy soldier. I grabbed the first plane north, that one-time Army plane, the Caribou, which had been transferred to the Air Force inventory in 1966. With its high wing and short runway requirement, the Caribou had been much loved in the Army. During my first Vietnam tour, it had been used in a Medevac role during the Operation White Wing/Masher battles at Bong Son. The Caribou would take me only as far north as An Khe, my base camp of some three years earlier, where I was told I could get a hop to Phu Bai the next day.

Compared to the robust activity there in 1965-66, An Khe was now a mere ghost camp. Its lodgings, which were modernized with hootches after my tour there, were now mostly deserted. First Cav had moved out shortly after my first tour, and the hustle and bustle of the time when it was First Cav's base camp was ancient history to and practically incomprehensible to the soldiers who were now there. I found a deserted hootch with cots and prepared for the night. It was a little difficult falling asleep in that absolute quiet of the camp. After awhile some troopers began H&I fires with grenade launchers. Back when it was my home base, we conducted H&Is with our impressive eight-inch howitzers. The area obviously had been pacified to a great extent since 1965-'66. The more active war had moved to other locations. As darkness descended, I looked for the full moon, an old An Khe habit. We had commonly believed that the VC or NVA would never attack when they were illuminated by that heavenly spotlight. Near midnight, as a sort of reminder of that earlier time, every soldier on the perimeter joined the grenade launchers with his M-16 or .45 pistol in a "mad minute" of firing, directing a deadly curtain of lead outward from the camp, an old First Cav routine that had been set off at random times to keep the NVA or VC off guard and also to celebrate the coming of that Year of the Horse, at midnight of New Year's Eve, three years ago. In the morning, I caught a C-123 for Phu Bai, happy to leave ghostly An Khe. On the way, we stopped at Danang to drop off a soldier who had been on the flight with us. The only thing we knew about him was that he was headed toward home in a black rubber bag.

While I was gone, Sergeant Adams provided the Marine section refresher training in ballistic met calculations. They had been trained in ballistic meteorology procedures at our school at Ft. Sill, but the lessons had faded somewhat during that long down period with an inoperable GMD-1. Under Adams' excellent instruction, they regained their wind and air density calculation skills in a short time. When I returned with the repair part, Sergeant Adams installed it, checked out the crew on their now-operational GMD-1, and the Marines had a running meteorological section in time to support the operation.

The Commanding General 3rd Marine Division, R. G. Davis, forwarded through General Pixton a letter from the CO of 12th Marines,

Peter J. Mulroney, who wrote we had "rendered invaluable assistance in reestablishing the 12th Marines Meteorological Section....Their knowledge, selfless devotion to duty and willing spirit of cooperation has contributed materially to the successful accomplishment of the mission of this organization." I appreciated the kind words but what we had done was only what I envisioned our job to be, though that kind of service was not anticipated when the concept of a Meteorological Quality Control Team was set forth. With the Met Section now running at Vandegrift Combat Base, I visited them a number of times to see if they needed any additional help. They were doing fine. On one of the visits, the sergeant in charge told me the camp had been attacked and an ammunition dump situated close by had gone up. "But," the sergeant proudly explained, "we stayed with the GMD and except for shrapnel hits in the dish, it wasn't hurt at all, gunner." (Marines call their warrants "gunner.") Those hits in the GMD's antenna would not affect its performance. I thought of the contrast between those Marines and the anti-US-military protesters back home. In my opinion, which was supported by everyone I knew in-country, the protesters were prolonging the war. The Marines were fighting it. And some of these Marines would die in that extended period of the war made almost inevitable by the encouragement the protesters gave the North Vietnamese. At about this time I began to call the protesters the NVA auxiliary for they were effectively as much a part of that army as was its leader, General Vo Nguyen Giap. But in fairness to them, most of them probably knew too little of US history or had not given the issue enough thought to realize the real effect of their actions. They could have been sincere in thinking they were ending the war, but the war had been won militarily by that year of 1969 and *the only reason for its continuance was the hope the protesters were providing to Ho Chi Minh.* If I were not in the military, could I have been a protester? It was not possible. I was raised in rural Michigan where people cherished the Land of the Free. That citizens might refuse when asked to fight for America was a thought they couldn't entertain. I had boyhood chums killed in World War II battles they probably didn't want to go to. They went and they won their war. Now, I was in a war doomed to be lost even though our military performed far better than any objective observer would have believed possible, winning, in fact, all the battles. It was true that the rules of engagement, such as the one giving the enemy safe haven across the Vietnam border, were important factors, but those decisions, which flowed, at least in part, from the shrill voices of the protesters, were not why we were going to lose.

In may, I took an R&R to Hawaii. As she had during my first Vietnam tour, Marie had been staying in our house near the back gate of Selfridge Air Force Base where she could be within driving distance of her dad and my mother. She took a flight to L.A. and, from there, another to Honolulu. Five glorious days together on the beach at Waianae. Five days respite from rocket attack. Then back to Vietnam.

Once more at Phu Bai, we settled in for the siege. The rockets of each salvo would be fired one at a time. There'd be as few as four in one salvo and as many as 13 in another. Compared to the shelling endured by the defenders of Khe Sanh, these were a small number, but regardless of how few, every rocket was a potential killer. Unofficially, I was given the toll. During an attack on June 18, two were killed and 21 wounded; on July 7, eight were killed; on July 11, the rockets claimed as victims our Vietnamese allies — three were killed and 52 wounded.

Weathermen find it difficult to refrain from practicing what seems to be a natural bent for statistics. The enemy's firing schedule was recorded in my little pocket notebook to the nearest five minutes. (Only to the nearest five minutes because even the most dedicated data gatherer, who is imprinting the ground with his body to create as slim a silhouette as possible, will not usually consult his watch until he is separated from the last explosion by a discernible, however little, increment of time.) This was the schedule:

time	date	number of rockets
0535	Sat 22 March	13
0005	Fri 28 March	8
2210	Sat 26 April	7
0115	Fri 9 May	6
2100	Sun 1 June	5
0115	Tue 10 June	4
2045	Mon 16 June	8
2010	Wed 18 June	4
1930	Thu 26 June	4
0130	Mon 7 July	12
0205	Mon 7 July	4
2230	Fri 11 July	5
2110	Mon 21 July	7

As can be seen, with the exception of the attack on 26 June which took place in fading daylight, the tactic was to launch the rockets during the night to reduce the likelihood that those enemy gunners would be pursued; and presumably with the goal of keeping us off balance, they were scheduled at random times. I did not know until the moment of copying this list from my notebook that the number of salvos totaled 13 — another good luck indicator even if only in retrospect.

The worst part of the shelling was the utter helplessness you felt. I thought I would rather have been in a drop zone where the enemy might show himself, allowing you the opportunity to return fire. Few Americans knew war as General Sherman did. In his memoirs, in a sentence that seems most apropo, he wrote this: "The hardest task in war is to lie in support of some position or battery, under fire without the privilege of returning it...."[2] When you were waiting for the rockets, knowing that the incoming missile you are listening to might be the very last one you can ever hear, to a surprising extent fright seems buried in the anger generated by your helpless plight.

After the first rockets hit, there was a flurry of bunker-building. All over the compound soldiers of nearly all ranks could be seen filling sand bags. The walls of the bunkers were made of stacked sand bags. Additional sand bags, piled atop a layer of logs, made up the roofs. A few bunkers had PSP roofs, the perforated steel plate that would be used as a covering for unimproved aircraft runways. Sand bags were placed over the PSP. Someone with the sense of humor of the soldier who had painted PHU BAI IS ALL RIGHT on the headquarters jeep, attached this sign to his completed bunker: ELLSWORTH (after Ellsworth Bunker, the U.S. ambassador in Saigon).The bunkers were good protection against mortars, but we weren't being hit with mortars. In fact, as I saw it, the bunkers then being built at Phu Bai, tall enough for a man standing, increased your exposure, they were just catcher's mitts for rockets. They wouldn't stop one, and if you were inside when a rocket slammed into it — it was *Good Bye, Irene.*

I discussed the rocket situation with Jerry Tabb, a specialist in S-2 who frequently rode shotgun for me when I went by jeep to the Met Sections. I didn't want to raise any concerns he might have, nevertheless I did let slip my evaluation of the bunkers. Jerry had a degree in education. If the bunkers were no good, Jerry wanted to know what one could do to protect himself from the rockets. Just as there is often no perfect answer in weather forecasting, so there was no perfect protection from the rockets. If that rocket had your name on it, it was going to get you. You could minimize the hazard by diving into as small a foxhole as would accommodate your body, but if the rocket made a direct hit on your chosen spot of earth — it was all over. In spite of that gloomy assessment, I explained that the foxhole was your best protection. The most important thing was to dive into it as soon as you heard a rocket coming. Since the rockets were being launched at night when the troops would most likely be in their bunks, the hootch itself, standing up there like another catcher's mitt primed to snag a low flying rocket, was a part of the problem. None of the hootches that had been hit had caught fire, although a corrugated steel building used as a warehouse had. When it was hit, the hootch would become a pile of splinters and curled steel roofing.

Given that information, the best protection, as I explained to Jerry, would be a foxhole that would be dug directly under one's bunk — this would

---

[2] Sherman, p. 898.

eliminate the time lost in scrambling to the screen door, opening it, and running around to the side of the hootch in search of your hole, only to find, perhaps, that someone else was already in it and there was no way you were going to evict him. And all this, of course, was going to be done in the dark.

For access to the foxhole, I further explained, one should cut a hole in the plywood floor of the hootch and close the hole with a trapdoor. Then, at the first sound of a rocket, the routine would be a roll-out from the bunk, a drop through the trapdoor and a slide into the hole. In addition to having had teacher's training, Jerry was accomplished in woodworking skills. He also had a fine sense of humor and told me he did just as I said and that his first sergeant liked it so much he said he thought he'd build one like it for himself.

In the years that followed, I would remember this with a slight feeling of guilt. I had advised a young soldier to modify government property, and as far back as my basic training days, that had been a definite no-no. Years later I learned that Jerry was only joking about cutting the hole in the floor of his hootch. Regardless, that would have been the best nighttime protection from the shelling.

Back when I was a weather forecaster, it was not unusual to be crowded into a decision corner. I am referring to the occasions where you didn't have as much data as you thought you needed and what data there were would be pointing toward a decision somewhat contrary to the norm. In the end, you had to go with the data regardless of the consequences. Something like that happened one day when I inspected Firebase Tomahawk, south of Phu Bai and close to the Gulf of Tonkin.

Tomahawk was being occupied by C Battery of the 2nd of the 138th Artillery, one of the most interesting artillery units in Vietnam. It was a National Guard outfit, a microcosm of the small Kentucky town of Bardstown from which it came. Unlike other artillery units which drew its men from all parts of America, the soldiers of 2/138th Artillery were essentially a cross section of that one town. When you talked to them, their voices telegraphed the warmth of their distant Kentucky homes.

We needed to move the 138th's battalion meteorology section from its site near Camp Eagle (at the entrance to the "Yellow Brick Road") to a position from which it could provide close support to the 138th's artillery at Tomahawk. Tomahawk was situated in a wooded, hilly area close to the Gulf of Tonkin's sandy beach. With the battalion's warrant, cool, competent CWO Hancock, I surveyed a site on the beach and another somewhat wooded location at Tomahawk. We discussed the pros and cons of each site. In terms of the meteorology, either location would have served. The Tomahawk site would have been directly at the guns (howitzers), the beach site a short distance away but still close enough to either hand carry or radio-transmit the ballistic messages. If I had followed the routine I had established on my first tour, I would have sited the Met Section directly at the guns. But I didn't do it that time, concluding that, of the two sites, Tomahawk, with its hilly, wooded terrain, seemed more vulnerable to a VC or NVA attack. Not long after the 138th's Met Section began operating at the beach site, Tomahawk was infiltrated by NVA soldiers armed with satchel charges and rocket-propelled grenades. During the attack, five of those Kentucky troopers were killed. If the Met Section had been sited at Tomahawk, being preoccupied with their ballistic calculations, those ballistic meteorologist-troopers of the 138th would have been easy targets. To have the required balloon-launching space, they would have been located somewhat removed from the firing battery they were supporting. They could have contributed little if anything toward thwarting the attack and, being up and about at all hours with their attention riveted on the radiosonde run, they would have been seen as an easy target to be quickly silenced. I believed they would have been included in the casualties. In that decision, as with a number of those weather forecasting decisions of the distant past, someone must have been watching over me.

At about that time, I was called on to make a big career decision. When my Air Force and Army service was combined, I was approaching the 20-year mark, a departure point for many career soldiers. I recalled that if he hadn't called off the bet, I could have had a bit of fun with Andy Anderson, my buddy during basic training and weather school days (he had advised me to enter the meteorology field). I would have called Andy and demanded my winnings — that whole twenty-dollar bill he had bet against my reenlisting for three years, not to mention sticking it out, albeit in a wavering sort of way, for 20 years.

As I prepared to submit retirement papers, Colonel McLeod, the Corps Artillery XO (executive officer) called me in. He had helped me in the past by suggesting I write meteorological newsletters to get "the word" to the sections which, along with some studies I made of the Met data, were sent to Ft. Sill for their "Lessons Learned" (in Vietnam) compilations. If I would agree to stay in the Army for 30 years which entailed, I believed, acceptance of a Regular Army warrant officer commission, he said my future in the Army would be assured. He reminded me that I was on the published list for promotion to CWO-3. I thanked the colonel for his thoughtfulness but declined, telling him I wanted to pursue a graduate degree and, fully aware that I had an almost identical decision to make some 15 years in the past, one that did not turn out as I thought it would, I told him I believed I could do better on the outside. Marie and I were then in the process of putting three daughters through college, not a very comfortable undertaking on a warrant officer's pay. And, if I stayed in, I'd go back to Ft. Sill for another stint as a meteorology instructor. That was a nice job, particularly after Major

Mollichelli had conceived of the idea of a meteorology elective to the Artillery Officers' Advanced Career Course, had produced several topics to be included in the course and had then turned the development and expansion of that course over to me. I then constructed the complete curriculum and, having that bias for forecasting, included more meteorological theory as well as practical aspects not before taught in artillery meteorology such as the character of pressure systems, analysis of weather charts and the uses of the Skew T, Log P diagram. The major had me teach those classes, handle the VIP briefings, and generally improve the instructional materials for all courses taught in the Meteorology Division. A truly good job. But, as the saying goes, "I'd been there, done that." As a warrant officer, that would be as much as I'd be allowed to do. I needed more responsibility. In my Air Force and Army tours, I had been sent to meteorology's great challenges, to the search for the unknown, had consistently been taken to the center of action. I saw no other hills to climb. My 20 years in the military, coming after a somewhat uncertain start, had been immensely satisfying. Like my friend at Long Binh, I mused, it seemed I too had not signed on as Christmas help. I did give the 30-year question a second thought. But the conclusion came out as before. Of no little import was the realization that the anti-US-military protesters, that NVA auxiliary, were increasingly successful in influencing the political process. The war, therefore, had to continue and with no end in sight. I'd then return to the rice paddies for a third tour, then a fourth, etc. The statistician part of my meteorological makeup argued that I had already used up all the near misses I'd likely be granted in that sad country.

So I took a charted jet from Danang, touched down for a refueling stop in Tokyo, and landed at the Seattle-Tacoma Airport, returning from Vietnam for the last time. It was a day I could not ever forget. The date was 20 July 1969, and Neil Armstrong on the Apollo 11 mission spoke to Earth from a place called "The Sea of Tranquillity."

Outside the airport, I stepped into a rain shower, looking for a cab to take me to Ft. Lewis for my retirement processing. It was nighttime. A cab's headlights, reflecting off the wet street, came into view. "Where you coming from?" the driver asked, smiling. I could see he was one of those cheerful sorts, no doubt was going to bore me with chit chat, including the obligatory remarks about the weather. I hoped that wouldn't include the usual cab-driver weather prediction.

"Vietnam," I said. There was no response. The drive to Ft. Lewis took about a half hour. A strange sort of ride, it was completed with not another word spoken. I got out of the car at Lewis' main gate, gave the cabbie a bill, told him to keep the change. He grunted.

****

Top: In the distance, a stratocumulus deck bumps into Diamond Head as vacationers enjoy carefree Waikiki Beach, Honolulu.
Bottom: Saying good bye at Honolulu Airport after a much too short five-day R&R. May 1966

Top: The roof and sides of this Phu Bai hootch disintegrated when a VC or NVA 122 millimeter rocket exploded. The piece of corrugated metal is a small part of the roof.
Bottom: A soldier digs out the casing of another 122 millimeter rocket. This one buried itself in Phu Bai's sandy soil as it exploded.

Top: The author's smartly styled brief case, a Christmas present that seemed incongruous in Vietnam, rests on a skid of the Artillery Commanding General's OH-6A "Loach," a chopper he was privileged to borrow.
Bottom: Long before the home office became common, the ballistic meteorologists at Camp Evans dug themselves an underground ballistic computation "office" with sand-bagged sides and roof in which they also installed their bunks. Not a *Better Homes and Gardens* edifice, perhaps, but it provided some comfort when mortar shells and rockets were incoming.

Two Met Sections in the XXIVth Corps. At LZ Nancy, the ballistic meteorologist manually aims the GMD-1 at the radiosonde. The radiosonde and balloon have just been released from the balloon inflation shelter. After locking on the radiosonde's signal, the GMD-1 will automatically track it, even through the cloud ceiling. At LZ Roy, on the Gulf of Tonkin, a CH-47 Chinook shares the real estate with a GMD-1 that is protected with sand bags.

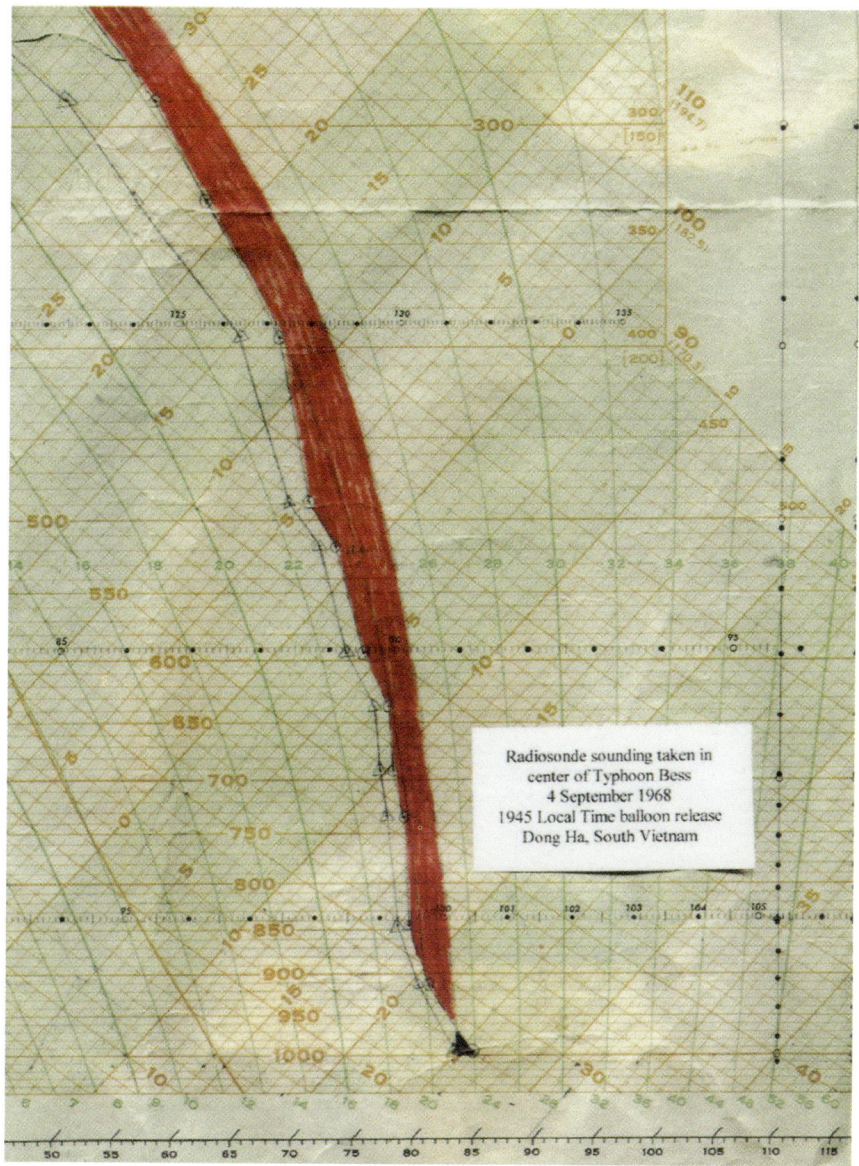

Radiosonde sounding taken in center of Typhoon Bess
4 September 1968
1945 Local Time balloon release
Dong Ha, South Vietnam

This thermodynamic diagram, the "Skew T, Log P," shows the temperature and dew point plots of a radiosonde that was released in the middle of Typhoon Bess. The area shaded in red displays the amount of energy available for free convection, depicting, as would be expected, a very unstable atmosphere.

Top: This XXIV Corps U-21 is about to take off at Phu Bai, near the old imperial city of Hue, for Tan Son Nhut, at Saigon, carrying the author in search of a critical GMD-1 part. Much faster than the helicopters he often rode in, this aircraft had a cruising speed of 240 mph.
Bottom: A gun crew prepares to fire a 105 millimeter howitzer.

Top: Saigon, South Vietnam in 1969.
Bottom: On a morning that is typically wet, a patrol clears a road in the South Vietnamese I Corps area. They check for signs indicating that mines have been planted by the VC or NVA during the night.

The interior of the van that was the office for the XXIV Corps Artillery Met Quality Control Team at Phu Bai.

Epilogue

After leaving the military for the last time, I enrolled in graduate school of one of our respected universities. In my first class the professor seated us in a circle, facing each other. After asking we identify ourselves and say a few words about what we had been doing for the past few years, he turned the conduct of the class over to us. Most of the students were just out of undergrad school. Like me, they were there to gain the credentials that might lead to a job. If any had been in the military, and I suspected some of the males had, they didn't say so. There were polite questions, pleasantly worded answers. At my turn, I opened with: "I just returned from Vietnam." And that was all I was given the opportunity to say. Hostility became the order of the day. How many Vietnamese houses had I burned? How many children had I killed? Incredibly, the rest of the class period was taken up with similar discourse. All my attempts to answer questions were cut short with more accusations or insinuations. Several were flung at me at the same time. Out in the jungle and rice paddy country, I deplored the activities of the protesters, but they were then distant. I assumed they'd fade away. Now, what bothered me most was that I was seated in a classroom in an institution of higher learning where discussion open to all points of view ought to be the norm. But the vocal ones, who were in the minority, didn't want anyone to interfere with their conclusions. The others merely sat in silence. As I would later learn, the latter were as nice a group as anyone would want to meet. They wanted to support me, but like President Nixon's "silent majority," they couldn't quite bring themselves to perhaps be embroiled in controversy. So the vocal ones carried the day. Here, in this ordinary classroom, one could see a microcosm of the national dilemma. Like a scene in a nightmare, Americans had gone from cheering troop trains to where it became acceptable to castigate returning soldiers as a mere reflex action. I thought I must have dozed in those 20 odd years in the military and had lost touch with America. With all my heart I hoped that when the Vietnam period was relegated to the history books, the silent ones would "get over it" as the saying goes and understand that it was not inappropriate to make their voices heard when the future of their country was at stake. This was my Rip Van Winkle wake-up, and it seemed that Rip's reception had been better than mine.

As I write this, it is Veteran's Day, 2001. Today, while waiting in a mall parking lot for Marie to finish shopping, a 50ish man walked up to my car window. "I see you have a veteran's license plate," he said. "And I just wanted to thank you for what you did for my country." I thanked him for his thoughtfulness, thinking — he's about the right age to have been an anti-US-military protester in that long ago war; and I saw that it really didn't matter if he was one of them or one of the silent ones, for that aberrational time was past and could not return.

Abbreviations

AAC — Army Air Corps
AAF — Army Air Force
Acc — altocumulus castellanus
ADC — Air Defense Command
ANAPQ/13 — From a B-29 bomber, the first weather radar
AO — Airdrome officer
ARVN — Army of the Republic of Vietnam (South Vietnam)
ATC — Air Transport Command
Cb — Cumulonimbus
Col. — Colonel
CMSgt. — Chief Master Sergeant
Corps AO — Corps area of operations
Cpl. — Corporal
CPS-9 — The first fully functional operational weather radar
CQ — Charge of Quarters
CWO — Chief Warrant Officer
DMZ — Demilitarized zone (which it was not); in Korea, at the 38th parallel; in Vietnam, at the 17th parallel
ITCZ — Intertropical convergence zone
Lt. Col. — Lieutenant Colonel
MAC — Military Airlift Command
MATS — Military Air Transport Service
Met Section — Army ballistic meteorology section
Metro Section — Same as Met Section
ML-574 — Artillery altitude-pressure-density chart
MQCT — Meteorological Quality Control Team
M/ Sgt. — Master Sergeant
NVA — North Vietnamese Army
Ob — Weather observation
Prog — Prognostic chart or forecast map
Pvt. — Private
Pfc. — Private first class
Raob — Radiosonde observation
Rawin Section — Air Force upper air sounding section
R&R — Rest and relaxation
SAC — Strategic Air Command
SFC — Sergeant First Class
Sgt. — Sergeant
Skew T , Log P — Thermodynamic diagram
SMSgt. — Senior Master Sergeant
S/Sgt. — Staff Sergeant
St — Stratus
TIROS — the first family of weather satellites
T/Sgt. — Technical Sergeant (also Tech Sergeant)
USAF — United States Air Force
VC — Viet Cong
WO — Warrant Officer
WOJG — Warrant Officer Junior Grade

Index (* denotes picture or map)

AAF shoulder patch 16*
AAF Technical Training Command 17, 29
AAF Weather Service 17, 119
AACS (Airways and Air Communications Service) 59, 60
Abke, Albert 4
Abke, George 2, 4, 5, 20
Abrahamson, M/Sgt. Karl 81, 82
Adams, SFC Forrest W. 319, 320
ADC (Air Defense Command) 124, 187, 193
Admiral 33
Advection fog 127
Africa 154, 172, 218
Air and Space Museum, Washington, D.C. 39
Air Force ii, 10, 83, 203, 205, 208, 209
Air Force Association 195
Air Force Climatic Center 265
Air Force One 109*, 231*, 264*
Air Force Reserve 210
Air France 159, 171, 175
Air-sea rescue 73
Air Weather Association 91, 198
Air Weather Service 17, 25, 33, 89, 101, 116, 117, 119, 141, 153, 178, 204, 205, 210, 257, 276
Albiston, Sgt. 298
Albuquerque, New Mexico 172
Altocumulus 68, 189, 190, 282
Altocumulus Castellatus 99, 100*
Ambassador Bridge 253
American International College 216
AN/APQ-13 130, 212
Anderson, Donald C. "Andy" 1, 9, 14, 15, 23, 24, 57, 119, 147, 148, 325
Andrews AFB, Maryland 153, 269
Angus, Major Frank 82, 83
An Khe, South Vietnam 282-285, 287, 289, 290, 296, 298, 310, 320
An Khe Pass 282, 290
An Lao River 294
Apollo 11 326
ARAMCO 164, 171
Arctic Circle 55
Armor 10, 82
Armstrong, Neil 166, 326
Army Air Corps 1, 83
Army Air Force (AAF) i, ii, 1, 7, 11, 12, 51, 82, 83
Army Air Service 83, 216
Army Artillery ii, 10, 66, 69, 75, 82, 83, 257, 275, 287*, 332*
Army pay 10, 18, 19, 32, 33, 325
Army-Navy E-flag 39
Arsenal of Democracy 38

Index (continued)

Artillery and Missile School 276, 311
Artillery Officers' Career Course (Meteorology Elective) 304
Artillery song 285
ARVN (Army of Vietnam) 281, 282, 288*, 297
Asmara, Eritrea 172, 173
AT-6 (North American Texan) 74*, 121*, 130, 131, 144-146, 157
AT-11 (Beech Kansan) 65-68, 70, 72-74*, 251
ATC (Air Transport Command) 23, 101, 257
Athens 171
Atlantic 115, 179, 215, 216, 219, 227, 233, 251, 255, 306
Austria 3, 96
Avianca 218
AWOL 21
Azores 55, 57, 101, 134, 218, 220, 223
B-17 (Flying Fortress) 7, 8* 73, 83, 86, 115, 122*, 154
B-24 (Liberator) 38, 86
B-25 101, 201
B-29 (Superfortress) 11, 23, 38, 39, 48*, 69, 86, 130, 140, 172, 212, 231
B-47 (Stratojet) 254, 258, 260, 261, 264*
B-50 231*, 243
B-52 186
B-377 226, 249
B-707 251, 252, 264*, 311
Bahrain Island 164, 165
Baffin Island 55
Ball lightning 208, 209
Balloon 137, 150*, 170, 330*
Bardstown, Kentucky 324
Barkley, Vice President Alben 108
Barracks 53, 54, 82, 84*, 120*
Base Hospital 201, 202, 269-273
Base Ops 65, 68, 107, 118, 123, 127, 131, 186, 190, 217
Base Weather 55, 57, 64*, 65, 68-70, 81, 83, 101, 119, 124, 147, 187, 193, 211, 212, 217, 218, 255, 266, 311
Battalion Information Officer 310
Bataan death march 290
Battle of the Bulge 277
Battles in the Monsoon 299
Bay City, Michigan 94
Beckham, Lt. Colonel Charles 206, 207
Behind the Eight Ball 61
Beirut, Lebanon 172
Belleville, Illinois 36*
Beneke, Capt. Laverne 55
Bengtsen, Captain 312
Benjamin, SMSgt. 256

337

Bennett, F/O Lawrence C. 40
Berkshires 66, 68
Berlin 101, 102
Berlin, Irving 108
Berlin Airlift *ii,* 101, 102, 108, 128, 172, 210
Bermuda *ii,* 55, 134, 215-230*, 232-248, 249*, 250*, 252, 267
Bermuda triangle 224
Biloxi, Mississippi 2
Bladensburg High School 276, 303
Blanchard, Captain (later Lt. Colonel) L. Dayton 186, 198
Blimps 218
Blue Star 30
Blue uniform 134, 180
BOAC 218, 226, 248
Bohr, S/Sgt. Wencel J. 40
Bolling AFB 157, 268
Bombay, India 171, 172
Bong, Major Richard 133
Bong Son, South Vietnam 294-298, 320
Bootstrap Program 276
Boxing 44, 66
Bracci, Eugene 95, 96
Briggs, Spike 212
British West Indies Airways 218
Brown, CWO Albert 313
Bryant, Pamela (nee Cogut) *vii,* 128, 141, 181*, 208, 215, 249*, 276, 304
Buddhists 270, 285
Bunker, Ellsworth 323
Bunkers 323
Buntyn, Colonel 277, 312
Burgess, Capt (later Lt. Colonel) James 186
Burrows, Mr. and Mrs. Ben 95
C-45 66, 124, 126
C-46 (Curtiss Commando) 71, 72, 76*
C-47 (also DC-3, the Navy's R4D, and "Gooney Bird") 65, 71, 72, 74*, 115, 124, 157, 164,173, 176, 179, 191, 297
C-54 (Douglas Skymaster, also DC-4 and the Navy's R5D) 57, 104, 107, 115, 116, 118, 122*, 124, 155, 172, 178, 179, 215, 251, 252
C-74 (Douglas Globemaster) 153- 155, 167*, 179
C-97 231*, C-97*
C-118 109*
C-123 320
C-124 221, 230*
C-rations 282, 286, 311
Cairo, Egypt 171, 172
Calculators 266
Camp Carroll, South Vietnam 313, 316, 317
Camp Custer, Michigan 75
Camp Eagle, South Vietnam 324

Camp Evans, South Vietnam 318, 329*
Camp Skeel, Michigan 202
Canada 67, 70, 73, 138, 194, 218, 252, 253
Cape Harrison, Labrador 55
Capitol 267
Caribbean 88, 216, 218
Caribou (DeHavilland DHC-4) 320
Carlsbad Caverns 304-307
Carmel, California 306
Caro High School 4
Caro, Michigan *iii,* 2, 3, 5-8, 73
Carpenter, 2nd Lt. Jonas 40
Carswell AFB, Texas 177
Casaday, Anne Cogut 48*
Cass, Lewis (Michigan Territorial Governor) 5
Cass City, Michigan 5
Cass River 5, 6, 93
Catton, Brigadier General Jack 257, 262
CAVU 125, 126, 132, 198
Cedar Lake, Michigan 197
Central Coastal Plain of Vietnam 280
Central Highlands of Vietnam 281, 282
CH-47 (Chinook) 285, 293, 295, 330*
CH-54 (Flying Crane) 285, 302*
Champaign, Illinois 29, 90
Chemical Corps 82
Chanute Field (also AFB), Illinois 17-24, 81-83, 107, 142, 157, 218, 265, 296
Chaplain 185
Charleston, South Carolina 222, 225
Cheverly, Maryland 265, 267-273, 276, 303
Chicago 1, 13, 21
Chicopee Falls, Massachusetts 52, 78, 79, 260
Christmas 44, 83, 108, 179, 180, 255-257, 263, 290, 291
Churchill, Prime Minister Winston 101, 216
Civil War 1, 6, 93
Class A uniform 67, 80
Climatic Center (also "Clime Center") 265, 267, 276, 303
Climatology Course 265, 266
Climatology vs. meteorology 265
Cloud Chasing With Weather 177
Cobb, Lt. Colonel Melvin 266, 267
Cogut (the pronunciation of) 2
Cogut Creek 5, 20, 93, 98
Cogut, Bill 7, 8*
Cogut, Louis (Dad) *iii*,* 3, 4, 11, 23, 37-39, 69, 70, 93, 94, 96-99, 142, 178, 179, 180, 279
Cogut, Grandfather 15, 94
Cogut, Grandmother Elizabeth 15
Cogut, Marie (nee Nordstrom) *vii*
Cogut, Mary (Mother, nee Evanish) *iii*,* 3, 4, 96
Cole, Colonel Frederick J. 55

Cold War 257, 259
Computers in forecasting 117, 224, 225, 253
Conger, Joan *vii*
Conger, William *vii*
Connecticut River 116
Connor School 6, 95
Constellation, Lockheed 167*, 171, 175, 221, 231*
Continental polar air mass 215
Continental Weather Wing 34, 37, 51
Corona 189, 190
CPS-9 130, 212, 222, 229
CQ 53, 54
Crombie, Mr. Howard 266
Cruthirds, Captain 72
Cumulonimbus 99, 125, 145, 147, 150*, 188, 219, 259
Cumulus (and towering cumulus) 68, 93, 125, 147, 188, 259, 301*
Curtis racing plane 202
Custer, Brigadier General George A. 291
Dammam, Saudi Arabia 175
Dana, Richard Henry 11
Danang, South Vietnam 309, 320, 326
Davis, General R.G. 320
Day, Doris 285
Delta H chart 217, 218
Department of Defense 83
Detachment 10, 9th Weather Group 233
Detroit, Michigan 2-4, 11, 19, 20, 30, 38, 42, 44, 54, 58, 65, 66, 69, 99, 133, 160, 210, 252
Detroit City Airport 121
Detroit News 3, 7, 131, 141, 142, 299
Detroit River 253
Detroit Tigers 203, 212
Detroit Wayne Major (DTW) 22
Dewey, Tom 5
D-day *i*
Dhahran, Saudi Arabia 48*, 122, 153-166, 169-180, 182*-184*, 186, 265
Dhahran Duster 166, 177
Diamond Head, Oahu 327*
Diem, Ngo Dinh 266
Division artillery 283
DMZ 317, 318
Dong Ha, South Vietnam 312, 313, 316, 317, 319, 331
Doolittle, Lt. General Jimmy 17, 123
Doppler radar 147
Doty, Airman Ray 151*
Dover, Delaware 222
Downey, T/Sgt. Paul E. 136
Draft 44, 178
Dropsondes 223, 243
Dry adiabatic lapse rate 172, 173
Durand, Michigan 2, 147
East Orange, New Jersey 32

Eberle, Pfc Ralph F. 37
Eckrem, Captain (later major) S.P. 168*
Edsel Ford Expressway 208
Eighth Air Force 261
Eighth Weather Group 37, 50, 52, 62
Eighth Weather Squadron 52, 55, 56*, 57-63, 81, 115
Eisenhower, General of the Army and President *i,* 13, 109, 216, 231, 239, 275, 298
Eisenhower High School 304
El Khobar 164
Ellis Island 37
Enemy artillery 313, 314, 316, 317
England 7
Enola Gay 39
Erwin, SMSgt Carl 172
ETAC (Environmental Technical Applications Center) 265
ETO (European Theater of Operations) 2, 50, 51
Europe 101, 104, 108, 140, 218
F-84 Thunderjet 124
F-86 Sabrejet 186, 187, 193, 195, 200*, 251
FAC 172
Fagan, Pfc. 32
Falkenberg, Jinx 108
Farm 66, 81, 93, 99*
FAX charts 151*, 224
FAX machine 28, 29, 145, 152*
Ferrying Command 257
Fetherolf, Pfc 30
Field Force I 313, 319
Field Force II 312, 319
Field jacket 67
Fifty mission crush 47, 118
Fifty-third Weather Reconnaissance Squadron 221
Fifty-sixth Fighter Interceptor Wing (and Group) 124
Firebase Tomahawk, South Vietnam 324, 325
Firebase Vandegrift, South Vietnam 313, 318, 321
First Cavalry Division (also 1st Air Cavalry Division) *ii,* 97, 275-287, 289-302, 311, 318, 320
First Infantry Division 277, 280
First Marine Division 160
First Weather Squadron 186
Fisher, Helen 93
Fisher, Silas 6
Fitzsimmons Hospital 211
Flint, Michigan 2, 130, 208, 213
Florida 218
Focke Wulf 190 7, 73
Ford, President Gerald 5
Ford, Henry 4, 23, 178
Forecast study 255

339

Form 175 212, 227-229, 232, 261
Fort Shawnee, Ohio 8
Fort Wayne, Indiana 21, 145
Fourth Infantry Division 278, 280
Fourth of July 7
France 13, 295
Francis, Connie 285
Franz Joseph, Emperor of Austria 15
French (in Vietnam) 281
French fighter pilots 66
Ft. Benning, Georgia 277
Ft. Chimo, Quebec 30, 55
Ft. Lewis, Washington 278, 280, 284, 326
Ft. Meade, Maryland 268
Ft. Myer, Virginia 276, 317
Ft. Sheridan, Illinois 1, 12, 13, 31, 42, 54, 119, 153
Ft. Sill, Oklahoma 276, 304, 305, 310-312, 316-319, 325
Ft. Totten, New York 37, 49, 50
G2 section 14
Gabreski, Colonel Francis 133
Gander, Newfoundland 103
Garry Owen 291
General Gordon 278
General of the Allied Armies 32
Geostrophic wind scale 106
German POWs 12, 28
German soldier-bakers 95
German weathermen *i*
Germany 13, 14, 95
GI Bill 205
GI party 10
Giap, General Vo Nguyen 299, 321
Gibson, Capt (later Colonel) Ralph "Hoot" 195
Gilbert, lst Lt. Lester E. 40
Given, CWO 319
Glen Miller 30
Gliders 115
GMD-1 89, 213, 214*, 279, 280, 283, 284, 289-292, 297, 300*, 316, 319, 320, 321, 330*
Golden Gate Bridge 278
Golden, Martha 11
Gone with the Wind 3
Goose Bay, Labrador 30, 55, 103, 140
Grand Army of the Republic 7
Grand Canyon 306, 308*
Grand Central Station 45, 46
Grand Rapids, Michigan 5, 179
Gray, Lieutenant 291
Great Britain 216
Great circle 103, 104
Great Depression *ii*, 4, 9, 15, 98, 179
Great Lakes 125, 195, 207, 208, 253
Greater Pittsburgh Airport 71
Green Beret 281
Greenland ice cap 115

Greenwich (also Greenwich time) 145, 169
Griffith, S/Sgt. (later CWO) Henry 179
Grondal, Greenland 55
Ground fog 126
Group Mobile 281
Gulf of Mexico 2
Gulf of Tonkin 316, 318, 324, 330*
Gulf Stream 107, 215
Gulf War 164
Gullion, T/Sgt. (later CWO) Jim 270, 271, 276
Gun (artillery) 277, 284, 285
H&I (harass and interdict) fires 283, 284, 320
Hancock, CWO 324
Hanoi Hannah 285, 292
Happy Valley 298, 299
Hastings, Michigan 99
Harmon Field (and AFB), Newfoundland 55, 103
Hartel, M/Sgt. (later Lt. Colonel) Bud 90
Hartford, Connecticut 52, 116
Hastings, Michigan 11
Hatteras Wave 255
Hawaii 321
Hazel Park, Michigan *iii*, 142
Helium 170
High Altitude Course 141-143
Highway 19, South Vietnam 281, 285
Hitler *i*, 102, 295
Ho Chi Minh 321
Hoag, Mr. and Mrs. Walter and son Thurlow 94
Holloman AFB, Alamogordo, New Mexico 24
Honolulu, Oahu 321
Honolulu Airport 327*
Hope, Bob 108, 291
Hopkins, Capt Joe 177
Horse blanket (GI overcoat) 67, 163
Horses 99*, 138
Howitzers 277, 284, 285, 287*, 332*
Hue, South Vietnam 309, 332
Hurley, 2nd Lt. Kenneth D. 40
Hurricane 78, 79, 88, 221-223
Hurricane Hunters 221, 223
Hydrogen 170
I Corps 310, 333*
Ia Drang Valley, South Vietnam 282-284, 290, 291, 293
Icing 137, 139
Idlewild Airport 251, 252
IFR (instrument flight rules) 70, 72, 126, 127, 129, 267, 270
Ikateq, Greenland 55
Inchon 160
Independence (President Truman's C-118) 107
Indian House Lake, Labrador 52, 55

Indianfields Township 6
Indians 6
Infantry 10, 82
Instability line 258
Instrument shelter 149*
Island water works 222, 230*
Italian POWs 30
Italians at Dhahran 162, 163
ITCZ (intertropical convergence zone) 89, 168
J.L. Hudson department store 20
Jackson, SMSgt Ronald 219
Japan 14, 39, 140
Japanese Zero 39
JATO (jet assisted takeoff) 116, 124
Jet stream 140, 144, 146
JFK Airport 251
John J. Pershing High School 30, 32, 66
John Lodge Expressway 142
Johnson, General Harold K. 290
Johnson, President Lyndon Baines 290
Joy, Henry B. 123
Joseph, Joe 8*
Kaline, Al 203
Karachi 171
Katherine, Aunt 15
KC-97 218, 226, 238, 266
Keesler Field, Mississippi 2, 9-15, 42
Keith, Sgt. James B. 40
Kennedy, President John F. 267
Khe Sanh, South Vietnam 318, 322
Khobar Towers 164
Kiel Auditorium 32
Kilroy 60
Kindley, Captain Field 216
Kindley Field (and AFB), Bermuda 55, 103, 216, 218, 220, 222, 225, 227, 233, 251
King Ibn Saud 158
Kinnard, Major General Harry W.O. Jr. 277, 289, 290
Klier, Ardis 48*
Klier, Sgt. Howard "Henry" 48*, 56, 62
KLM (Royal Dutch Airline) 159, 171, 218
Knudsen, William S. 203
Korean stage show 313
Korean War 85, 89, 91, 133, 141, 153, 159, 172, 213, 266, 279, 283, 284
KP (Kitchen Police) 7, 10, 13, 28, 31, 33, 37, 38, 86, 154, 202
Kwajalein 95
Lafferty, M/Sgt. Robert 166
Lagens (and Lajes), Azores 55, 57, 103, 104, 118, 154, 179, 220, 221
Lake Erie 253
Lake Huron 180, 185, 209
Lake Michigan 1, 54, 130, 132, 194
Lake St. Clair 66, 68, 69, 73, 123, 127, 134, 138

Landon, Alf 15
Lansing, Michigan 2, 122, 130
Lawrence of Arabia 155
Lawton, Oklahoma 276, 284
Layne, CWO Bruce 316, 317
Le Bon, S/Sgt. Joseph W. 40
Legion of Merit 128
LeMay, General Curtis 123, 254
Lemberg, Austria 96
Lenticular clouds 192, 200*
Lewis, M/Sgt. Theodore A. 40
Lindbergh, Charles 66, 103, 123, 140, 146, 202, 227
Lightning 193
Litzenberg, George 39
Lockbourne Army Air Field 71
London 103, 169, 226, 248
Long Binh 312, 319
Long Island Railroad 42, 49, 50
Luftwaffe 7
Luty, S/Sgt. Paul 65, 77
LZ Bastogne 318
LZ Nancy, South Vietnam 318
LZ Roy, South Vietnam 330*
M-14 rifle 280, 281, 290, 299
M-16 rifle 280, 281, 320
M-60 machine gun 290
MAC (Military Airlift Command) 257
Macon, Georgia 73
Mach, Darrell 250
Mach, Leta (nee Cogut) vii, 65, 77, 79, 83, 90, 92, 99*, 162, 181*, 208, 215, 276, 303, 304
Mad minute 320
Madison Heights, Michigan 65, 121
MANOPS 62
Marine Corps 177, 312, 318-320
Marshall, Brigadier General S.L.A. 299
MATS (Military Air Transport Service) 57, 101, 122, 124, 218
Mauldin, Bill 314
Maximum temperature 317
May flies 134, 135*, 158
McArthur, General of the Army Douglas 160
McAuliffe, Major General Anthony 277
McClung, S/Sgt. Tom 61, 62, 83
McCurley, Specialist 6 280, 298
McDaniel, S/Sgt. Eddie 227, 238, 239, 246
McGlew, M/Sgt. Robert 255
McGuire, Major Thomas 133
McGuire AFB, New Jersey 216
McLeod, Colonel 325
McMillan, Prime Minister Harold 216, 239
Mecatina, Quebec 55
Medals 7, 8, 73, 128, 254, 312
Medevac 320
Medical Corps 82
Mediterranean 155

341

Meeks Field, Iceland 55
Memorial Day parade 7
Met Quality Control Team 309, 312, 313, 321
Meteorology Division 318
Michigan Central Station 30, 71, 77
Microburst 147
Middle East Airline 171
Middletown, Pennsylvania 215
Mid-Ocean Golf Club 217
Mid-west 203
Militia forces, Vietnam 282
Miller, CWO Chuck 316
Milwaukee's Billy Mitchel Field 196
Mines 275, 318, 333
Mississippi 2, 33
Mitchel Field, New York 120
ML-574 316
Mleziva, Dorothy 128
Mleziva, Matt 128, 129, 141
Mollichelli, Major Edward V. 318
Monsoon 172, 285, 296
Mortar 277
MOS (Military Occupational Specialty) 9, 28, 34, 57, 81, 102, 103
Most hazardous occupations 313
Moxon, Lt. Colonel George 268
MP's 46, 47
Mt. Clemens, Michigan 120, 126, 133
Mulroney, Peter J. (CO 12th Marines) 321
Mussolini 31, 162
Napoleon i
National Guard 324
National Hurricane Center 222
National Meteorological Center 29, 224, 253, 255
National Weather Service 117
Navigators 173, 225-227
Navion 171
Navy 12
NCO Academy 289
New Baltimore, Michigan 128
New England 252
New York 5, 46, 50, 103, 218, 251, 252, 285, 296, 297
New York Central Railroad 42
New York World's Fair 14
Newman, Ange and Bob 207, 252, 253
Nha Trang 313, 319
Niagara Falls Air Force Station (also airport) 180, 187, 196
Nicosia, Cyprus 155
Nixon, President Richard 335
Nobel, Mr. 88, 194, 296
Nordstrom, Alexander 186
Nordstrom, Martha (nee Golden) 66, 70, 99
Nordstrom, Robert 23, 84*, 142
Nordstrom, Walter 11, 38, 69, 70, 75*, 77, 141, 142, 144

North Africa 30
Northern Pike 5, 6, 93, 94
NOTAM 192
NVA (North Vietnamese Army) 281, 290-295, 297, 299, 313, 314, 316, 317, 325
Objective (also semi-objective) forecasts 266
O-Club (Officers' Club) 317, 319
OCS 203, 206
O'Diorn, Major John 198
O'Donnell, General Emmett "Rosy" 123
O'Dwyer, William J. 57
Okinawa 280
Oklahoma City 40-44
OH-6A Cayuse LOH (or "Loach") 313, 329*
OH-13 Sioux 285, 302*
O'Hare International Airport 196
Old Country 3, 4, 15, 94
Omaha, Nebraska 169
Ontario 138
Operation Blue Book 190
Operation Black Horse 292, 293, 297
Operation Clean House I, II, III 292
Operation Crazy Horse 292, 298
Operation Flying Tiger 292
Operation White Wing/Masher 292, 297, 301, 302, 320
Orchard Place (ORD) 22
Oscoda AFB (later Wurtsmith AFB), Michigan 180, 185-198, 199*, 200*, 201, 209, 210, 251, 258, 296
Owosso, Michigan 5
P-38 (Lockheed Lightning) 83
P-47 (Republic Thunderbolt) 66, 83, 133
P-51 (North American Mustang, also F-51) 66, 84*, 124, 185, 199*
P-80 (Lockheed Shooting Star, also F-80) 24*, 50, 124, 139, 140, 147
Pacific 279, 306
Padloping Island 52, 55
Pan Am 218
Parachute 65, 67, 68, 146
Paris 103, 106, 175, 251
Patton, General George 13
Paulus, Field Marshal Friedrich 102
Pay 10, 18, 19, 32, 33, 325
Pease, Capt Harl Jr. 254
Pease AFB, New Hampshire 252, 254-262, 263*, 267
Penn Station 49
Pentagon 7, 276
Period of National Emergency 2, 11
Persian Gulf 155, 162
Phu Bai, South Vietnam 309-313, 315, 319, 322, 323, 332*, 334*
Phu Cat, South Vietnam 294, 300*
PIBAL (pilot balloon) 59, 63, 169-171, 178, 182, 282, 283

342

Pilot-to-forecaster radio 186
Pima Air and Space Museum, Tucson, AZ 39, 48*, 73, 74*, 76*, 109, 167*, 231*, 264*
Pistol (.45) 320
Pittsburgh Pirates 266
Pixton, Brigadier General Allan G. 318, 319, 320
Pleiku, South Vietnam 282
Polk, Lt. Colonel Gilbert 318
Popadines, Sgt. George 64*, 133, 149*
Port Lyautey, French Morocco 154
Porter, CWO 317
Potsdam Conference 101
Precipitation 317
Presque Isle, Maine 55
Pressure pattern (flying the) 104
Pressure surfaces and maps:
   200 millibar 105, 140;
   300 millibar 105, 114*, 140, 143, 195, 217, 228;
   500 millibar 104, 105, 107,113*, 140, 143, 195, 217, 218, 228, 253, 255, 258, 296;
   700 millibar 105, 107, 112*, 173, 173, 195, 217, 218, 228;
   850 millibar 105, 107, 111*, 145, 217, 228;
   1000 millibar 104
Prim, M/Sgt. William T. 40
Prince (the black horse) 3, 97
Prognostic charts 218, 253, 255
Project Vittles 101
Protesters 278, 295, 315, 321
Pseudo-Adiabatic Chart 89
Puff the Magic Dragon 74, 297
Pulitzer-Schneider Race 202
Pusan 160
Putnam, Major 212
PX (post exchange) 70, 130, 131, 166,
Quang Tri, South Vietnam 317
Queen of Bermuda 235
Qui Nhon, South Vietnam 281, 285, 288, 292, 298
Quonset hut 155-159, 161-164, 168*, 176, 177, 183
Queen of Bermuda 251
R&R 172, 173, 321
Radar 130, 131, 188, 193, 222
Radiation fog 126, 127
Radio City Rockettes 108
Radio Moscow 160, 161
Radio Station KMOX 32
Radiosonde 59, 89, 150*, 145, 223, 279, 283, 294, 317, 330*
Radiosonde Course 17, 28
RAF 218, 220
Rantoul, Illinois 22
Rawinsonde 137, 138, 140

RCAF 177, 218
REA 14
Red Ball 298
Red Cross 179, 180
Refugees 295
Rendulic, Pfc Anthony Jr. 37, 41
Rexall Drug Store 3, 7
Rhumb line 103, 104, 179
Rice, Grantland 141
Rickenbacker, Captain Eddie 123
Riverside, California 219
Rockets 314, 315, 322-324 328*
ROK (Republic of Korea) Capital Division 292, 297
Roketa, Anthony 95
Rome 96
Rooney, Andy 7
Roosevelt, President Franklin D. 216
Royal Bermuda Meteorological Office 223, 229
Royal Oak (city of), Michigan 12, 51
Royal Oak Township, Michigan *iii*, 20, 65, 121
Rub al Khali 162, 169
Ryals, CMSgt Jim 78
S-2 309-312, 323
SAC (Strategic Air Command) 254, 257, 259, 261
Saigon, South Vietnam 296, 311, 319, 332, 333*
Sandburg, Carl 21
Satellites 193, 224, 282
Saturation vapor pressure 127
Saudi Arabia *ii*, 153-180, 182-184, 186, 208, 215, 283, 296
Schilling, Colonel David 140
Schuffert, M/Sgt. Jake 118
Scott Field, Illinois 18, 29, 31, 34*, 50
SCR-658 59, 89, 137, 213, 214
Scroll 303-306, 307*
Seabees 318
Seattle-Tacoma Airport 326
Second Battalion, 138th Artillery 324, 325
Selfridge, Lt. Thomas 117
Selfridge Field (also AFB), Michigan 41, 65-73, 120, 123-152, 157, 158, 172, 177, 179, 180, 186, 196, 202, 204, 210- 213, 269, 273, 321
Sentimental Journey 39, 48*
Setla, Mr. and Mrs. Rudolph and family 94
Seventh Cavalry 291
Seversky P-35 69
SFERICS 28
SHAFT (Specialized High Altitude Forecasting Training) 141, 218
Shamaal 170, 171
Shamaal Talk 177
Shannon, CWO Tom 318, 329

343

Shark oil 223, 229
Shepherd, Mr. and Mrs. Eugene and son Chloral 94
Sherman, General William T. 309
Sidi Slimane, Morocco 219
Signal Corps 82, 83, 316
Silent majority 335
Simuitak, Greenland 55
Single-station analysis 169, 296
Sixteenth Weather Squadron 136
Sixty millimeter mortar tube 294
Sixty-third Fighter Interceptor Squadron 187, 195, 198, 200
Skeel, Capt Burt E. 202
Skew T, Log P Diagram 89, 316, 317, 331*
Slide rule 266, 274*
Smith, Capt William E. 136, 140
Smith, S/Sgt. Maynard 7, 8, 73
Snipers 275
Snow 116, 252, 253, 255, 256, 257
Sokol, Austria 96
Sondrestromfjord, Greenland 55
Song Con Valley 298
Sonic boom 202
South America 218
South China Sea 280
Soviet Union 159, 161, 173
Spaatz, General Carl "Tooey" 123
Spider web weather predictor 223
Spirit of Caro 8
Spirit of St. Louis 66, 140
Springfield, Massachusetts 52, 116
Springfield Art Museum 81
Squadron (423rd) 7
Stalin, Joseph 101, 102, 159, 161
Stalingrad *i*, 102
Station Chief 81, 136, 187, 204
Station Weather Officer 116
Steam fog 127
Stewart AFB, New York 204, 205, 209
Stimson, Henry L. (Secretary of War) 7
St. Elmo's fire 220
St. Louis, Missouri 21, 22, 31, 32, 43, 44, 46, 198
St. Louis and San Francisco Railroad 42
Stock Market Crash 4
Strand Theater 3
Stratocumulus 125, 188
Stratosphere 282
Stratus 72
Suitland, Maryland 224
Sunspots 218, 219
Surface inversion 127
Surface weather chart 110*, 195, 217, 258
Swartz, Willa (nee Cogut) *vii*, 73, 119, 121*, 181*, 208, 215, 216, 249*, 256, 276, 304
Synoptic code 59
T-29 226

Tabb, Jerry 323, 324
TAFORS 219
Tan Son Nhut Airfield, South Vietnam 311, 319, 332
Tapline 171
Teheran, Iran 172-174
Telegraph Road 58
Teletype machine 59, 63, 64*, 65, 72, 152*, 188, 193, 260
Television 14
Temple, T/Sgt. Harry L. 40
Tents 154
Tesla, Nikola 14
Tet "offensive" 291
Thailand 276
Thanksgiving 20, 285, 286
The Rockpile, South Vietnam 313
The Sea of Tranquility 326
Theodolite 59, 170, 182*, 282, 283
Third Army 13
Third Battalion, 18th Artillery 277, 278
Third Marine Division (also Third Marine Amphibious Force) 297, 318, 321
Third Weather Group 204
Thomas, Lowell 102
Thumb section (of Michigan) *iii*, 97
Thunderstorm(s) 145, 157-159, 193-198, 219, 224, 258-261
Tibbets, General Paul 39
Times Square 50
Tinker Field, Oklahoma 34, 37-42, 50, 51
Tokyo, Japan 326
Toledo, Ohio 20, 88
Tolono, Illinois 29, 42
Tornado 177, 208
Transportation 82
Traverse City, Michigan 130-132, 134, 224
Tripoli, Libya 154, 155, 167, 172, 174, 179, 249*
Troop train 1
Troy, New York 253
Truman, President Harry S. 2, 11, 72, 101, 108, 160
Turbulence 144, 215, 219-221, 259
Tuscola County 5
Tuy Hoa, South Vietnam 292
TWA 159, 160, 167, 171, 218
Twain, Mark 218
Twelfth Marines 319, 321
Twelfth Weather Squadron 119, 120, 186, 204-206, 209
Twenty-ninth Weather Squadron 122, 154, 172, 174
Twenty-seventh Pursuit Squadron 202
Typhoon Bess 316, 331
U-21 319, 332*
UFOs (also flying saucers) 188-192
UH-1 (Bell Iroquois, also "Huey") 282, 285, 294, 301*, 313

Ukrainians 2, 101
Union Station, St. Louis 33, 43, 44
University of Maryland 268, 276
University of Michigan 206
Upper Frobisher, Baffin Island 55
US Air Force museum 74*, 231*
US Lake Survey 207, 208
US Navy 154
USO 32,33
US Weather Bureau 116, 213
Van Dyke, 2nd Lt. Orley W. 40
Van Ettan Lake 202, 203
VC-118A (Douglas Liftmaster, also Air Force One and DC6A) 109*
Vertical weather cross sections 115, 179
VFR (visual flight rules) 70, 126, 129, 130, 132, 133, 270
Viet Cong (also VC) 279, 281, 282, 290, 292, 293, 313, 314, 318, 325
Viet Minh 281
Vietnam ii, 83, 97, 128, 172, 214, 266, 270, 275-302, 309-334
Vinh Thanh Valley 298
Virginia (Virginia Elizabeth, nee Cogut) Martin iii*
Vorticity 144
Voss, Capt Julian 203
Waianae, Oahu 321
Waikiki Beach, Honolulu 327*
War Department 41, 42, 83
Warrant Officers ii, 86, 87, 103, 118, 141, 276, 279, 309
Warrant Officer casualties 279
Wartime weather forecast curtailments 98
Washam, Sgt. Robert W. 40
Washington, DC ii, 78, 157, 169, 210, 211, 265, 289, 303
Water wheel 294
Wayne County Road Commission 207
WB-50 218, 221, 223, 224, 243
WBAN (Weather Bureau, Army and Navy weather observations form) 63, 100, 126, 255, 262*, 267
Weather badge 100*
Weather balloon 59
Weather Detachment 1-18L 186
Weather Detachment 12-8 186
Weather Detachment 12-23L 119
Weather Detachment 29-2 168*
Weather Equipment Teletype Technician Course 17, 18, 27-34, 43
Weather Forecaster Course 17, 18, 83, 85, 107, 169, 170, 296
Weather Observing Course 17, 18, 22, 23, 44, 147
Weather pin 25*, 26*
Weather recon 172
Weather sleeve patch 25*, 26*

Weatherman's Song 27, 50, 143
Weather Wizard 223
Weekend warrior 65, 67, 68, 70, 211, 212
Western Allies 101, 102
Where Did All the Years Go? 40
Westover Field, Massachusetts 52-55, 57-64, 65, 67, 72, 77, 78, 83, 101, 104, 106, 108, 115-118, 120, 124, 128, 144, 153, 157, 179, 191, 192, 217, 225, 260, 261
Wheelus Field 122, 154, 167, 249*
Whipple, S/Sgt. Norman A. 40
Wiggins, CWO Chuck 318
Wind shear 144, 147
Windsor Locks, Connecticut 65
Windsor, Ontario 253
Wilbur Wright Field 202
Willimansett, Massachusetts 72, 79, 80
Willow Run 23, 196
Willys jeep 12
Wimbledon, England 11
Wind factor 226, 227
Wisconsin 131
Witherington, CWO J.D. 276, 277, 317, 318
Wolfe, Captain (later Colonel) Donald J. vii, 186-188, 204, 206, 209, 210
Wolfe, Lucy 209
Wood, Pfc Glen S. 37
Woodward Avenue, Detroit 20
World War I 12, 69, 94, 155, 216
World War II (also "the big one") 2, 11, 14, 15, 45, 94, 123, 133, 160, 162, 169, 172, 186, 212, 293
Wright, 1st Lt. Clayton 180
Wright-Patterson Field (also AFB) 71, 74, 77, 116, 179, 187, 190, 191, 196
Wurtsmith, General Paul 201
Xuan Loc, South Vietnam 311
XXIV Corps and XXIV Corps Artillery ii, 309, 311-313, 318, 334
Yates, Brigadier General Donald 116, 117
Year of the horse 292, 320
Yellow Brick Road 324
Youngstown Airport 196
Zoot suiter 44

345